INDUSTRIAL SAFETY
Management
and Technology

David A. Colling

College of Engineering
University of Lowell

PRENTICE HALL, Englewood Cliffs, New Jersey 07632

Library of Congress Cataloging-in-Publication Data

Colling, David A.
   Industrial safety : management and technology / David A. Colling.
     p. cm.
   ISBN 0-13-457235-1
   1. Industrial safety—Management. 2. Industrial safety.
I. Title.
HD7262.C617   1990
658.3'82—dc20   89-22766
                                                                      CIP

Editorial/production supervision: *Raeia Maes*
Manufacturing buyer: *Gina Brennan*
Cover design: *Bruce Kenselaar*

TO: SUZANNE, in trying to help her, I have learned so much how to help others,

and to all students of safety, who face the challenge of providing a truly safe work environment. May they succeed!

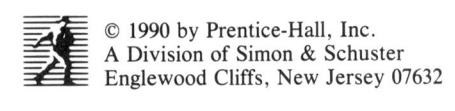

© 1990 by Prentice-Hall, Inc.
A Division of Simon & Schuster
Englewood Cliffs, New Jersey 07632

All rights reserved. No part of this book may be
reproduced, in any form or by any means,
without permission in writing from the publisher.

Printed in the United States of America

10 9 8 7 6 5 4 3 2 1

ISBN 0-13-457235-1

Prentice-Hall International (UK) Limited, *London*
Prentice-Hall of Australia Pty. Limited, *Sydney*
Prentice-Hall Canada Inc., *Toronto*
Prentice-Hall Hispanoamericana, S.A., *Mexico*
Prentice-Hall of India Private Limited, *New Delhi*
Prentice-Hall of Japan, Inc., *Tokyo*
Simon & Schuster Asia Pte Ltd., *Singapore*
Editora Prentice-Hall do Brasil, Ltda., *Rio de Janeiro*

Contents

PREFACE xi

1 SAFETY TRADITIONS AND REGULATIONS 1

    The History of the Safety Movement, 2

    The Development of Safety Programs, 5

    The 1970 Occupational Safety and Health Act, 5

    The Occupational Safety and Health Administration (OSHA), 6

    The Right-to-Know Laws and SARA-III, 10

        Case History 1.1   More Than Conformance Is Needed, 17

    Further Reading, 26

    Questions, 26

2 ACCIDENT CAUSATION 27

    Heinrich's Domino Theory, 28

    Human Error Model, 31

    Petersen's Accident/Incident Model, 32

    Epidemiological Models, 33

    Systems Models, 33

    Multiple Causation, 35

    Summary, 35

Case History 2.1   Pointing the Way, 36

Further Reading, 39

Questions, 39

3 SAFETY: A PEOPLE PHENOMENON 40

The KISS Principle, 41

Perception: A Key Process, 42

Worker Attitudes and Perceptions of Safety, 44

Human Error, 45

Case History 3.1   Feelings and Business, 46

Case History 3.2   Choosing the Wrong Carpenter, 49

Ergonomics or Human Factors in the Workplace, 55

Case History 3.3   Anthropometry in a Hospital Chemistry Laboratory, 56

Case History 3.4   A Simple On/Off Switch Choice, 65

Case History 3.5   Who Controls? 67

Further Reading, 72

Questions, 73

4 SYSTEMS SAFETY: MANAGEMENT 74

The Tasks of Management, 75

Managerial Roles and Skills, 76

Management by Objectives, 77

Case History 4.1   High-tech Industrial Safety Management, 84

Transition in Management Theory, 86

Case History 4.2   The Alcoa Experience, 88

Behavior Management, 90

Training, 94

Case History 4.3   Training for Supervisors: The Federal Hazard Communication Law (OSHAct 1910.1200), 95

Contents

Safety Committees and Safety Circles, 97

Risk Management, 99

Further Reading, 101

Questions, 102

5 SYSTEMS SAFETY: HAZARD IDENTIFICATION 103

Case History 5.1 Safety in Research, 104

Preliminary Hazards Analysis (PHA), 107

Case History 5.2 A Model for Analysis, 108

Case History 5.3 A Case for Preliminary Hazard Analysis, 109

Failure Mode and Effects Analysis (FMEA), 109

Case History 5.4 A Case for FMEAs, 110

Hazard and Operability Review (HAZOP), 110

Case History 5.5 A Case for HAZOP, 114

Human Error Analysis in Combination with Hazard Analysis, 116

Fault Tree Analysis (FTA), 116

Risk Analysis, 118

Summary, 122

Further Readings, 123

Questions, 123

6 HAZARD CONTROL 124

Case History 6.1 No Change: The Right Decision, 125

Decisions for Action, 127

Case History 6.2 The Wrong Remedy, 127

Engineering Versus Management Controls, 128

Case History 6.3 Multiple Engineering Remedies, 129

Case History 6.4 Combined Engineering and Management Remedies, 131

Implementing Hazard Control Measures, 131

Monitoring Remedies, 132

Evaluating Effectiveness, 133

Putting It Together, 133

Wholistic Safety Programs, 134

A Last Word, 136

Further Readings, 137

Questions, 137

7 MATERIAL HANDLING 138

Lower Back Pain, 138

The Relationship Between Lower Back Pain and Manual Handling, 140

Manual Handling Programs to Limit Lower Back Injuries, 141

Case History 7.1 A Packaging Problem, 143

Mechanically Assisted Material Handling, 144

Case History 7.2 Better to Miss the Brass Ring, 148

Palletizing the Workload, 148

Powered Industrial Vehicles, 150

Case History 7.3 A Case for Training, 152

Case History 7.4 An Unplanned Task, 155

Automated Material Handling, 157

Case History 7.5 Automation Needs Ergonomics, 159

The Future of Material Handling, 160

A Last Word on Material Handling, 160

Further Reading, 162

Questions, 163

Contents vii

8 MACHINERY SAFEGUARDING *164*

Hazards and Machine Motion, 164

Principles of Safeguarding, 167

Task Factors, 169

Case History 8.1 Press Brake Accident, 169

Operator Factors, 171

Case History 8.2 Press Brake Accident, 171

Learning to Recognize Hazards, 171

Case History 8.3 Leather Band Buffer, 172

Some Realities in Hazard Recognition, 174

Case History 8.4 A Case of Similar, But Different Machines, 175

Safeguarding, Automation, Robotics and Factories of the Future, 175

Further Reading, 178

Questions, 179

9 FALLS *180*

Slips and Falls, 181

Trips and Falls, 183

Prevention of Falls Caused by Slips or Trips, 183

Case History 9.1 Slipping Injury in Print Room, 184

Case History 9.2 Tripping Incident on Freight Elevator, 185

Stairway Falls, 187

Falls from Ladders, 189

Case History 9.3 An Unnecessary Ladder Fall, 191

Free Falls, 192

Case History 9.4 Elevator Shaft Fall, 192

Further Reading, 193

Questions, 193

10 FIRE PREVENTION AND PROTECTION   195

Chemistry of Fire, 196

Products of Combustion, 199

Case History 10.1   Propane Fire in Warehouse, 201

Fire Development: Severity and Duration, 202

Case History 10.2   Fire Spread in Cold Storage Warehouse, 203

Effect of Enclosure and Heat Transfer in Fire Development, 206

Case History 10.3   A Challenge to the Professional Fire Investigator, 206

Fire Prevention and Protection, 210

Case History 10.4   A Hydrogen Explosion, 210

Early Detection, 211

Extinguishing Fires, 211

Case History 10.5   A Case of Liability, 213

Disaster Preparedness, 216

Electrical Safety, 217

Case History 10.6   What Can You Suspect? 220

Case History 10.7   The Wrong Wire, 221

Further Reading, 222

Questions, 223

11 INDUSTRIAL HYGIENE   224

Routes of Entry of Foreign Substances, 225

Long-term Medical Disorders and Epidemiology, 229

Industrial Solvents, 231

Contents

    Case History 11.1   A Case of Worker Comfort, 234

    Case History 11.2   A Matter of Intake, 238

Stress and the Workplace, 240

    Case History 11.3   Distress at Evaluation Time, 244

Industrial Noise, 245

    Case History 11.4   A Punch Press Noise Problem, 253

Hazardous Waste, 254

    Case History 11.5   Know Your Waste, 259

Further Reading, 260

Questions, 261

12 PRODUCT SAFETY 262

The Legal System, 263

The Origins of Products Liability Laws, 264

Status of Products Liability Laws, 266

Statutory Law and Products Liability, 267

The Expert Witness in Products Liability, 269

    Case History 12.1   Kales v. Gem Wheels, Inc., 270

    Case History 12.2   Mahoney v. Bernardston Chair, Inc., 273

    Case History 12.3   Berger v. Middlesex Conveyor, 276

    Case History 12.4   Pottle v. Foreign Motors, Ltd, 278

Products Liability Prevention Programs, 280

Management of Product Safety Programs, 281

Technical Requirements of Product Safety Programs, 282

Planning for Warnings and Instructions, 287

Recall Planning and Customer Relations, 289
Summary, 290
Further Reading, 290
Questions, 291

APPENDIX A: SAFETY INFORMATION *292*

APPENDIX B: PROFESSIONAL ORGANIZATIONS *294*

APPENDIX C: BIBLIOGRAPHY *296*

APPENDIX D: GOVERNMENT AGENCIES *301*

INDEX *303*

Preface

Safety is an exciting field! It has all the elements of science fiction, mystery and drama, combined with the satisfaction derived in knowing how people and things work. In addition, it is always changing—dynamic in all respects. Yet this concept is somehow lost, lost because of outdated tradition, lack of innovative thinking and lack of commitment. It seems strange that safety is naively considered a common-sense subject while society is demanding safer workplaces, safer highways and safer products, goals readily achieved if dependent on common sense.

The primary purpose of this book is to convey the real excitement of safety to those who have an interest in or direct responsibility for safety or are simply studying the subject because it is part of their required college curriculum. The subject matter has been designed to stimulate the reader, yet provide the necessary information for understanding basic safety engineering and management principles. Any record-keeping or checklist formats have been minimized or eliminated, because proper attention to safety will point the way to what is needed. The same is true of traditional safety remedies, including training methods and procedures all the way through personal protective equipment. It is not the intent to reduce these areas to insignificance, but to permit room for creativity and innovation. The appendices, which are devoted to answering the question of how to find safety information, are an important part of the book.

This book is intended both as an introductory textbook for undergraduate technical, business or engineering courses and as a reference book for manufacturers, executives, labor leaders and interested employees. It is organized into two major sections reflecting the importance both individually and collectively of safety management and safety engineering. The significance of management understanding technology and engineers understanding management concepts is nowhere more evident than in industrial safety. The organization of the text is derived from teaching industrial safety to engineering and business students. My students have been required to conduct an evaluation of industrial safety programs and report their findings to the classroom. Because of the student motivation value, I urge all teachers who use this text to initiate such studies which are modeled on safety internships. I also urge these teachers to develop small discussion groups to address safety topics that cannot be fully exploited in the classroom; members of these groups are excellent prospects for facilitators in successful industrial safety circles. The questions at the end of each chapter have been developed by one discussion group who first analyzed how we learn and then looked at each subject. My thanks to the members of this group.

The first six chapters deal with general safety principles and describe some historical perspectives and legislative regulations. Accident causation models are presented in Chapter 2 and they lay the foundation for the importance of studying about people, management and hazards in the next three chapters. Safety is about people, about how they behave, how they interact with others and with machines and about why and how they commit errors. Human factors, people and perception all dictate how effective a safety program can be. That program has to be managed effectively as well and Chapter 4 on safety management incorporates all the attributes of successful business management principles.

Nowhere are creativity and innovation more important than in identification of hazards. Chapter 5 focuses on effective utilization of present knowledge of how to study and identify hazards in the workplace. Conventional methods are described, but effective studies of machine design and process sequencing will help remove some of the mystery of how things work and how workers might interact. Chapter 6 completes the first section and covers remedies, their selection and application and monitoring and feedback. Engineering remedies, training and personal protective equipment are discussed in a general way, as are the importance of feedback information and communication to the workers.

The second section is devoted to safety engineering principles, industrial hygiene and product safety. In selecting the chapter topics, I considered frequency and severity of injuries as well as demonstration of safety engineering principles. Therefore, Chapters 7 through 10 describe safety in material handling, using both manual and powered handling, machinery safeguarding principles, falls on the same level and falls to other levels, and fire protection and prevention, including disaster planning and electrical safety. Industrial hygiene is becoming increasingly important to safety professionals because of societal pressure to reduce long-term health problems. For this reason, Chapter 11 covers anatomy and physiology of the human body, with example sections describing the effects of the four industrial hygiene stressors: chemical (solvents); physical (noise); ergonomic (psychological stress); and biological (hazardous wastes). Chapter 12 on products liability is included because of potential third-party action in many industrial accidents. Such actions affect employee morale and require administrative and legal expenses. The purpose of these actions, the evaluation of products liability law and products liability prevention programs are therefore topical.

A teacher has the responsibility not only to disseminate information but also to stimulate the appropriate application of the knowledge in solving real problems. One of the more effective teaching methods used in business schools, medicine and other professional fields is the case history. One of my purposes in preparing this textbook was to apply the lessons learned in specific cases to prevent others from making the same or similar mistakes. Of course, case histories have been fictionalized to protect personal and corporate privacy.

I would like to thank my students for their inspiration, my colleagues at the University of Lowell and the members of the Boston Chapter of ASSE for their encouragement and support, Dr. Justin Newmark and Dr. Jane Dreskin who have helped me better understand the people part of safety, Ralph Berger and the staff of Skinner and Sherman Laboratories, Inc., for providing much of the material for case histories, Dr. Paul Specht, Dr. Louis Trucks and Patrick Jones, Esq., for reviewing the manuscript, Jacquie Nicholas for typing the manuscript and Jeanne Maider for her administrative help.

David A. Colling

1

Safety Traditions and Regulations

A safe working environment is expected in modern society. That has not always been the case, however, with workers assuming all risks while working in intolerable conditions. If we trace the history of industrial safety, we find first a few concerned inventors who patented safety devices. For example, the first barrier safeguard was patented in 1868. Movement toward a safer workplace closely followed the stormy worker movement for unionization and legal efforts which formulated the first Workmen's Compensation Laws. At about the same time, i.e. at the beginning of this century, the first systematic studies of efficiency in manufacturing were being conducted by Fredrick Taylor, the Father of Scientific Management. While Taylor's interests were in efficiency, productivity and profits, his emphasis on lost time, lost personnel and lost material led to an early understanding of the important interrelationship between safety and management that we recognize today. Those safety programs which are in existence today have evolved by the combined endeavors represented in these historical roots—the safety inventor, the concerned legislator, and sound management. This book is written on the premise that optimum industrial safety programs can and will result from the creative, innovative continuation of these dedicated groups.

There is no doubt that there has been steady progress in providing safer places in which to work. For example, recent data show the death rate from work-related injuries to be 4.4 per 100,000 workers, about a third of that 50 years ago.[1] However, safety programs cannot be judged on death rates alone. Another val-

---

[1] *Accident Facts,* 1987 Edition, National Safety Council.

uable indicator is the cost of injuries. In 1982, the total cost of benefits paid to disabled workers was $16.1 billion, while the National Safety Council estimated the *total* cost of disabling work injuries, including production losses and other damages, to be $31.4 billion. This amounts to more than $200 for every worker in the entire country. Cost is almost insignificant, however, when we consider that 16.5% of the population is disabled, nearly half of whom are unable to work at all.

Much of what is faulty or lacking in our approach to safety is rooted in the history of the safety movement. There are many economic, political, social and cultural aspects which are intimately interrelated. Improvements in safety are most likely to come in the future when the interests of the workers are integrated with those of management and when cultural division, whether it be based on racism, sexism or ethnicity, is eliminated. However, let us examine how far we have come to date, what legislation is involved and how today's safety programs have developed. Only then can we begin to understand how safety programs come about, something that will be examined in detail in later chapters.

THE HISTORY OF THE SAFETY MOVEMENT

One of the major handicaps we encounter in safety is the attitude that safe behavior is based on common sense. The term common implies "gained by experience" and sense refers to the ideas of right and wrong. If safety was based on common sense, we would encounter the same or similar conditions throughout the world. However, this is simply not the case. In underdeveloped countries, it is as true today as it once was everywhere in the world, that an individual's rights and safety are subordinate to the preservation of the community or civilization.

With this in mind, we are able to view safety as being rooted in the industrial revolution of the late eighteenth century, the main development of which was the introduction of the steam engine for manufacturing. The early factories developed alongside apprenticeship programs, with child labor therefore becoming commonplace. Child labor became the focus of public outrage with an outbreak of fever in the cotton mills near Manchester in England. Societal pressures led the government to act, and in 1802 the Health and Morals of Apprentices Act was introduced, the aim of which was to combat the exploitation of children, the excessive hours they worked, and the wretched conditions within which they worked. This legislation marked the entry of government into work safety.

The industrial revolution reached the U.S.A. in the nineteenth century, but it was not until the rapid expansion of factories in the latter half of the century, that labor became strong enough to organize itself. The upper classes, i.e. management, had effectively treated the workforce as slave labor, with working conditions that did not take account of human values, or health and safety. The weakly organized workforce fought for the elimination of hazards introduced with large moving machinery parts and for compensation when workers were injured or killed in the workplace. Government was again made to intervene after the public and humanitarian community leaders were made aware of the workers' plight.

Massachusetts, a textiles center, introduced factory inspection in 1867 and established what was perhaps the first safety committee when the Bureau of Labor Statistics was formed in 1869 to study the kinds and causes of accidents in factories. In 1877, the first legislation requiring the use of guards for hazardous machinery was passed. These advances are particularly noteworthy, for they also inspired safety inventors and led to the earliest reconstructions of accidents to analyse how accidents happen. However, the contributions of these innovators were not to be recognized for many years.

Labor's other demand, i.e. compensation for injuries or loss of life suffered at work, was much more difficult to achieve. Societal differences were more likely to be settled over the back fence than in a court of law and workers were still viewed as extensions of independent craftsmen or family-run businesses. Although it was agreed that risk of injury should be reduced, it was still the individual worker who was liable. Common law heavily favored the employer and an injured worker's chances of winning compensation was practically nil. This was the case due to three defenses available to lawyers representing the employer:

- assumption of risk;
- contributory negligence; and
- the fellow servant rule.

Assumption of risk was based upon voluntary actions and the consequences of those actions. When a worker accepted a job, he or she voluntarily accepted the risks associated with that employment and, if they later incurred an injury, how could the employer be found at fault? Contributory negligence placed the blame for an accident firmly upon the shoulders of the individual whose carelessness caused the injury. Once again, how could an employer be held responsible for the negligence of an employee?

Assumption of risk and contributory negligence covered most employers for litigation regarding worker injury; however, if they did not, the fellow servant rule could be invoked. In this case, a fellow worker such as a maintenance man could be held responsible, not the employer. The fellow servant rule is still valid today in New Hampshire, where injured workers can sue individual corporate employees if their actions are held to contribute directly to the injury.[2]

[2] 498 Atlantic 2d 741, 127 New Hampshire 162.

Workmen's Compensation Laws

Public concern about work-related injuries in the late nineteenth century led legislators in some jurisdictions to pass what became known as employers' liability acts. Although litigation was still required in these jurisdictions, assumption of risk and the fellow servant rule were virtually eliminated and contributory negligence became comparative negligence. For a jury to award damages, they had to find the employer guilty of negligence; however, the amount of damages could be reduced by an amount equal to the worker's comparative negligence.

Although employers' liability acts were an improvement on common law practices, workers were still faced with the task of finding a lawyer, proof of negligence, finding fellow workers to testify for them, the length of time before damages would be awarded, and the likelihood that damages would be reduced. On the other hand, employers had to hire defense lawyers, who did not work on contingency fees but who were paid on an hourly basis, and had to pay any damages awarded. Although the practice of settling small amounts with injured workers was commonplace, it was viewed as rather reprehensible, both for the injured employee and for the corporate image.

While these problems were being worked on in the U.S.A., a system known as workmen's compensation was developed by Bismarck in Germany. By 1903, most European countries had adopted some form of compensation for injuries arising from employment, regardless of who was deemed to be at fault. This concept was first introduced in the U.S.A. in 1908, but the law was limited to specific classes of federal workers and benefits were minimal. The first state workmen's compensation law which was upheld was passed in 1911 in New Jersey. Other states followed, rapidly at first, but Mississippi did not do so until 1948.

With the development of such laws over such a long period of time and by all 50 states individually, we cannot but be surprised by the variation in legislation from one state to another. Nevertheless, these laws do have the same objective, i.e. to compensate a worker for injury or loss of health associated with his/her employment. In all cases, the question of fault is eliminated with respect to compensation and employers are required to provide for medical expenses and disability benefits as well as compensation for time lost and for specific injuries or health issues. In exchange for these benefits, workers are universally barred from suing their employer.

Workmen's compensation laws have done more to promote safety than all other measures collectively, because employers found it more cost-effective to concentrate on safety than to compensate employees for injury or loss of life. However, many employers took up insurance to cover compensation claims (in no case did an employee ever have to pay toward workmen's compensation coverage). Over the years, insurance companies have been a driving force in establishing and maintaining effective safety programs, supporting research in safety and employing safety specialists.

THE DEVELOPMENT OF SAFETY PROGRAMS

The first recorded safety program began in response to the explosion of a flywheel at the Joliet plant of the Illinois Steel Company in 1892. A committee of plant executives was formed to evaluate the problem, and their first action was to inspect and test all flywheels in the plant. Gradually, other high-injury industries, spearheaded by railroads and steel, took similar action, but it was not until the workmen's compensation laws became general that accident prevention programs received serious attention. As interest in safety grew, a national conference was convened in 1912. Chief among the accomplishments of that conference was the establishment of what is now the National Safety Council (NSC). The NSC promotes safety chiefly through compilation and dissemination of safety information and the promotion of safety consciousness.

As industry became more aware of safety, methods to prevent accidents were developed and engineered, efforts to train and educate workers were undertaken and safety rules were established and enforced. Safety programs were built upon the principles of the "Three E's of Safety" — Engineering, Education and Enforcement. There is no doubt as to the success of these traditional safety programs, but we need to build upon them. One continuing flaw is the tendency to correct hazards *after* an accident has occurred. Other problems, notably the tendency to find fault and place blame on the injured worker, still persist.

Recommendations for changes in safety programs, intended to continue and improve upon the successes to date, fall into two categories – federal legislation, such as the Occupational Safety and Health Act, and improved safety methodology, perhaps best exemplified by systems safety concepts (the subject of Chapters 4 and 5). Legislative efforts provide direction by mandating the reduction of fatalities, injuries and diseases associated with the workplace. However, such mandates require funding for administration and enforcement, a politically dependent process. The law also requires competent and well-informed administration, plus a reliable basis for standards. These are also expensive and their funding is also politically dependent. Thus, the effectiveness of safety legislation is limited, and we must therefore look toward improved safety methodology for future improvements in safety. The rest of this chapter will look at the status of safety legislation and at safety programs which have evolved to date. The rest of the book will examine directions for future safety programs.

THE 1970 OCCUPATIONAL SAFETY AND HEALTH ACT

The most far-reaching safety legislation to date is the Occupational Safety and Health Act (OSHAct), which was passed on December 29, 1970, and became effective in 1971. Whereas earlier laws were in general designed to provide compensation for occupational injuries and illnesses, the focus of the OSHAct was *prevention*. The major goal of the Act was to insure "so far as possible every working man and woman in the nation safe and healthful working conditions and to preserve our human resources." Everyone involved with safety must be familiar with the impact of OSHAct and compliance procedures.

A number of ideas were proposed in the OSHAct to achieve its fundamental aim, and although they are all worthy, the following have proved to be the most important:

- separate but dependent responsibilities and rights of employers and employees with respect to achieving safe and healthful worker conditions;
- mandatory occupational safety and health standards;
- effective enforcement;
- uniform record-keeping and reporting procedures; and
- the establishment of the Occupational Safety and Health Administration (OSHA) and the National Institute of Occupational Safety and Health (NIOSH).

THE OCCUPATIONAL SAFETY AND HEALTH ADMINISTRATION (OSHA)

OSHA Standards

The responsibility of employers to provide safe and healthful workplaces is based primarily on the standards established by OSHA, a responsibility given by the OSHAct. OSHA standards have been categorized in several ways. Perhaps the most common distinction is between "safety" standards which are intended to protect against traumatic injury and "health" standards which deal with toxic substances and long-term health effects. Another distinction is based upon the scope of standards. "Horizontal" standards apply to a wide variety of operations in virtually all industries, whereas "vertical" standards are developed for a specific type of employment, such as construction or telecommunications.

Many OSHA standards are "consensus" standards which have been adopted from nationally recognized organizations, notably the American National Standards Institute (ANSI), the National Fire Protection Association (NFPA) and the American Society of Mechanical Engineers (ASME). Others have been issued through specific rule-making procedures, the major steps of which are as follows:

1. OSHA proceeds on the basis of its own information, petitions from interested parties, and recommendations from other government agencies. Recommendations from NIOSH form an important basis for OSHA health standards.
2. OSHA may establish an advisory committee to make recommendations for the development of a standard. Requirements are laid down for the com-

position of the advisory committee and for the time periods within which it must act.

3. If OSHA decides that a standard should be issued, it must publish a proposed standard and give the public at least 30 days to comment in writing. If objections to the proposal are filed and a public hearing is requested, such a hearing must be held.
4. On the basis of the entire record, OSHA must either promulgate the standard or determine that no standard is needed and must publish a statement outlining its actions.
5. Certain prescribed time-frames for most stages of rule-making must be followed.

There are many types of standards, all of which involve control, a management function described in Chapter 4. For example, company and industry standards exist for specific purposes, which differ from the horizontal and vertical standards of OSHA. We must remember, however, that no standard can be relied upon which is incomplete or obviously self-serving. There is no substitute for thorough understanding of a system. The goal of a good standard should be to coordinate work on the same problems in order to generate routine solutions which can guide those who face similar problems in the future.

There are particular characteristics which apply to good standards. These are as follows:

- it must suggest something which can be attained;
- it should be economically feasible;
- it should be meaningful, and applicable to the situation in which it is to be used;
- it should be understood by its users;
- it should be consistent in its interpretation; and
- it should be both stable and maintainable.

When OSHA first implemented safety standards, many deficiencies were encountered. Many of them were not exclusive to worker safety, were not binding or enforceable, lacked specificity and were even conflicting and inconsistent. For these reasons a Presidential Task Force was established in 1976 to develop a model approach to safety standards. The task force approached its assignment with the following objectives in mind: standards directly related to worker safety; a better interpretation of safety standards, so as to make them binding and enforceable; and the maintenance of safe working conditions when introducing new technology.

Today's OSHA standards are performance standards which state obligations in terms of ultimate goals, allowing employers flexibility to select the means of achieving these goals. They are very comprehensive, and the following are just those which are covered in 29 CFR Part 1910 of the OSHA standards:

Subpart	Subject
A	General
B	Adoption and extension of established federal standards
C	General safety and health provisions
D	Walking-working surfaces
E	Means of egress
F	Powered platforms
G	Health and environmental control
H	Hazardous materials
I	Personal protective equipment
J	General environmental control
K	Medical and first aid
L	Fire protection
M	Compressed gas and equipment
N	Materials handling and storage
O	Machinery and machine guarding
P	Hand and portable power tools
Q	Welding, cutting and brazing
R	Special industries
S	Electrical
T	Commercial diving
Z	Toxic and hazardous substances

OSHA Enforcement

One major provision of the OSHAct was to give meaning to the purpose of insuring "so far as possible every working man and woman in the nation safe and healthful working conditions." The Act enforced OSHA standards by allowing surprise workplace inspections and, if violations were found, citations could be issued and civil penalties proposed. These aspects of enforcement have been upheld by the courts.

There are two ways that OSHA enforcement methods are intended to promote a safe workplace. First, for those workplaces inspected, it is thought that the imposition of penalties or even possible imposition of penalties, promotes conformance at that particular workplace. However, the discrepancy between the number of workplaces covered by OSHA (approximately 5 million) and the number of inpectors or compliance officers (about 2500) means that many workplaces are inspected infrequently or not at all. Quite often, OSHA responds to serious injuries of fatalities, thereby inspecting workplaces *after* an accident has occurred. To counteract the potential non-conformance of unconscientious employers,

The Occupational Safety and Health Administration (OSHA)

OSHA ENFORCEMENT ACTIVITY

	FY80	FY87	Percent Change
Inspections			
Establishment	63,363	59,071	−7
Health	11,871	9,722	−18
Accident	2,286	1,458	−36
Complaint	16,093	9,743	−39
Manufacturing	27,224	17,384	−36
Construction	26,315	35,456	+35
Violations			
Willful	1,238	529	−57
Serious	44,645	36,982	−13
Penalties	$25.5 million	$24.5 million	−4
Employees covered by inspections	3.67 million	2.46 million	−33
Time spent on inspections			
Safety	16 hours	11 hours	−31
Health	41 hours	32 hours	−22

Source: U.S. Senate Committee on Labor and Human Resources, OSHA

Figure 1.1

OSHA enforcement can also include meaningful first-instance penalties or sanctions issued after surprise OSHA inspections.

Available statistics do not prove or disprove the effectiveness of OSHA enforcement. During 1981 Senate oversight hearings, OSHA was criticized for overregulation and onerous enforcement practices; however, in similar 1988 hearings, OSHA was criticized for becoming a "non-entity." These two divergent descriptions originate more from political sources than OSHA enforcement activity (see Fig. 1.1).

Such arguments as these only point to the wisdom of multiple approaches to safety—improved methodology as well as legislated programs. A 1974 amendment to the OSHAct recognized this when it allowed on-site consultations by federal compliance officers. On-site consultation can be undertaken with enforcement or as an alternative to it. Typically, the consultation involves a walk-through inspection, a discussion of violations and a written report containing recommendations for conformance. This program has been supplemented by voluntary programs where compliance officers give general assistance, upon request, to employers in identifying methods for correcting safety and health hazards in the workplace. Today, we find that OSHA is one of the first sources we can turn to for assistance in safety and health matters.

THE RIGHT-TO-KNOW LAWS AND SARA-III

In recent years, a great deal of time and money has been spent complying with hazard communication laws, better known as the Right-to-Know Laws. Although there are many state and local laws, the oldest and best known is the OSHA Hazard Communication Standard enacted in 1983. This Standard requires all companies and employers handling any hazardous substance in any form to assess the hazards associated with the substance in the work area, to inform workers of those hazards and to train them in safe handling procedures. The Environmental Protection Agency (EPA) has also become involved with the Right-to-Know Laws as a result of the Superfund Amendments and Reauthorization Act of 1986 (SARA-III), which requires companies to provide authorities with information concerning toxic chemical releases and other potential chemical hazards to the community. SARA-III also encourages emergency planning efforts in response to emergencies arising from chemical incidents.

All of the Right-to-Know Laws have three primary objectives. First, they require that the dangerous properties of materials used or produced in the workplace are determined; secondly, employees are to be trained in the recognition of and safe handling of these materials; and, finally, the laws force companies to disclose the presence of hazardous substances, thereby making employees and others aware of the possible dangers.

Compliance to the OSHA Hazard Communication Standard or to SARA-III should take precedence whenever contradiction or conflict with local or other various Right-to-Know Laws exist. It is known that the task of compliance is expensive, sometimes confusing and seemingly endless, but nevertheless, steps should be taken to meet the requirements and do it on a cost-effective basis by using the following as building blocks:

- hazard determination — the Material Safety Data Sheet;
- the written program; and
- training.

Material Safety Data Sheets (MSDS)

A MSDS (see Fig. 1.2) is a key to communication and compliance to the Right-to-Know Laws. Each MSDS must identify the particular material or mixture, its physical and chemical characteristics, any health hazards and other identifying criteria. Every company should develop an internal review process for MSDS forms received with orders to insure that it contains the following *minimum* information:

- The chemical and common name of the hazardous substance, and for a mixture the proportion of each chemical and its hazardous ingredients.

The Right-to-Know Laws and SARA-III

UNIVOLT N 61

EXXON COMPANY, U.S.A.
A DIVISION OF EXXON CORPORATION

DATE ISSUED: 10/24/88
SUPERSEDES DATE: 03/02/88

MATERIAL SAFETY DATA SHEET
EXXON COMPANY, U.S.A.   P.O. BOX 2180   HOUSTON, TX 77252-2180

A. IDENTIFICATION AND EMERGENCY INFORMATION

PRODUCT NAME
UNIVOLT N 61

PRODUCT CODE
331831 - 01831

CHEMICAL NAME
Petroleum Electrical Insulating Oil

CAS NUMBER
Complex Mixture
CAS Number not applicable

PRODUCT APPEARANCE AND ODOR
Clear water-white liquid
Mild, bland petroleum odor

EMERGENCY TELEPHONE NUMBER
(713) 656-3424

B. COMPONENTS AND HAZARD INFORMATION

COMPONENTS	CAS NO. OF COMPONENTS	APPROXIMATE CONCENTRATION
Hydrotreated light naphthenic distillate, petroleum	64742-53-6	100%

This product, as manufactured by Exxon, does not contain polychlorinated biphenyls (PCB's) as per ASTM D 4059.

See Section E for Health and Hazard Information.

See Section H for additional Environmental Information.

HAZARDOUS MATERIALS IDENTIFICATION SYSTEM (HMIS)

Health	Flammability	Reactivity	BASIS
1	1	0	Recommended by Exxon

EXPOSURE LIMIT FOR TOTAL PRODUCT   **BASIS**
5 mg/m3 for oil mist in air   OSHA Regulation 29 CFR 1910.1000

5 mg/m3 for oil mist or fumes   Recommended by the American Conference of Governmental
(10 mg/m3 STEL)   Industrial Hygienists (ACGIH)

5 mg/m3 for mist in air   Recommended by Exxon

C. PRIMARY ROUTES OF ENTRY AND EMERGENCY AND FIRST AID PROCEDURES

EYE CONTACT
If splashed into the eyes, flush with clear water for 15 minutes or until irritation subsides. If irritation persists, call a physician.

SKIN
In case of skin contact, remove any contaminated clothing and wash skin thoroughly with soap and water.

945-0277(MWH001)

Figure 1.2 (Reprinted by permission of Exxon Co., U.S.A.)

12   Safety Traditions and Regulations   Chapter 1

UNIVOLT N 61

INHALATION
Vapor pressure is very low. Vapor inhalation under ambient conditions is normally not a problem. If overcome by vapor from hot product, immediately remove from exposure and call a physician. If breathing is irregular or has stopped, start resuscitation; administer oxygen, if available. If overexposed to oil mist, remove from further exposure until excessive oil mist condition subsides.

INGESTION
If ingested, DO NOT induce vomiting; call a physician immediately.

D. FIRE AND EXPLOSION HAZARD INFORMATION

FLASH POINT (MINIMUM)   **AUTOIGNITION TEMPERATURE**
145°C (293°F)   Greater than 204°C (400°F)
ASTM D 92, Cleveland Open Cup   ASTM E 659

NATIONAL FIRE PROTECTION ASSOCIATION (NFPA) - HAZARD IDENTIFICATION
Health   Flammability   Reactivity   BASIS
  1   1   0   Recommended by Exxon

HANDLING PRECAUTIONS
Use product with caution around heat, sparks, pilot lights, static electricity, and open flame.

FLAMMABLE OR EXPLOSIVE LIMITS (APPROXIMATE PERCENT BY VOLUME IN AIR)
Estimated values: Lower Flammable Limit 0.9%   Upper Flammable Limit 7%

EXTINGUISHING MEDIA AND FIRE FIGHTING PROCEDURES
Foam, water spray (fog), dry chemical, carbon dioxide and vaporizing liquid type extinguishing agents may all be suitable for extinguishing fires involving this type of product, depending on size or potential size of fire and circumstances related to the situation. Plan fire protection and response strategy through consultation with local fire protection authorities or appropriate specialists.

The following procedures for this type of product are based on the recommendations in the National Fire Protection Association's "Fire Protection Guide on Hazardous Materials", Eighth Edition (1984):

Use water spray, dry chemical, foam or carbon dioxide. Use water to keep fire-exposed containers cool. If a leak or spill has not ignited, use water spray to disperse the vapors and to provide protection for men attempting to stop a leak. Water spray may be used to flush spills away from exposures. Minimize breathing gases, vapor, fumes or decomposition products. Use supplied-air breathing equipment for enclosed or confined spaces or as otherwise needed.

DECOMPOSITION PRODUCTS UNDER FIRE CONDITIONS
Fumes, smoke, carbon monoxide, sulfur oxides, aldehydes and other decomposition products, in the case of incomplete combustion.

"EMPTY" CONTAINER WARNING
"Empty" containers retain residue (liquid and/or vapor) and can be dangerous. DO NOT PRESSURIZE, CUT, WELD, BRAZE, SOLDER, DRILL, GRIND OR EXPOSE SUCH CONTAINERS TO HEAT, FLAME, SPARKS, STATIC ELECTRICITY, OR OTHER SOURCES OF IGNITION; THEY MAY EXPLODE AND CAUSE INJURY OR DEATH. Do not attempt to clean since residue is difficult to remove. "Empty" drums should be completely drained, properly bunged and promptly returned to a drum reconditioner. All other containers should be disposed of in an environmentally safe manner and in accordance with governmental regulations. For work on tanks refer to Occupational Safety and Health Administration regulations, ANSI Z49.1, and other governmental and industrial references pertaining to cleaning, repairing, welding, or other contemplated operations.

945-0277(MWH002)   PAGE: 2   DATE ISSUED: 10/24/8
                                                                                                      SUPERSEDES DATE: 03/02/8

Figure 1.2 (continued)

The Right-to-Know Laws and SARA-III 13

UNIVOLT N 61

E. HEALTH AND HAZARD INFORMATION

VARIABILITY AMONG INDIVIDUALS
Health studies have shown that many petroleum hydrocarbons and synthetic lubricants pose potential human health risks which may vary from person to person. As a precaution, exposure to liquids, vapors, mists or fumes should be minimized.

EFFECTS OF OVEREXPOSURE (Signs and symptoms of exposure)
Prolonged or repeated skin contact may cause skin irritation.

NATURE OF HAZARD AND TOXICITY INFORMATION
In accordance with the current OSHA Hazard Communication Standard criteria, this product does not require a cancer hazard warning. This is because the product is formulated from base stocks which are severely hydrotreated, severely solvent extracted, and/or processed by mild hydrotreatment and extraction. Alternatively, it may consist of components not otherwise affected by IARC criteria, such as atmospheric distillates or synthetically derived materials, and as such is not characterized by current IARC classification criteria.

Prolonged or repeated skin contact with this product tends to remove skin oils possibly leading to irritation and dermatitis; however, based on human experience and available toxicological data, this product is judged to be neither a "corrosive" nor an "irritant" by OSHA criteria.

Product contacting the eyes may cause eye irritation.

Product has a low order of acute oral and dermal toxicity, but minute amounts aspirated into the lungs during ingestion or vomiting may cause mild to severe pulmonary injury and possibly death.

This product is judged to have an acute oral LD50 (rat) greater than 5 g/kg of body weight, and an acute dermal LD50 (rabbit) greater than 3.16 g/kg of body weight.

PRE-EXISTING MEDICAL CONDITIONS WHICH MAY BE AGGRAVATED BY EXPOSURE
None Recognized

F. PHYSICAL DATA

The following data are approximate or typical values and should not be used for precise design purposes.

BOILING RANGE
IBP Approximately 238°C (460°F)
by ASTM D 2887

SPECIFIC GRAVITY (15.6 C/15.6 C)
0.88

MOLECULAR WEIGHT
Approximately 255

pH
Essentially neutral

POUR, CONGEALING OR MELTING POINT
-45°C (-50°F)
Pour Point by ASTM D 97

VISCOSITY
55 SSU @ 100°F

VAPOR PRESSURE
Less than 0.01 mm Hg @ 20°C

VAPOR DENSITY (AIR = 1)
Greater than 5

PERCENT VOLATILE BY VOLUME
Negligible from open container
in 4 hours @ 38°C (100°F)

EVAPORATION RATE @ 1 ATM. AND 25 C (77 F)
(n-BUTYL ACETATE = 1)
Less than 0.01

SOLUBILITY IN WATER @ 1 ATM. AND 25 C (77 F)
Negligible; less than 0.1%

Figure 1.2 (continued)

UNIVOLT N 61

G. REACTIVITY

This product is stable and will not react violently with water. Hazardous polymerization will not occur. Avoid contact with strong oxidants such as liquid chlorine, concentrated oxygen, sodium hypochlorite or calcium hypochlorite.

H. ENVIRONMENTAL INFORMATION

STEPS TO BE TAKEN IN CASE MATERIAL IS RELEASED OR SPILLED
Recover free product. Add sand, earth, or other suitable absorbent to spill area. Minimize skin contact. Keep product out of sewers and watercourses by diking or impounding. Advise authorities if product has entered or may enter sewers, watercourses, or extensive land areas. Assure conformity with applicable governmental regulations.

THE FOLLOWING INFORMATION MAY BE USEFUL IN COMPLYING WITH VARIOUS STATE AND FEDERAL REGULATIONS UNDER VARIOUS ENVIRONMENTAL STATUES:

REPORTABLE QUANTITY (RQ), EPA REGULATION 40 CFR 302 (CERCLA Section 102)
Not applicable

THRESHOLD PLANNING QUANTITY (TPQ), EPA REGULATION 40 CFR 355 (SARA Sections 301-304)
Not applicable

TOXIC CHEMICAL RELEASE REPORTING, EPA REGULATION 40 CFR 372 (SARA Sections 311-313)
Not applicable

EPA HAZARD CLASSIFICATION CODE:	Acute Hazard	Chronic Hazard	Fire Hazard	Pressure Hazard	Reactive Hazard	Not Applicable XXX

I. PROTECTION AND PRECAUTIONS

VENTILATION
Use local exhaust to capture vapor, mists or fumes, if necessary. Provide ventilation sufficient to prevent exceeding recommended exposure limit or buildup of explosive concentrations of vapor in air. No smoking or open lights.

RESPIRATORY PROTECTION
Use supplied-air respiratory protection in confined or enclosed spaces, if needed.

PROTECTIVE GLOVES
Use chemical-resistant gloves, if needed, to avoid prolonged or repeated skin contact.

EYE PROTECTION
Use splash goggles or face shield when eye contact may occur.

OTHER PROTECTIVE EQUIPMENT
Use chemical-resistant apron or other impervious clothing, if needed, to avoid contaminating regular clothing which could result in prolonged or repeated skin contact.

WORK PRACTICES / ENGINEERING CONTROLS
Keep containers and storage containers closed when not in use. Do not store near heat, sparks, flame or strong oxidants.

In order to prevent fire or explosion hazards, use appropriate equipment.

Information on electrical equipment appropriate for use with this product may be found in the latest edition of the National Electrical Code (NFPA-70). This document is available from the National Fire Protection Association, Batterymarch Park, Quincy, Massachusetts 02269.

PERSONAL HYGIENE
Minimize breathing vapor, mist or fumes. Avoid prolonged or repeated contact with skin. Remove contaminated clothing; launder or dry-clean before reuse. Remove contaminated shoes and thoroughly clean before reuse; discard if oil-soaked. Cleanse skin thoroughly after contact, before breaks and meals, and at end of work period. Product is readily removed from skin by waterless hand cleaners followed by washing thoroughly with soap and water.

Figure 1.2 (continued)

The Right-to-Know Laws and SARA-III

UNIVOLT N 61

J. TRANSPORTATION AND OSHA RELATED LABEL INFORMATION

TRANSPORTATION INCIDENT INFORMATION
For further information relative to spills resulting from transportation incidents, refer to latest Department of Transportation Emergency Response Guidebook for Hazardous Materials Incidents, DOT P 5800.3.

DOT IDENTIFICATION NUMBER
Not applicable

OSHA REQUIRED LABEL INFORMATION
In compliance with hazard and right-to-know requirements, the following OSHA Hazard Warnings should be found on a label, bill of lading or invoice accompanying this shipment.

(OSHA Hazard Warnings not applicable for this product; therefore, no OSHA warnings would appear on the label.)

Note: Product label will contain additional non-OSHA related information.

The information and recommendations contained herein are, to the best of Exxon's knowledge and belief, accurate and reliable as of the date issued. Exxon does not warrant or guarantee their accuracy or reliability, and Exxon shall not be liable for any loss or damage arising out of the use thereof.

The information and recommendations are offered for the user's consideration and examination, and it is the user's responsibility to satisfy itself that they are suitable and complete for its particular use. If buyer repackages this product, legal council should be consulted to insure proper health, safety and other necessary information is included on the container.

The Environmental Information included under Section H hereof as well as the Hazardous Materials Identification System (HMIS) and National Fire Protection Association (NFPA) ratings have been included by Exxon Company, U.S.A. in order to provide additional health and hazard classification information. The ratings recommended are based upon the criteria supplied by the developers of these rating systems, together with Exxon's interpretation of the available data.

FOR ADDITIONAL INFORMATION ON HEALTH
EFFECTS CONTACT:
DIRECTOR OF INDUSTRIAL HYGIENE
EXXON COMPANY, U.S.A.
P. O. BOX 2180 ROOM 3157
HOUSTON, TX 77252-2180
(713) 656-2443

FOR OTHER PRODUCT INFORMATION CONTACT:

MANAGER, MARKETING TECHNICAL SERVICES
EXXON COMPANY, U.S.A.
P. O. BOX 2180 ROOM 2355
HOUSTON, TX 77252-2180
(713) 656-5949

945-0277(MWH002)   PAGE: 5   DATE ISSUED: 10/24/88   SUPERSEDES DATE: 03/02/88

Figure 1.2 (continued)

- The physical and chemical characteristics of the hazardous substance, such as vapor pressure and flash point.
- The hazards posed by the substance, including potential for fire, explosion and reactivity.
- Health hazards, including symptoms of exposure and medical conditions aggravated by exposure.
- Precautions for safe handling and use, including procedures for the cleaning up of spillages and leaks.

Written Programs

All employers using hazardous substances, regardless of size, must comply with the written program requirements of OSHA's Hazard Communication Standard. This entails developing, implementing and maintaining a written hazard communication program. A standard form describing the program must be clearly posted and the program itself must describe the presence and location of hazardous substances in the workplace as well as identify the location and availability of the written program and where to find and how to use the MSDS, labeling procedures and all other matters related to hazardous substances.

Training

Training is required by OSHA's Hazard Communication Standard, other OSHA standards and generally by other Right-to-Know Laws. It is left to the discretion of individual companies to decide what training program to introduce. The net result of this policy has been to yield excellent, well thought out, expensive training and retraining programs, which have been developed internally and therefore are well tailored to individual companies. However, in some instances, it has led to poorly prepared and poorly understood training programs which hardly qualify as training at all. Such companies are able to benefit from outside consultation where at least a standardized but general training format is available.

In general, the minimum training program should alert employees to the hazardous materials that they may come into contact with, emphasize the written program, thoroughly explain the information given on an MSDS, and then describe the procedures for safe handling. Training should be approached in a general fashion on the first day of employment, and then for specific assignments at the time of assignment and annually thereafter. Training programs should include lectures, discussion groups and prepared presentations, each of which can be supplemented by the use of slides or videos, produced in-house or purchased from outside. The choices are numerous, but the goal is the same — the safe handling of hazardous substances by employees.

The Right-to-Know Laws and SARA-III

Case History 1.1   More Than Conformance Is Needed

When we hear about high-technology industries, we think of computers and electronics. However, "high tech" would not exist if it were not for the miniaturization made possible by the development of semiconductor devices, most importantly the silicon chip. High-purity silicon is extremely important for performance and one way we can prepare it is by epitaxial deposition of a film from a vapor chemical reaction of hydrogen and trichlorosilane ($SiHCl_3$). An intermediate product used in this purification reaction is tetrachlorosilane, TET ($SiCl_4$). The MSDS for TET, given in Fig. 1.3, clearly describes its reactivity with water and the hazards of hydrochloric acid which is formed by the reaction.

Silectronics, Inc., a small electronics manufacturing company, is one of many which has grown up in southern New Hampshire, not far from Lowell. They use about 100 lb of TET annually in the production of epitaxial silicon for their products, purchasing 55-lb stainless steel containers. The company has established a training program in conformance with the Right-to-Know Laws and had the MSDS for TET on file.

However, the safety office was not informed that a vent tube handle was missing from a new container that was to be connected. Arrows on the handle indicate the direction of flow, and without it the valve is able to rotate continuously and the open and closed positions cannot be distinguished. Nevertheless, there was no replacement TET, and so the container was installed and used successfully.

Normally, containers remain in place until the TET is completely consumed. However, after 3 months' use, the system was shut down to relocate it. In doing so, the system is purged with nitrogen (see Fig. 1.4) and the fittings are removed. However, because the vent tube handle was missing, the vent tube valve was thought to be closed when it was actually open. When the technician first cracked the fittings, he heard a hissing sound, and then liquid TET spilled out of the container under pressure, causing him minor burns, but extensive corrosive damage to Silectronics' property.

Silectronics' safety department became involved because of their record-keeping responsibilities. They immediately arranged a meeting with the production personnel to determine why they were not consulted when the faulty container was first used. The outcome of this meeting was to agree upon the needs for better communication, better training (beyond conformance) and better hazard identification. ∎

OSHA Record-Keeping Requirements

The keeping of records is as important for safety as it is for any other business undertaking. The OSHAct mandates, however, that all companies with more than 10 employees prepare and maintain records of occupational injuries and illnesses. This responsibility has, of course, been expanded by OSHA's Hazard Communication Standard and by SARA-III. Required record-keeping in itself is not com-

SECTION I

MANUFACTURER'S NAME	EMERGENCY TELEPHONE NO.
ADDRESS (Number, Street, City, State, and ZIP Code)	
CHEMICAL NAME AND SYNONYMS	TRADE NAME AND SYNONYMS Silicon Tetrachloride
CHEMICAL FAMILY Chlorosilane	FORMULA $SiCl_4$

SECTION II - HAZARDOUS INGREDIENTS

PAINTS, PRESERVATIVES, & SOLVENTS	%	TLV (Units)	ALLOYS AND METALLIC COATINGS	%	TLV (Units)
PIGMENTS			BASE METAL		
CATALYST			ALLOYS		
VEHICLE			METALLIC COATINGS		
SOLVENTS			FILLER METAL PLUS COATING OR CORE FLUX		
ADDITIVES			OTHERS		
OTHERS					

HAZARDOUS MIXTURES OF OTHER LIQUIDS, SOLIDS, OR GASES	%	TLV (Units)
Tetrachlorosilane	100	5 ppm*

*Ceiling value based on hydrochloric acid.

SECTION III - PHYSICAL DATA

BOILING POINT (°F.)	135 °F	SPECIFIC GRAVITY (H_2O = 1)	1.48
VAPOR PRESSURE (mm Hg.)	200	PERCENT, VOLATILE BY VOLUME (*.)	Greater than 75%
VAPOR DENSITY (AIR = 1)	>1	EVAPORATION RATE (ETHER = 1)	<1
SOLUBILITY IN WATER	Reacts		
APPEARANCE AND ODOR	Colorless, clear liquid with sharp, acid odor.		

SECTION IV - FIRE AND EXPLOSION HAZARD DATA

FLASH POINT None	FLAMMABLE LIMITS None	Lel	Uel

EXTINGUISHING MEDIA
Should not use water.

SPECIAL FIRE FIGHTING PROCEDURES
None

UNUSUAL FIRE AND EXPLOSION HAZARDS
None

Form OSHA-20
Rev. May 72

Figure 1.3

The Right-to-Know Laws and SARA-III

THRESHOLD LIMIT VALUE
5 ppm in an 8 hour day.

EFFECTS OF OVEREXPOSURE
Capable of causing skin and eye burns. Vapor irritates mucous membranes of respiratory tract and delayed burns to the eyes.

EMERGENCY AND FIRST AID PROCEDURES
Remove contaminated clothing and subject to drenching shower with generous amounts of water.
Irrigate eyes with generous quantities of running water. Seek Medical attention for burns resulting from hydrochloric acid.

SECTION VI - REACTIVITY DATA

STABILITY	UNSTABLE		CONDITIONS TO AVOID
	STABLE		

INCOMPATIBILITY (Materials to avoid)
Water and water vapor.

HAZARDOUS DECOMPOSITION PRODUCTS
Hydrochloric acid.

HAZARDOUS POLYMERIZATION	MAY OCCUR	X	CONDITIONS TO AVOID
	WILL NOT OCCUR	X	Water and water vapor in air.

SECTION VII - SPILL OR LEAK PROCEDURES

STEPS TO BE TAKEN IN CASE MATERIAL IS RELEASED OR SPILLED
Minimize quantity spilled. Control small spills by blanketing with inert material.

WASTE DISPOSAL METHOD
Per applicable EPA regulations.

SECTION VIII - SPECIAL PROTECTION INFORMATION

RESPIRATORY PROTECTION (Specify type)
Self-contained breathing appratus.

VENTILATION	LOCAL EXHAUST	SPECIAL
	Not recommended	Silicon halide scrubber.
	MECHANICAL (General)	OTHER
	Exhaust	

PROTECTIVE GLOVES	EYE PROTECTION
Rubber gloves.	Safety face shield or goggles.

OTHER PROTECTIVE EQUIPMENT
Plastic apron and sleeves.

SECTION IX - SPECIAL PRECAUTIONS

PRECAUTIONS TO BE TAKEN IN HANDLING AND STORING
Store in mild steel containers in complete absence of water. Should water enter the container, hydrochloric acid will attack the container.

OTHER PRECAUTIONS

Figure 1.3 (continued)

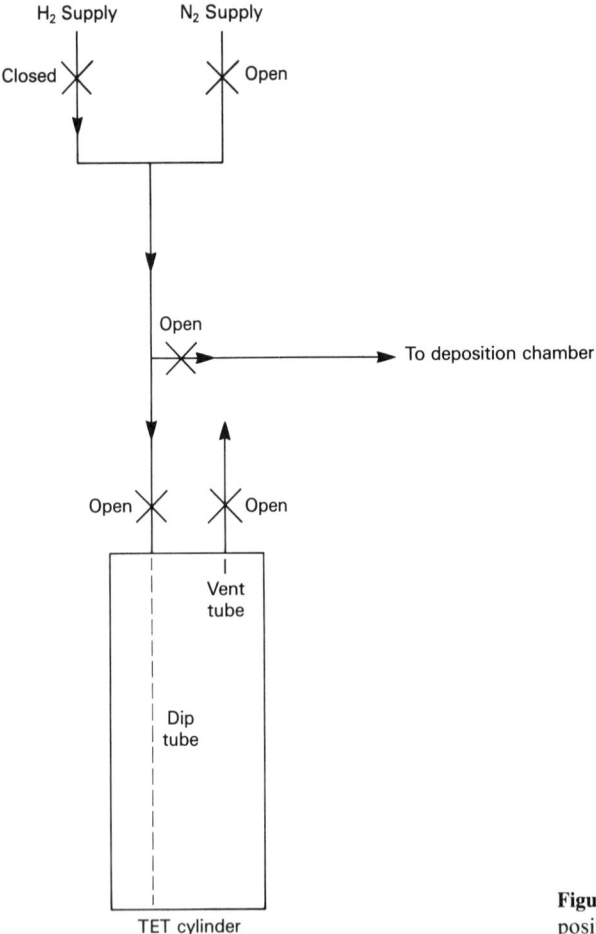

Figure 1.4 Schematic of valve positions during purge cycle.

plicated and Fig. 1.5 illustrates how easy it is to decide whether or not a record should be kept. When a case is to be recorded, only two forms need to be completed and retained, and made available for examination by an OSHA inspector. The log and summary, better known as OSHA Form 200, classifies injury and illness and records the extent and outcome of each (see Fig. 1.6). A supplementary record which delves into the nature of the cause of the accident which resulted in injury or illness is also necessary. An example of OSHA Form 101 is shown in Fig. 1.7; however, workmen's compensation or insurance forms, which provide the same information, can be used instead.

Records can provide much more than the information required by OSHA—they are able to provide the database for decision making for safety programs and the easy retrieval of hazard communication data, all made possible by computers. Several large companies began using computers for data management and anal-

The Right-to-Know Laws and SARA-III

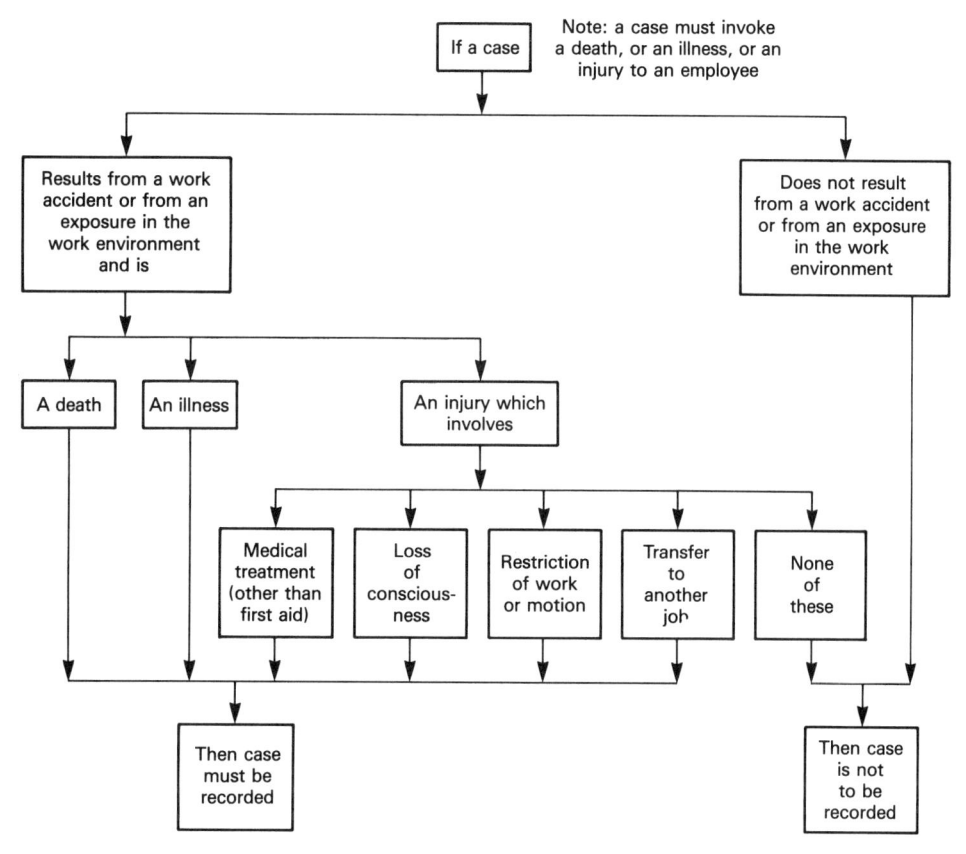

Figure 1.5 Guide to recordability of cases under the Occupational Safety and Health Act.

ysis, tasks that were virtually impossible using manual methods in companies of their size. More and more smaller companies have automated their occupational safety and health information in the 1980s, using both in-house programs and commercial packages. Many of these programs not only serve to keep records, but they can analyze information statistically and provide guidance for the direction of safety programs. However, caveats are in order because most safety workers are not well versed in computer technology and many have wasted time and money purchasing programs which are incompatible with their systems.

Progress in record-keeping computer applications (and safety in general) have reached the point where integrated information systems mean that workers and the environment are better protected. Justification for such systems is based upon the millions of different material safety data sheets in circulation and the obvious overlap between hazardous chemical data collection, retention and the reporting that is necessary to both OSHA and EPA because of the Right-to-Know and SARA-III laws (see Fig. 1.8).

Safety Traditions and Regulations — Chapter 1

Bureau of Labor Statistics
Log and Summary of Occupational Injuries and Illnesses

NOTE: This form is required by Public Law 91-596 and must be kept in the establishment for *5 years*. Failure to maintain and post can result in the issuance of citations and assessment of penalties. (See posting requirements on the other side of form.)

RECORDABLE CASES: You are required to record information about every occupational *death*; every nonfatal occupational *illness*; and those nonfatal occupational *injuries* which involve one or more of the following: loss of consciousness, restriction of work or motion, transfer to another job, or medical treatment (other than first aid). (See definitions on the other side of form.)

Case or File Number (A)	Date of Injury or Onset of Illness (B)	Employee's Name (C)	Occupation (D)	Department (E)	Description of Injury or Illness (F)
Enter a nonduplicating number which will facilitate comparisons with supplementary records.	Enter Mo./day.	Enter first name or initial, middle initial, last name.	Enter regular job title, not activity employee was performing when injured or at onset of illness. In the absence of a formal title, enter a brief description of the employee's duties.	Enter department in which the employee is regularly employed or a description of normal workplace to which employee is assigned, even though temporarily working in another department at the time of injury or illness.	Enter a brief description of the injury or illness and indicate the part or parts of body affected. Typical entries for this column might be Amputation of 1st joint right forefinger; Strain of lower back; Contact dermatitis on both hands; Electrocution--body.

PREVIOUS PAGE TOTALS ➤

TOTALS (Instructions on other side of form.) ➤

OSHA No. 200

Figure 1.6

The Right-to-Know Laws and SARA-III

Figure 1.6 (continued)

```
OSHA No. 101                                          Form approved
Case or File No. _____                          OMB No. 44R 1453
              Supplementary Record of Occupational Injuries and Illnesses
EMPLOYER
   1. Name _____
   2. Mail address _____
                 (No. and street)        (City or town)       (State)
   3. Location, if different from mail address _____
INJURED OR ILL EMPLOYEE
   4. Name _____ Social Security No. _____
          (First name)   (Middle name)   (Last name)
   5. Home address _____
                 (No. and street)        (City or town)       (State)
   6. Age _____     7. Sex: Male_____ Female_____ (Check one)
   8. Occupation _____
        (Enter regular job title, not the specific activity he was performing at time of injury.)
   9. Department _____
        (Enter name of department or division in which the injured person is regularly employed, even
        though he may have been temporarily working in another department at the time of injury.)
THE ACCIDENT OR EXPOSURE TO OCCUPATIONAL ILLNESS
  10. Place of accident or exposure _____
                 (No. and street)        (City or town)       (State)
        If accident or exposure occurred on employer's premises, give address of plant or establishment in which
        it occurred. Do not indicate department or division within the plant or establishment. If accident oc-
        curred outside employer's premises at an identifiable address, give that address. If it occurred on a pub-
        lic highway or at any other place which cannot be identified by number and street, please provide place
        references locating the place of injury as accurately as possible.
  11. Was place of accident or exposure on employer's premises? _____ (Yes or No)
  12. What was the employee doing when injured? _____
                        (Be specific. If he was using tools or equipment or handling material,
        _____
        name them and tell what he was doing with them.)
  13. How did the accident occur? _____
                 (Describe fully the events which resulted in the injury or occupational illness. Tell what
        _____
        happened and how it happened. Name any objects or substances involved and tell how they were involved. Give
        _____
        full details on all factors which led or contributed to the accident. Use separate sheet for additional space.)
OCCUPATIONAL INJURY OR OCCUPATIONAL ILLNESS
  14. Describe the injury or illness in detail and indicate the part of body affected. _____
                                                 (e.g.: amputation of right index finger
        _____
        at second joint; fracture of ribs; lead poisoning; dermatitis of left hand, etc.)
  15. Name the object or substance which directly injured the employee. (For example, the machine or thing
        he struck against or which struck him; the vapor or poison he inhaled or swallowed; the chemical or ra-
        diation which irritated his skin; or in cases of strains, hernias, etc., the thing he was lifting, pulling, etc.)
        _____
        _____
  16. Date of injury or initial diagnosis of occupational illness _____
                                                 (Date)
  17. Did employee die? _____ (Yes or No)
OTHER
  18. Name and address of physician _____
  19. If hospitalized, name and address of hospital _____
        _____
        Date of report _____ Prepared by _____
        Official position _____
```

Figure 1.7

Chapters 4 and 5 will deal more fully with systems safety, but let us look here at one practical application of occupation safety, health and environmental information systems. The first and basic step is to organize users and systems representatives into a study group to define needs and objectives, data gathering and dissemination techniques, and establish a logical requirements definition. This

The Right-to-Know Laws and SARA-III

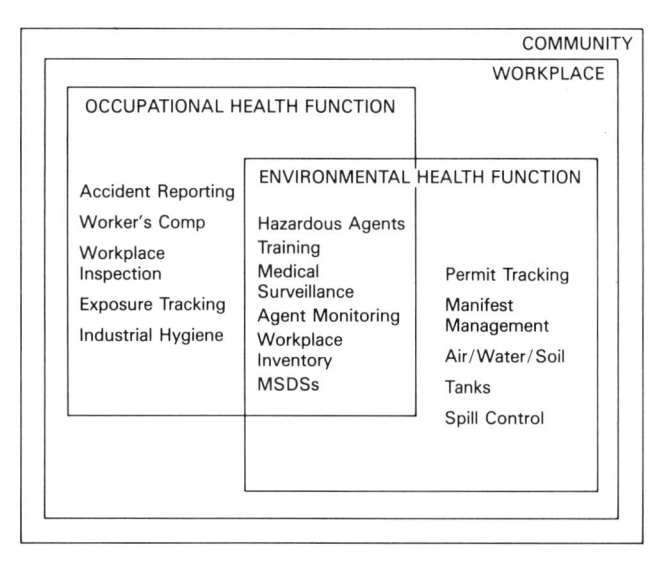

Figure 1.8 The domain of occupational health and environmental information. (Reprinted from W. Rapperport and M. J. Alken, *Occupational Health and Safety*, 57 (1), 52, 1988.)

Logical Definition Phase
— Identify the functions of each involved interest group
— Identify interrelationships and overlaps among all interest groups
— Formulate system objectives
— Formulate required system functions
— Define outputs (existing reports/requirements for new reports)
— Define specific data item needs

Physical Definition Phase
— Identify underlying assumptions and constraints:
  — scope of the system: volume of data to be managed
  — required response time: online and batch reporting
  — sensitivity of data
  — required levels of hardware, software and data security
  — data validation methods for required accuracy/reliability
  — acceptable/affordable/manageable hardware environments
  — database management characteristics to support requirements
  — communications requirements among functions/facilities
  — personnel requirements for data entry, system development and system support
  — training requirements
— Identify alternatives:
  — Maintain a manual system
  — Automate:
    — Build in-house
    — Buy a commercial package
— Evaluate the risks/benefits of each alternative and potential for success or failure
— Select the course of action that best meets logical and physical requirements and provides highest probability for success and acceptance.

Figure 1.9 The requirements definition procedure. (Reprinted from Rapperport and Alken.)

procedure is detailed and at times complicated (an outline appears in Fig. 1.9). Not all attempts to integrate all these functions have been successful, but safety professionals have identified flexibility as the key to success.

FURTHER READING

Best's Safety Directory, A. M. Best, published annually.

R. H. ELLING, *The Struggle for Workers' Health: A Study of Six Industrialized Countries,* Baywood Publishing, Farmingdale, N.Y., 1986.

R. J. FIRENZE, *The Process of Hazard Control,* Kendall/Hunt Publishing, 1978.

F. E. MCELROY, ed., *Accident Prevention Manual for Industrial Operations: Administration and Programs,* 8th Edition, National Safety Council, Chicago, Ill., 1981.

B. W. MINTZ, *OSHA, History, Law and Policy,* Bureau of National Affairs, 1984.

W. RAPPERPORT and M. J. ALKEN, "Comprehensive 'OH & EIS' Systems Help Firms Combine Related Data," *Occupation Health and Safety,* p. 52, June 1988.

S. H. SNOOK and B. S. WEBSTER, "The Cost of Disability," *Clinical Orthopedics,* 1987.

QUESTIONS

1.1. Explain how the origin of safety programs is directly tied to the Industrial Revolution.

1.2. Explain the OSHA recordkeeping requirements and the importance of carefully filling out each section of OSHA Form No. 101 (Figure 1.7).

1.3. Why has the effectiveness of OSHA been debated?

1.4. Conformance to safety standards is not only the beginning of a safety program but also a very important beginning. Explain how you would determine the safety standards to which your industry should conform.

2

Accident Causation

"Safety Is No Accident" is a well-publicized slogan famous, perhaps, because of its multiple interpretation. It does, however, illustrate that both terms—safety and accident—are poorly defined. In fact, in this book we talk of industrial safety programs, industrial accident prevention programs and hazard prevention programs interchangeably. The purpose of this chapter, however, is to understand some of the underlying causes of accidents in order to prevent future accidents from happening and not to argue word definitions. The definition of accident used throughout this book is:

> *Accident:* Any unplanned and uncontrolled event caused by human, situational or environmental factors, or any combination of these factors which interrupts the work process, which may or may not result in injury, illness, death, property damage or other undesired events, but which has the potential to do so.

This definition is preferred because it encompasses the components of all accident causation models described in this book. On many occasions, the term incident is used almost interchangeably with accident, but such interchangeability is technically incorrect. An "incident" is a definite and separate act, occurrence or event, a definition which eliminates the unplanned, uncontrolled and potential consequences of an "accident."

In industrial safety, we accept the fundamental premise that accidents are preventable; in fact, many of our programs are termed accident prevention programs as the result of this tenet. In order to prevent accidents, though, we have to understand the sequential occurrence of accidents and the underlying causes of accidents. Many of us also refer to our safety programs as hazard control programs; this is not contradictory at all. A hazard, defined on p. 119, can result in an accident. We recognize that potential as well as existing hazardous conditions must be considered and that external variable factors influence these conditions.

HEINRICH'S DOMINO THEORY

In the early days of safety programs, control was emphasized—control which primarily involved the safeguarding of hazardous machinery. Herbert W. Heinrich's *The Origin of Accidents* (1928) is a milestone in our understanding of how accidents are caused. Working for the Travelers Insurance Company, Heinrich and his colleagues analyzed 75,000 industrial accidents, and discovered that 88% of them could be attributed to the unsafe acts of workers, 10% to unsafe conditions and that 2% were unavoidable. Heinrich utilized these data, combined with his "Axioms of Industrial Safety" (see Table 2.1), to formulate the first theory of accident causation. It has become known as the Domino Theory because of the analogy to a row of dominoes placed on end—toppling the first precipitates toppling all others.

TABLE 2.1 Axioms of Industrial Safety

1. The occurrence of an injury invariably results from a completed sequence of factors—one factor being the accident itself.
2. An accident can occur only when preceded by or accompanied and directly caused by one or both of two circumstances—the unsafe act of a person and the existence of a mechanical or physical hazard.
3. The unsafe acts of persons are responsible for the majority of accidents.
4. The unsafe act of a person does not invariably result immediately in an accident and an injury, nor does the single exposure of a person to a mechanical or physical hazard always result in accident and injury.
5. The motives or reasons that permit the occurrence of unsafe acts of persons provide a guide to the selection of appropriate corrective measures.
6. The severity of an injury is largely fortuitous—the occurrence of the accident that results in the injury is largely preventable.
7. The methods of most value in accident prevention are analogous with the methods required for the control of the quality, cost, and quantity of production.
8. Management has the best opportunity and ability to prevent accident occurrence, and therefore should assume the responsibility.
9. The foreman is the key man in industrial accident prevention.
10. The direct costs of injury, as commonly measured by compensation and liability claims and by medical and hospital expense, are accompanied by incidental or indirect costs, which the employer must pay.

Source: H.W. Heinrich, D. Petersen and N. Roos, *Industrial Accident Prevention*, 5th ed., 1980. Reprinted by permission of McGraw-Hill Book Company.

Heinrich's Domino Theory

TABLE 2.2 Accident Factors in Heinrich's Domino Theory of Accident Causation

Accident Factors	Explanation of Factors
1. Ancestry and social environment	Recklessness, stubbornness, avariciousness, and other undesirable traits of character may be passed along through inheritance.
	Environment may develop undesirable traits of character or may interfere with education.
	Both inheritance and environment cause faults of person.
2. Fault of person	Inherited or acquired faults of person; such as recklessness, violent temper, nervousness, excitability, inconsiderateness, ignorance of safe practice, etc., constitute proximate reasons for committing unsafe acts or for the existence of mechanical or physical hazards.
3. Unsafe act and/or mechanical or physical hazard	Unsafe performance of persons, such as standing under suspended loads, starting machinery without warning, horseplay, and removal of safeguards; and mechanical or physical hazards, such as unguarded gears, unguarded point of operation, absence of rail guards, and insufficient light, result directly in accidents.
4. Accident	Events such as falls of persons, striking of persons by flying objects, etc., are typical accidents that cause injury.
5. Injury	Fractures, lacerations, etc., are injuries that result directly from accidents.

Source: Heinrich, Petersen, and Roos.

In the Domino Theory, Heinrich considered five sequential accident factors which result in injury (these factors, as described by Heinrich, appear in Table 2.2). The occurrence of a preventable injury is the culmination of a series of events or circumstances. The key to accident prevention is to remove the central factor, i.e. the unsafe act or hazard, which underlies 98% of all accidents. This simple, but workable concept is presented in Fig. 2.1.

Heinrich's theory has become obsolete, primarily because of increased knowledge and sophistication about people and management (see Chapters 3 and 4). Nevertheless, several authors have incorporated the Domino Theory into their own sequences. For example, Bird and Adams have substituted management (its structure and control) for the social environment and ethnic background. Bird has also substituted basic origins and symptoms for personal fault and unsafe act; Adams, on the other hand, has labeled these dominoes operational (management) errors and tactical errors. The only change either of these authors has made to Heinrich's last dominoes is to include property damage as well as injury.

Weaver's updated theory has grouped the last three of Heinrich's dominoes into symptoms of operational error where emphasis is placed on WHY the unsafe act and/or condition occurred and WHETHER management had the knowledge to prevent the accident. Zabetakis has introduced a new concept which follows unsafe act or unsafe condition, i.e. unplanned energy release and/or release of

(a) The five factors in the accident sequence.

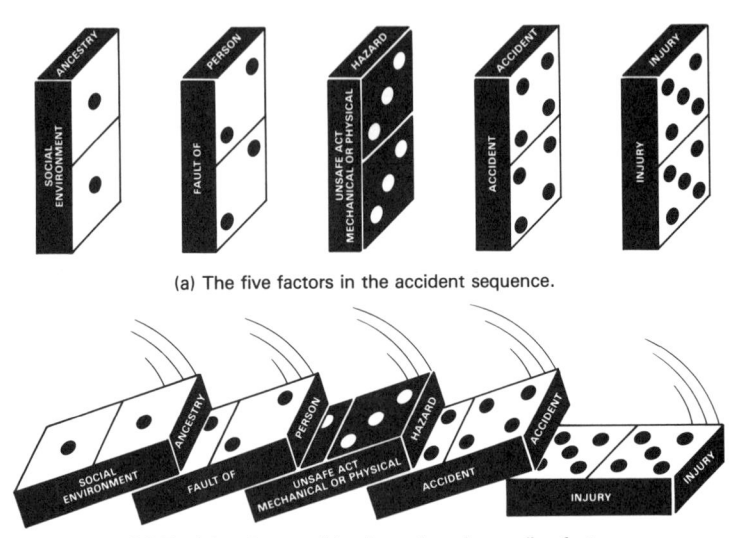

(b) The injury is caused by the action of preceding factors.

(c) The unsafe act and mechanical hazard constitute the central factor in the accident sequence.

(d) The removal of the central factor makes the action of preceding factors ineffective.

Figure 2.1 Heinrich's domino theory of accident causation. (From H. W. Heinrich, D. Petersen and N. Roos, *Industrial Accident Prevention,* 5th ed., 1980. Reprinted by permission of McGraw-Hill Book Company.)

hazardous material. In other words, most accidents are directly caused by the unplanned release of excessive amounts of energy (mechanical, electrical, chemical, etc.) or hazardous materials (chlorine, methane, etc.). In all but a few cases, however, these releases are in turn caused by unsafe acts and unsafe conditions.

Most safety programs which are built upon the principles of control have their roots in the original Domino Theory and its management-oriented updated forms. There exist many other models, though, which are based on human behavior (particularly human error), epidemiology (which we are more familiar with as occupational health workers), and systems concepts in safety (which recognize the interrelations of workers, their machinery and their work environment).

HUMAN ERROR MODEL

If we consider the early work of Heinrich, where 88% of all accidents were attributed to unsafe acts, it is logical that human errors which are the basis of unsafe acts are the foundation of an accident causation model. The Ferrell Human Factors Model is one such theory. Ferrell considers that accidents are the result of a causal chain of initiating incidents and that human error underlies all initiating incidents. Chapter 3 looks at human error in detail, so here it is sufficient to consider the three situations for human error proposed by Ferrell:

- overload—the mismatch between the load and the capacity of the person at the time of action;
- incorrect response by the person to a situation; and
- improper activity.

These causes of human error are outlined in more detail in Fig. 2.2.

OVERLOAD (a mismatch of capacity and load in a state)		
LOAD	CAPACITY	STATE
Task	Natural Endowment	Motivational Level
• Physical	Physical Condition	Arousal Level
• Information processing	State of Mind	
Environmental	Training	
• Light	Drugs – Pollutants	
• Noise	Pressure	
• Distraction	Fatigue	
• Stressors that require active coping	Stressors That Impair Ability to Respond	
Internal		
• Worry		
• Emotional stress		
Situational		
• Ambiguity of goals or criteria		
• Danger		
INCOMPATIBILITY		
Stimulus-Response	Response-Response	
• Control-display	• Inconsistent control types or locations	
Stimulus-Stimulus		
• Inconsistent display types	WORK STATION	
	• Size	
	• Force	
	• Reach	
	• Feel	
IMPROPER ACTIVITIES		
Didn't Know		
Deliberately Took Risk		
• Low perceived probability of accident		
• Low perceived cost of accident		

Figure 2.2 Human error underlying initiation of accidents. (From Heinrich, Petersen and Roos.)

PETERSEN'S ACCIDENT/INCIDENT MODEL

Dan Petersen has adapted the Ferrell model to include systems failure as well as human error, thus accounting for 98% of all accidents in the early Heinrich study. Petersen has termed his model the Accident/Incident Causation Model (Fig. 2.3). This model suggests three broad categories of human error: overload (similar to the Ferrell model), ergonomic traps (which we will learn more about in Chapter 3) and the decision to err.

A major contribution of Petersen's theory is the decision to err, a concept which suggests that workers often choose, consciously or unconsciously, to perform a task unsafely. It is simply more logical in a given situation to perform the job unsafely than it is to perform it safely. The major reasons behind the decision to err are the priority system we live under formed by the social, political and

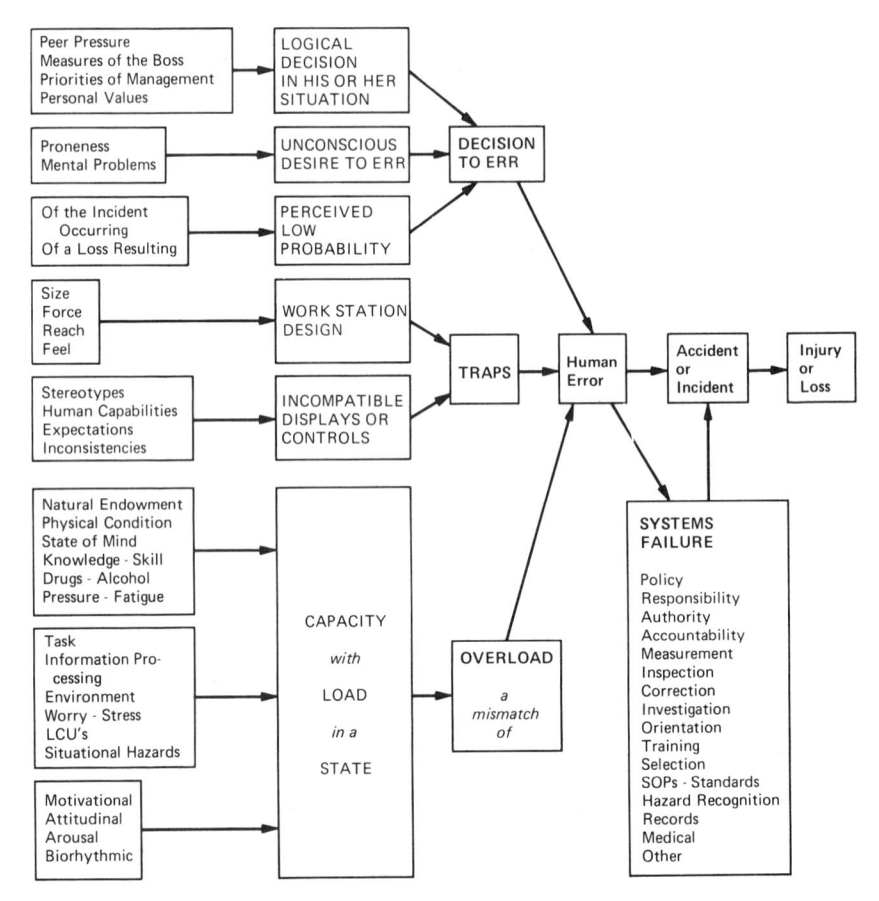

Figure 2.3 The Petersen accident-incident causation model. (From Heinrich, Petersen and Roos.)

economic forces in that system, the pressure for meeting production deadlines, and peer influences. The decision to err also notes that workers never think that they are going to be involved in an accident, regardless of the statistical probabilities.

EPIDEMIOLOGICAL MODELS

Epidemiology is the study of the causal relationships between disease and specific environmental factors. We will explore the subject of epidemiology in Chapter 11, but in the context of this chapter, we have to think of accidents and their consequences as the subject of an epidemic. An example of a simple epidemiological study would be the statistical analysis of accident and injury data used by insurance companies to establish workmen's compensation insurance rates.

An epidemiological model for accident causation has been proposed by Suchman and developed by Surry. In this model, predisposition characteristics which include worker susceptibility, perception, and environmental factors, combined with situational characteristics such as risk assessment through cognitive processes, can lead to or avoid accident conditions. Here the accident is "the unexpected, unavoidable, unintentional act resulting from the interaction of the host, agent (usually machine) and environmental factors."

SYSTEMS MODEL

Systems concepts are based upon the separation of a system into its components, and then examining those components and their interrelationships in detail. Systems models for accident causation treat an accident situation as a system, comprised of three interacting components: man (host), machine (agency) and environment. Changes in any of these components or their interrelationships can alter the probability of an accident occurring. When we think of systems approaches, we must think in terms of a team, because no one person will have the expertise to examine all of the details of the system and the interactions of the components. Thus we might need a safety officer, an industrial hygienist, a design engineer, a manufacturing engineer, a psychologist, a marketing manager and others on the team at different times.

The best-known systems model is that first described by Firenze, which he refers to as a risk-taking, decision-making model. In this context, risk can be defined as a product—in general, the product of the amount that can be lost and the probability of losing it, or, more specifically, the product of the frequency of an event and the severity of the consequences of that event. We all take risks and think little or nothing of it, e.g. when driving. That is true because the risks are reasonable and acceptable. The difference between a reasonable and unreasonable risk is where decision making comes into play in this model. In order to reduce risk, sound decisions are required and, in turn, sound information is required to base decisions on. The better the information, the better the decision, and subsequently the more estimable the risk. With information and calculable risk, the means of reducing that risk should become more apparent.

Firenze suggests that we must be aware of five factors in order to make sound decisions and take reasonable, calculated risks:

- job requirements;
- a worker's capability and his or her limitations relative to the job;
- what will be gained if we attempt the task and succeed;
- what are the consequences if we attempt the task and fail; and
- what we will lose if we do not attempt the task at all.

When we have applied the systems concept to a new task, and trained workers

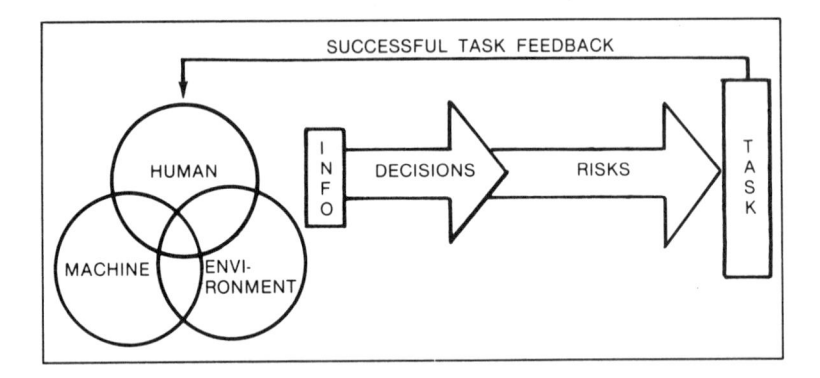

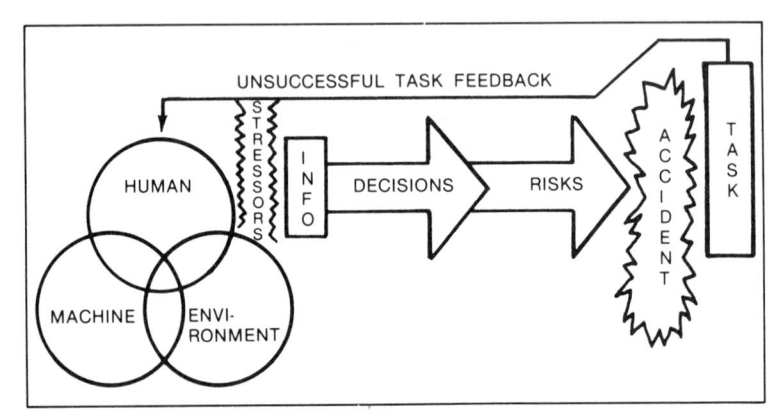

Figure 2.4 Systems model for accident causation. (From R. J. Firenze, *The Process of Hazard Control*. Copyright © 1978 by Kendall/Hunt Publishing Company. Reprinted with permission.)

fully as to what the hazards are and how to cope with them effectively so that their decisive actions are controlled, then the task should be successfully completed. Completion also provides new information for feedback to the system for future decisions. This successful process is shown schematically in Fig. 2.4.

Even when we have complete knowledge of a job, though, we can err in our decision making. Our capability is sometimes affected temporarily by stress and we can make poor decisions which can lead to the taking of unreasonable risks, and this can result in an accident (Fig. 2.4). And stress is not only experienced by man—machinery and the environment can also be affected. Man can experience psychological (see Chapter 11) and physiological (e.g. chemical substances) stress; glare and noise are environmental stresses; and machines are stressed when they are in need of lubrication.

MULTIPLE CAUSATION

Models for accident causation are essentially simplified patterns which outline the many factors which can cause accidents. Not one of the models is totally reliable for all accidents, nor are they totally unreliable. This is true because of the complexity of accidents. Multiple causation is where accidents occur because of sequential actions brought about by a combination of contributing factors. Although multiple causation is not in itself a model of accident causation, we should be aware of the phenomenon.

A real-world example of multiple accident causation is shown by the reconstruction of an accident involving a printing press. A dynamic adjustment of the tension of the print roll was periodically necessary and the hand position for the adjustment was 6 ft from the ground and 2 ft forward of the foot position. Near the hand position were unguarded in-running nip points (see Chapter 8)—an unsafe machine condition. To reach the adjustment position, the operator stood on an unstable stool—an unsafe act. This act was a repeated act, brought about because management did not provide a stable stool, nor did they attempt to improve working conditions in any general sense. On one occasion, as the machine operator attempted to make the necessary adjustment, his body weight shifted and the stool tipped over, causing his hand to be crushed in an in-running nip point. This accident was clearly caused through management's failure to provide the proper safety precautions, unsafe working conditions *and* an unsafe act.

SUMMARY

A model for accident causation is only significant if it helps to improve safety programs, thus preventing future accidents. Any model, in itself, is not the product but a guide to achieving the desired product, i.e. a safe workplace. No model is perfect, but each has its emphatic areas and each is simplistic, permitting us to customize our programs.

If we summarize the models and look at the areas we must emphasize and learn about in detail to establish meaningful accident programs, four topics emerge:

- people and their behavior;
- management;
- recognizing hazards; and
- systematic control.

These topics are covered in Chapters 3–6.

Case History 2.1 Pointing the Way

Katahdin Electric is a small rural power company in northern New England with a fleet of only 25 utility trucks. In 1984, they purchased eight new vehicles; only the chasses were purchased and standard utility bodies had to be fabricated to meet the special demands of the power company. The task of preparing the specifications for the modifications was assigned to Larry Moner, a mechanical engineer who had joined Katahdin Electric 6 months beforehand. Prior to that time, he had worked for a small steel company who specialized in steel treadplate and grating used primarily for exterior fire escapes.

The utility body standards prepared by Larry were 16 pages long, covering material compartments, superstructure, floor, roof and rear wheel fenders, as well as lighting and painting. On page 15 of the specifications, under a general miscellaneous heading, was the following:

> Insert tread plate in access step to chassis cab, right-hand side only (step type fuel tank will be furnished on left side).

In his research, Larry noted that the metal access step might be slippery when wet; the tread plates which he was familiar with would, in his opinion, improve worker safety. Nobody at Katahdin challenged Larry's idea and it became part of the complete specification.

The contract for the eight vehicles was awarded to A. C. Baxter Corporation. The shop drawing they prepared for the modification to be made to the access step is shown in Fig. 2.5. The A. C. Baxter shop then modified the eight vehicles, according to the shop drawing (Fig. 2.6).

Truck 18, which was one of those modified by A. C. Baxter Corporation, was put into service in 1985. In 1987, Bob McBoy, a Katahdin Electric employee who used the truck daily, noticed that two or three welds had broken; about 4 months after that, two more had broken. At that time, the grating would depress when it was stepped on. Bob reported the broken welds to his supervisor on both

Summary 37

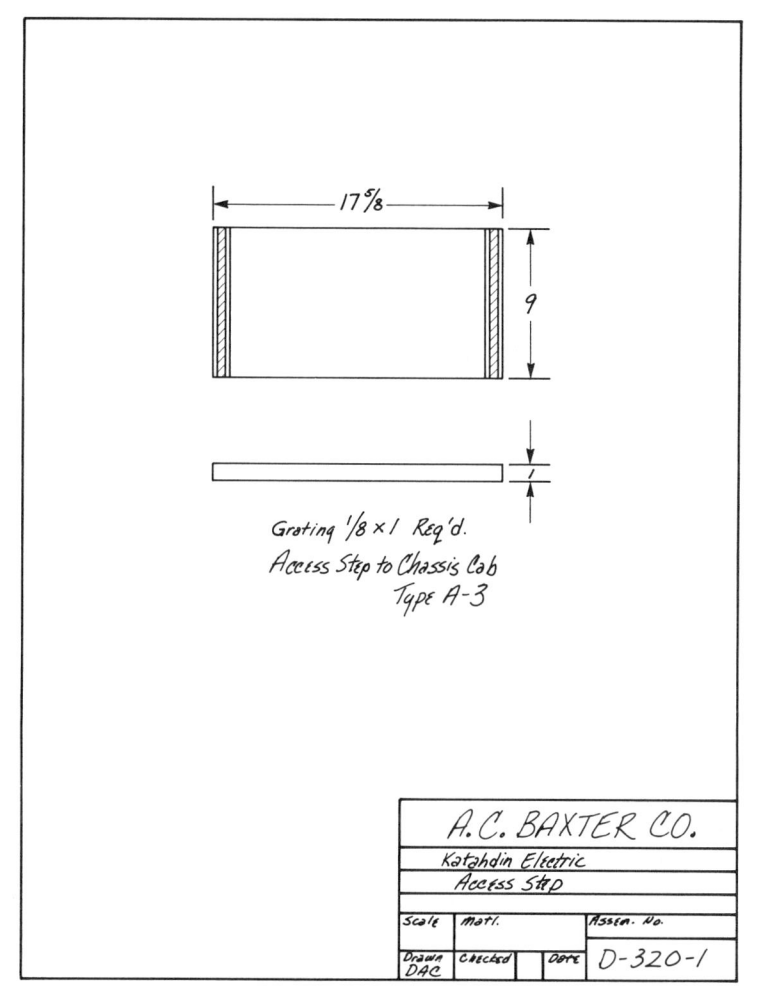

Figure 2.5 A. C. Baxter shop drawing.

occasions, but no repairs were made. In 1988, Bob went to enter the cab and stepped on to the insert grating; it collapsed under his weight and he fell to the ground, his foot became trapped, and he twisted his body such that his back has been permanently injured.

An examination of the broken access step showed that cutting the hole for the grating had eliminated the original rigid bracing beneath the step, that the grating was then inserted and tack-welded by arc-welding on three sides only. Only three welds near the front remained and these were bent.

Was this accident preventable? If so, do the accident causation models we have looked at point the way to prevention? Heinrich's original Domino Theory points to Bob as the cause of this accident, because he committed an unsafe act

Figure 2.6 Tread plate insert on right side access step.

by stepping on to the grating which he *knew* was weakened. However, the updated versions of the Domino Theory point to management—this includes the Katahdin management because they failed to understand the consequences of Larry's tread plate insertion and failed to respond to Bob's reports of broken welds, and A. C. Baxter's management who failed to select *and* specify exactly the method for insertion of the tread plate. Figure 2.5 is *not* an acceptable shop drawing, which is symptomatic of their management failure. A metallurgist also pointed out that arc-welding thin and thick metal sections is difficult, probably resulting in weak joints. The use of tack welding instead of continuous welding exacerbated the situation. We can therefore see the benefit of the updated domino theories in pointing the ways this accident could have been prevented.

Human factors and Petersen's Accident/Incident Model exculpate Bob for his unsafe act, and instead show that the accident could have been prevented by a correct management response to Bob's reports of broken welds, better perception by both Bob and management of improbable serious injury, and the elimination of ergonomic traps caused by the modification which led to Bob's unsafe act. Systems models point to the same areas which could have prevented Bob's injury. Here the bad information which led to poor decisions and increased risk began with Larry's original idea, followed by blind approval, poor shop drawings and the unsafe condition of the final insert. This model also points to the poor use of feedback information (Bob's reports of broken welds) by Katahdin Electric.

At first glance, an epidemiological model would be the only one which might not point the way to prevention of this accident because of the small number of vehicles that were modified and the learning experience of Bob's incident and injury. However, a basic cause of this incident is the original specification, which was both incomplete and not based on valid hazard analysis. Complete information and research into risk assessment are important issues raised by the epidemiological model.

Applicability of the multiple causation model to Bob's incident and injury is obvious, and therefore this case history clearly demonstrates that the models can work and should be used to our advantage. ∎

FURTHER READING

E. Adams, "Accident Causation and the Management System," *Professional Safety*, **22**, 1976.

F. Bird, *Management Guide to Loss Control*, Institute Press, Atlanta, 1974.

D. A. Gibson, "Herbert W. Heinrich, First Safety Engineer Elected International Insurance Hall of Fame," *Professional Safety*, **26**, (4), 20, 1980.

Herbert W. Heinrich, *Industrial Accident Prevention*, McGraw-Hill, New York, 1931.

H. W. Heinrich, D. Petersen and N. Roos, *Industrial Accident Prevention*, 5th Edition, McGraw-Hill, New York, 1980.

F. E. McElroy, ed., *Accident Prevention Manual for Industrial Operations: Administration and Programs*, 8th Edition, National Safety Council, Chicago, Ill., 1984.

D. Weaver, "Symptoms of Operational Error," *Professional Safety*, **17**, (10), 1971.

M. Zabetakis, *Safety Manual No. 4, Accident Prevention*, Mine Safety and Health Administration, Washington, D.C., 1975.

QUESTIONS

2.1. What are some of the incidental or indirect costs attributable to injury/accidents (Table 2.1)?

2.2. What are the three human-error situations proposed by Ferrell? Which is hardest to control? What type of action might control each?

2.3. Petersen's model states "It simply is more logical in a given situation to perform the job unsafely than it is to perform it safely." What does this indicate about workers' perception of the risk involved?

2.4. Petersen states that two of the causes for the decision to err are pressure to meet deadlines and peer influences. Explain how we can try to reduce these influences.

2.5. Compare the Systems Model with the updated Domino Theory. Which is more realistic? Why?

2.6. List the multiple causes of Bob's incident and injury in the case study.

3

Safety: A People Phenomenon

Being concerned about safety, we are automatically concerned about people, and therefore we have a responsibility to understand people. In this chapter, we explore what people are all about, why they behave the way they do, why human errors occur, and how humans relate to machines and the work environment. Perhaps the most important topic in this chapter is perception, because our safety programs are only as good as they are perceived by the workforce.

The best way to start learning about others is to begin learning about ourselves. Before reading on, why not answer the questions posed by Herzberg in his Two-factor Theory? First, remind yourself about a time when you felt exceptionally good about your work. Then, remind yourself about a time when you felt exceptionally bad about your work. Only you can benefit from your own answers, but Herzberg and his associates analyzed over 4000 responses to such questions. Their analysis identified different sources of satisfaction, which they termed motivators, and different sources of dissatisfaction, which they termed hygiene factors. Contributions to these two factors are summarized in Fig. 3.1.

In this context, safety comes under the hygiene factor "work conditions." In Two-factory Theory, hygiene factors affect only job dissatisfaction—job satisfaction is considered separately. In other words, no matter how effective our safety programs are, we cannot expect increased job satisfaction: only reduced job dissatisfaction will result.

There is a very important lesson here for our safety programs, because we cannot depend on poster campaigns or other motivational programs except on limited occasions. Improved hygiene can eliminate dissatisfaction, but cannot lead to satisfaction. To create job satisfaction, we must turn to the motivators. In other

The KISS Principle   41

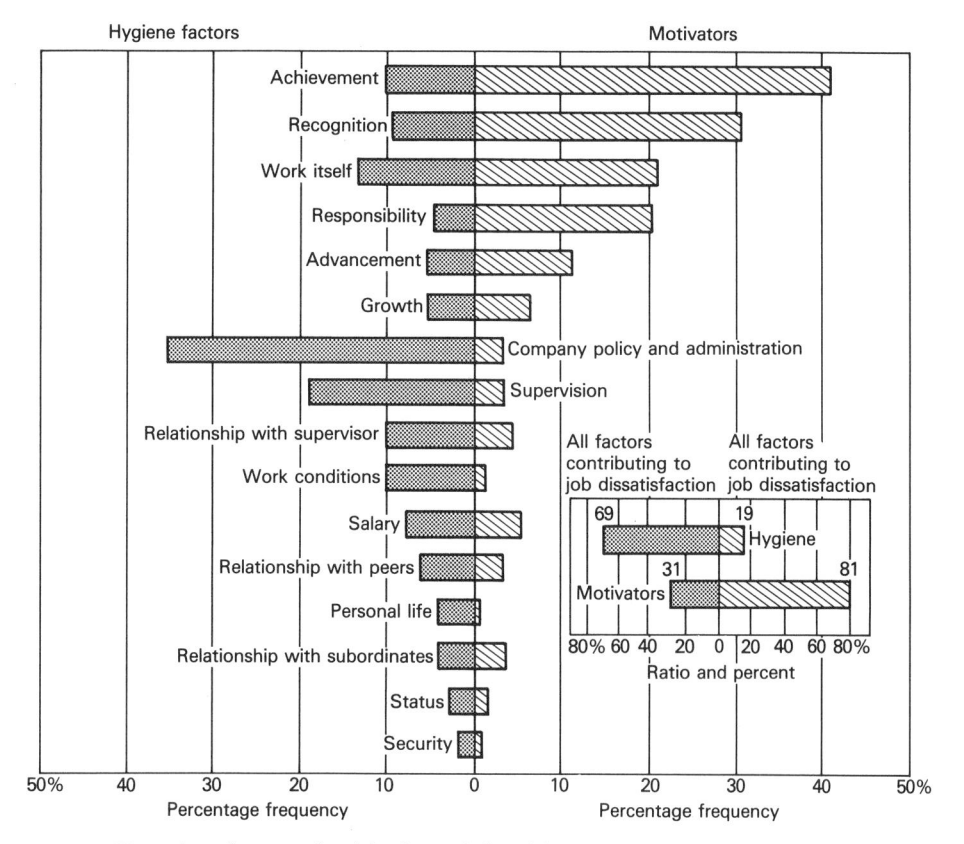

Figure 3.1 Sources of satisfaction and dissatisfaction in the workplace. (Adapted from Frederick Herzberg, "One More Time: How Do You Motivate Employees?" *Harvard Business Review,* Vol. 46, January–February 1968, p. 57. Copyright © 1968 by the President and Fellows of Harvard College. All rights reserved.)

words, we have to set up a competent safety program to eliminate dissatisfaction, and then determine what else our people need to provide satisfaction.

THE KISS PRINCIPLE

KISS is the acronym for "Keep It Simple, Stupid," which is much more than just a catchy slogan. Good leaders do keep things simple whenever possible and do *not* complicate matters unnecessarily. The KISS principle actually comes from the motivators. Achievement is the most frequent source of job satisfaction; people want to succeed, to be members of a winning group. We like to think of ourselves as winners, even if we practice a little self-deception to do it. Therefore, we should look for ways to make people succeed rather than fail. Recognition, growth and advancement are job satisfiers which are encompassed in the second KISS observation of our behavior—*nothing succeeds like success*. A little success makes us want to do more, to do better! Therefore, we should emphasize the successes, while playing down the failures. The other job satisfiers—the work itself and responsibility—fit the KISS tenet, control. While we want to feel secure in our jobs, we also want to exercise some control over our own destinies, and participate in decisions about our work.

The beauty of the KISS principle is its simplicity in understanding ourselves and thereby giving some direction to our efforts. In the past we have tried to improve people's attitudes in order to improve performance, i.e. we thought that attitudes preceded actions. Today, though, it is more common to think that actions precede attitudes and that actions can be observed and corrected. Behavioral management programs will be examined in Chapter 4.

PERCEPTION: A KEY PROCESS

From our simple beginnings of understanding human behavior, it is clear that we have to appeal to the workers who are the objects of our safety program. In other words, we have to carry the safety message. How it is received and interpreted, though, is of prime importance. Worker options include not only acceptance or rejection, but also revenge or even attack on the messenger! This process through which people select, receive, organize and interpret information is called perception. It is probably the most important process we have to consider in safety.

Anyone's perception process is affected by many factors, which are usually separated into three categories: the characteristics of the perceiver; the characteristics of the perceived; and the situational context. Figure 3.2 summarizes these multiple influences, but let us now look at some of the details.

Characteristics of the perceiver. How we feel, and what our habits, training, values and our personality are like, all enter into how we perceive. There is a lot of truth in the statement that we hear what we want to hear and see what we want to see. What often happens is that we select from any situation only those signals which tend to reinforce our prejudices. Realizing these limitations can be a beginning for more accurate perception *and* for better communication with others.

Characteristics of the perceived. Other people in a situation also influence how that situation is perceived. We can be attracted to or repelled by their appearance or conduct. How we dress and how we act can create quite different images to those people who do not know us. And one of the most important points to remember is that first impressions are very longlasting.

Characteristics of the situation. The physical, social and environmental

Perception: A Key Process

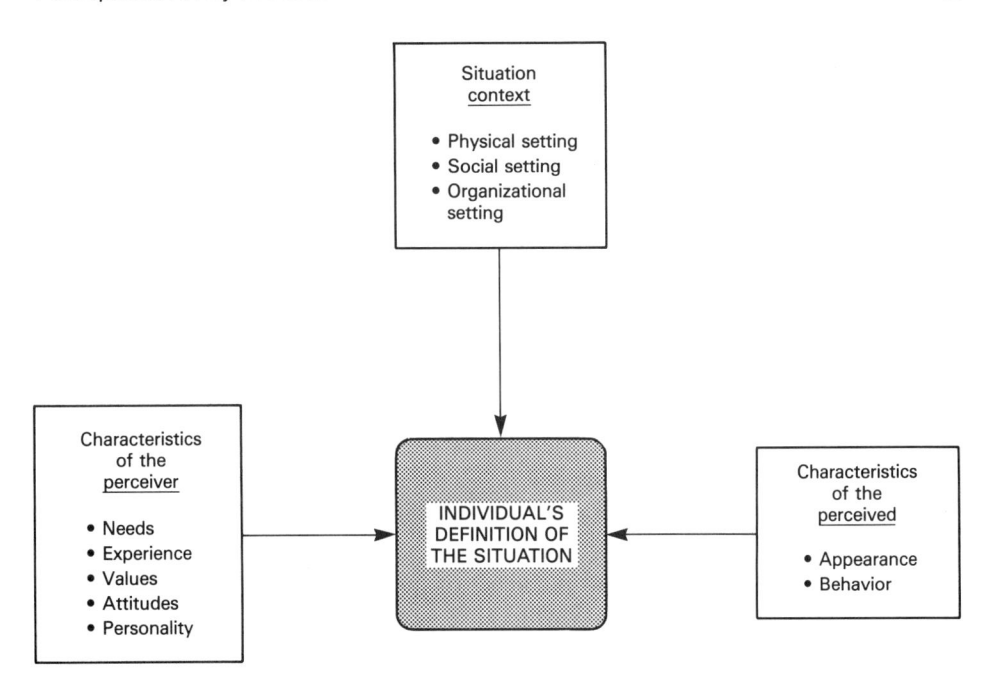

Figure 3.2 Multiple influences on the perception process. (From J. Schermerhorn, J. Hunt, and R. Osborn, *Managing Organizational Behavior*, 1985. Reprinted by permission of John Wiley & Sons, Inc.)

settings of a situation also influence perception. The same conversation can be viewed differently depending on whether it takes place in the office or at a ballgame. Even people can be perceived differently depending on the situation; for example, consider the case of a young learning-disabled woman who had strained the patience of many understanding employers because of her loud voice, but, when placed in the unique atmosphere of Filene's basement in Boston, her same loud voice became an asset!

Perceptions are also influenced by the way we organize information. Research in the 1960s on how we process information showed that we receive only about 4 bits of information out of the 10^5 bits provided each second from the environment. Perceptual organization is one means of coping with this discrepancy. There are four organizing tendencies which influence our perceptions:

- *Figure and ground:* the tendency to distinguish one outstanding feature or person in a maze or group.
- *Set:* the tendency to respond to predetermined anticipations rather than in terms of reality.
- *Gestalt:* the tendency to assign overall meaning to unorganized information.
- *Attribution:* the tendency to try to assign a cause for events or behavior.

Perception is not simple, and can often be distorted. Examples of common distortions include stereotyping, first impressions (either good or bad), selective perception which reinforces existing beliefs, the projection of one person's behavior to group behavior, and to search for behavior you expect. All of these can exaggerate what the situation and/or the behavior truly is and create incorrect perception.

We have to avoid the bias problems that incorrect or distorted perceptions can cause. In order to do so, we should be skilled in the following perceptual areas:

1. We should have a high level of self-awareness, and know and recognize when we are inappropriately distorting a situation.
2. We should seek the viewpoints of others—we have to develop a network of trusted colleagues.
3. We should attempt to see the situation in the same way as others, i.e. be empathetic.
4. We should point out incorrect or misleading impressions, i.e. influence the perceptions of others when they are distorting the situation.
5. We should avoid distortions which bias our views.
6. We should avoid inappropriate attributions, e.g. in accident investigations, we must determine what happened, not place blame.

WORKER ATTITUDES AND PERCEPTIONS OF SAFETY

In much of the safety literature, the worker has been the object of discussion, but the worker's own attitudes and perceptions of safety have been neglected. To fill this void, ReVelle and Boulton conducted a survey to determine how workers felt about government and company involvement in safety, their perceptions of management (particularly supervisors) and co-workers, and their individual attitudes toward safety. Half the workers felt government involvement in workplace safety was about right, but about 25% thought more intervention was necessary. Workers in large companies expected more from their employers in terms of safety than those in small companies. Supervisors who talked about safety were perceived to represent their company's attitude and co-workers were perceived as caring for personal safety and the safety of others.

The data clearly showed that workers who had received no safety training had experienced almost twice as many accidents as those who had received such training. Only 20% of the surveyed workers felt that their safety training was adequate, yet 75% felt comfortable about protecting themselves in the workplace.

We can learn a lot from these worker perceptions of safety. First, the role of the supervisor is critical in safety, as we will reiterate throughout this text. Secondly, the perception of workers' comfort in protecting themselves in the workplace is inappropriate and must be changed through proper safety training. And, finally, why should the distorted perception of safety expectations vary by size of company? The provision of industrial safety programs, therefore, can be understood in terms of people and their perceptions.

HUMAN ERROR

ReVelle and Boulton's survey showed that workers who had experienced accidents were generally candid and willing to accept responsibility for their part in the accident. Nevertheless, 85% said their accidents could have been prevented. This is close to the 88% suggested by Heinrich (Chapter 2) for accidents caused by human error. We should avoid the all-too-common connotation of human error being wrong, a source of blame or cause. It is much more productive if we consider human error simply as an event whose cause can be determined. Numerous definitions of human error can be found, but the definition used throughout this text is that of Sanders and McCormick: *Human error is an inappropriate or undesirable human decision or behavior that reduces, or has the potential for reducing, effectiveness, safety, or system performance.*

This definition points out the undesirable consequences and also includes those errors which have the potential for the undesirable consequences. We will not consider useful functions of human error, such as learning by trial and error or establishing limits of acceptable behavior. Although there is a tendency to think of human errors only as operator errors, this tendency is frequently incorrect. Errors are also made by designers, managers, supervisors and others involved in the design and operation of systems. Chapanis argues effectively that the only true errors are design errors, because the actions we call human errors were not anticipated, i.e. they did not conform to expected behavior. Thus, if we were smart enough and knew enough about human behavior, we could anticipate methods to forestall such behavior in our designs to ensure no serious consequences.

Classification of Human Error

Classifying the types of error which occur is important for our understanding of their causes and how they might be prevented. In their early work, Fitts and Jones classified "pilot error" experiences as:

- substitution errors;
- adjustment errors;
- forgetting errors;
- reversal errors;
- unintentional activation; and
- inability to reach.

We find these same errors being made today in many diverse situations—in the workplace, at home, and in public places. But these error classifications can also fit into Swain and Guttman's much simpler scheme of discrete actions:

- errors of omission;
- errors of commission;
- sequence errors; and
- timing errors.

Errors of omission involve the failure to do something, a classification which includes forgetting errors and the inability to reach, but it is more comprehensive than that, e.g. not wearing personal protective equipment because it is uncomfortable. *Errors of commission* involve incorrect actions, and therefore encompass substitution and adjustment errors. Many critical failures occur because of improper material selection, running equipment at the wrong speed, or by fixing equipment with "baling wire and chewing gum" so that production can continue temporarily. *Sequence errors* occur when a person performs some task out of turn. This includes reversal errors and is most common when the wrong control is activated or a step is "skipped." *Timing errors* occur when a person fails to perform an action within the allotted time—performing it too quickly or too slowly—or performs it at the wrong time. We can consider inadvertent activation as a timing error. Many timing errors occur because of poor judgment, e.g. overconfident people often underestimate the likelihood of an accident.

Overload

In Chapter 2, overload—the mismatch between the load and capacity of a person—was shown to be one source of human error in accident causation models. To understand how overload can occur, though, we have to understand the factors that contribute to capacity, load and state (see Fig. 2.2). Though some of these factors are easy to comprehend (e.g. differences in physical attributes, fatigue, training, and even chemical substance abuse), others are more abstract (e.g. feelings and thoughts). Psychological issues are often ignored, but they must be confronted if we are to make progress in industrial safety, just as in any other research.

Case History 3.1 Feelings and Business

There are almost 17 million small businesses in the United States, the majority of which are family owned and the dream of the inventor—entrepreneur. Some 70 percent of these businesses do not survive into the second generation. To combat such statistics, there is a growing network of support institutions trying to prevent a business from destroying relationships and to prevent relationships from destroying a business. Such was the case for TAC Laminates, a 35-year-old company that manufactures composite laminates for the electronics industry. The acronym stands for Thomas and Alfred Candrell, two brothers who began the company on a shoestring shortly after Al finished college where he studied plastics, a relatively new field at the time. Tom, who did not go to college, had developed a flair for sales and worked his way up to sales manager for a local brewery. TAC Laminates developed a dependable reputation for consistently high-quality, one-sided and two-sided copper-clad laminates and grew to a company with seven-figure annual sales. Much of TAC's success is attributable to the rapid growth of the electronics industry, particularly in the late 1960s when microelectronics were first introduced. But the company mainly flourished due to the salesmanship of Tom and the technical leadership of Al.

The company with 24 employees, was outwardly successful, but it had grown beyond the control of Al and Tom. Financial controls were virtually nonexistent, job descriptions were vague and confusing, and production costs were uncontrolled. Al has never married; Tom has three children, two daughters and a son, Tom, Jr., who is currently the production manager. In 1985, Tom went through a divorce that not only was personally stressful but also affected relationships within the company and the family. Al, by nature quiet and reserved, blamed Tom for the divorce and began criticizing Tom's business practices which he had overlooked all these years. Tom's behavior, in turn, deteriorated. When Tom, Jr., joined the company, he accepted his father's criticisms, but he did feel he was doing a good job as production manager. As Tom's criticisms became stronger, his son's feelings turned to anger. As a result, the business suffered.

It was the company's long-term accountant, Jack, who first recognized the problems that these family relationships were causing in the company. Jack's wife is a social worker who studied at Lowell's Institute for Psychotherapy. At her graduation in 1987, Jack learned of a new program in Organizational Development. This program is based on recognizing that psychotherapy and its understanding of human development readily apply to larger systems, especially the workplace. Jack urged Tom and Al to provide a grant to the Institute for internships at TAC Laminates. After six months of continued dissension, Tom and Al reluctantly agreed.

Mark, a licensed social worker, and Janis, a doctor of education, studying at the Institute under Dr. Peter Horowicz, were assigned to TAC Laminates. After a year-long study, Mark and Janis confronted Al, Tom and Tom, Jr., in a day-long meeting held at the Institute's Faculty Club. The purpose of this meeting was to achieve family consensus on the nature of their problems and to examine how irrational obstacles to change might be faced for *the benefit of the company*.

The psychological profile of the three family members showed that all three emphasized feelings over logic when weighing decisions. They did not make necessary decisions for fear of upsetting the family system and made unnecessary decisions in reaction to family problems. Their styles were different: Tom was extroverted, high-spirited and bored by detail, yet insecure because he lacked Al's education. Al was deliberate, persevering, dependable—the technical backbone of the company—but introverted by nature. Tom, Jr., like his mother, was very private, good at detail, but crushed by criticism. How these differences came about was explored in some difficult, heart-wrenching but also enlightening discussions.

To get the company moving, a breakthrough in the interpersonal relationships was essential. Mark and Janis asked them to select an issue over which they had recently clashed and asked each to give his version. Then they asked them to summarize what the other two had said. This would determine how well they listened to one another. The results showed that Tom was unreasonable and did not listen to Al or Tom, Jr.'s arguments and that Al let his feelings about the divorce affect his work with Tom and had never bothered to find out how Tom felt about his divorce.

A report and transcripts of the tape made at the intervention were issued. Al, Tom and Tom, Jr., are now talking out their differences more honestly, all have agreed to abide by decisions unless there is new information and Tom has even occasionally complimented his son's efforts. Many questions remain, but it is clear that an impressive transformation has taken place both in personal and business relationships at TAC Laminates. ∎

Psychological aspects are only a part of defining the capacity of a person, no matter how important a part it is. In a clinical sense, capacity comes down to how we sense, process and respond to the information we receive, as described by van Cott and Kinkade (see Fig. 3.3). Much can be said about each entry in this figure, but it is only important that we recognize that defects in any parameter

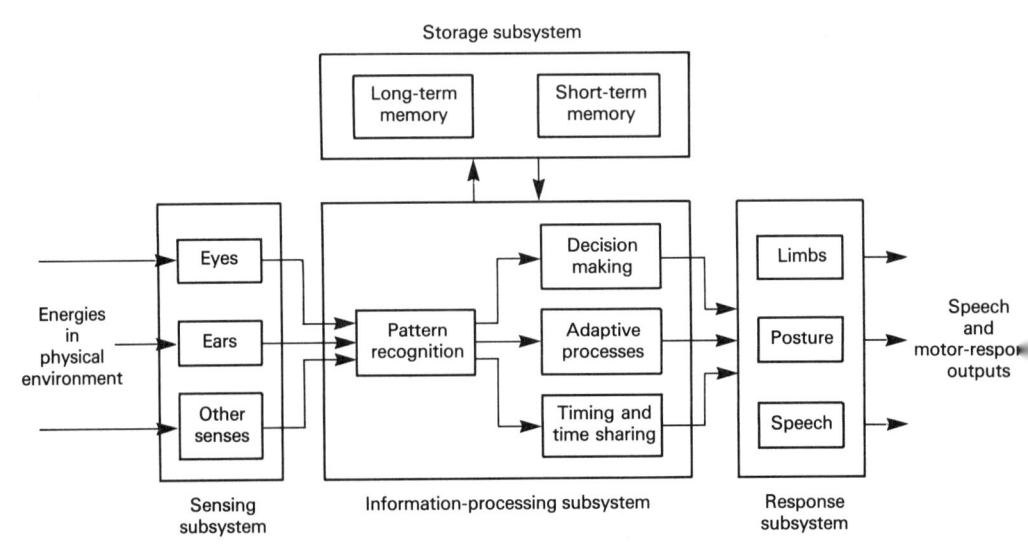

Figure 3.3 The human information-processing system. (From H. Van Cott and R. Kinkade, *Human Engineering Guide to Equipment Design*. Washington, D.C.: U.S. Government Printing Office, 1972.)

can alter the outcome. For example, Chapter 11 shows that noise can interfere with communication, thus changing the information we sense and therefore are able to process. Or, as in Case History 3.1, because Tom did not *listen* to what Tom, Jr. or Al were saying, he processed the wrong information.

Sensation and motor responses will be discussed in more detail later in this chapter, but for now let us look at the human information-processing subsystem. It is in this area that we can exercise some control over our safety programs. Except at the extremes, intelligence has little bearing on accidents and injuries; exceptions occur when judgment is involved. Personality, on the other hand, is known to be related to accidents, where behavioral traits such as aggression, impulsiveness, anxiety and mood changes represent a high risk of accidents. The controls which we can exercise include improving relationships (as in the case of Al and Tom), changing behavior (behavioral management programs are discussed in Chapter 4) and selecting and training employees.

We tend to think of selection and training together because of government civil rights legislation which has established selection procedures and because the labor pool for many jobs is limited. It is imperative that we provide the proper training if selection is not perfect! Nevertheless, we must strive to make the best selection possible. Selection begins with a job description of the vacant post, a description that must include *all* of the duties to be performed, including the training and experience required *before* employment and what tasks are to be performed once employed. As we will see in Chapter 4, the most important management decisions are people decisions, and therefore it is imperative that we strive to match the best applicant to the job description. If a job description is incomplete, then the selection process is made more difficult and it becomes more likely that the wrong applicant is selected.

There are two sources of information which we can use to base our selection of new workers: the biographical information contained on a résumé or a completed application form and the information gained by testing. Experience and past performance remains the best indicator we have for future performance. Tests, although of secondary importance, are used to determine the level of skills or pre-employment health of an applicant. Interviews are able to provide information in both these areas if they are structured properly; however, if they are not, they will be of limited use. Some of the pitfalls that should be avoided when interviewing include stereotyping applicants, relying on first impressions, asking questions which are designed to support or refute one's preconceived ideas, and overemphasizing unfavorable as opposed to favorable information.

Case History 3.2 Choosing the Wrong Carpenter

The small Billerica Mattress Company manufactures inexpensive generic box springs and mattresses near Lowell. Most of the 30 employees are involved in assembly or fabric cutting and stitching. All of the materials used in these processes are purchased. In 1988, an opening in the framing department was advertised. The advertisement, the only available job description, stated:

> Opening for trained carpenter in framing department. Duties include cutting and assembly of frames for box springs and maintenance of equipment. Billerica Mattress Company. Telephone 555-1234.

There were only three applicants for the job which had to be filled in order to continue box spring production. The first was a high-school dropout and not yet 18, and therefore barred from employment by law; the second was also a dropout, with a single high-school shop course, but no experience. The job was therefore given to Jose Santano, a 25-year-old Hispanic who had been in the U.S.A. for less than a year and who had no record of employment. He could speak no English and his interview was conducted using one of the stitchers as interpreter. He was hired because he had studied carpentry in his native Caribbean and had worked there as a carpenter for 2 years.

Jose reported to the general foreman, Joe Simons, who instructed him in his daily assignments, and he had a general helper, an older Portuguese man. As part of Jose's daily tasks, he had to cut the sides, ends and cross-braces of mattress frames, using a vertical band saw which had been purchased second-hand when the company began. Jose complained to the foreman that the bandsaw vibrated and had to be constantly adjusted, but his complaints went unheeded, partly because of the language barrier. In addition, the foreman was unsympathetic and knew Jose was responsible for maintenance.

Three weeks after Jose was hired, the blade started vibrating when he was cutting cross-braces, stacked four high. Jose's hands slipped off the wood and the moving blade caught his right hand, resulting in the amputation of four fingers. The Workmen's Compensation insurance agent interviewed Jose through an interpreter at the hospital, but he was unable to understand how the incident occurred, and so hired Dalzell Associates to examine the machine. After reviewing the transcript of the insurance agent's interview with Jose, Dalzell Associates examined the bandsaw. Their examination revealed that the lower blade guide (indicated by the arrow in Fig. 3.4) had loosened and had fallen out, causing the blade to vibrate. The upper blade guide and the guard, which was adjustable, was 3 in. above the stock height. If it had been properly adjusted, the injury would not have occurred, and if the lower guide had been properly sized as well as adjusted, the blade would not have vibrated.

As a result of the findings, a private investigator was hired and found that Jose did *not* possess the certification and experience that he claimed to have, an unforunate case of poor worker selection. ■

Load Considerations

Load considerations include task, environment, situational and internal factors (see Fig. 2.2). It is arguable whether or not internal factors such as personal problems affect capacity (reduce it) or load (increase it), but the net effect is the same, i.e. an increase in mismatch with an increased probability of human error.

Human Error 51

Figure 3.4 Vertical band saw (arrows indicate upper and lower blade guides).

Some situational factors, such as breaking-in periods for new equipment or distractions, can also be considered in the same way. But task factors and environmental factors are unequivocally load considerations.

There are many environmental factors which we may encounter in specific jobs (e.g. radiation, vibration or shock), but there are some which we find in almost all types of employment (e.g. heat or cold, noise and light). Noise is an important issue, and is covered in detail in Chapter 11. However, let us look briefly at light and heat to see how they affect us at work. It is not possible to look at these factors in an absolute sense because we each have different levels of sensitivity (or capacity) and different tasks present different levels of heat and light. Figure 3.5 illustrates how temperature and relative humidity affect our comfort when we are working at different tasks or activities. With illumination, the contrast of an object with its surroundings affects how visible it is because it is actually light that is reflected from the object that we see. The amount of light

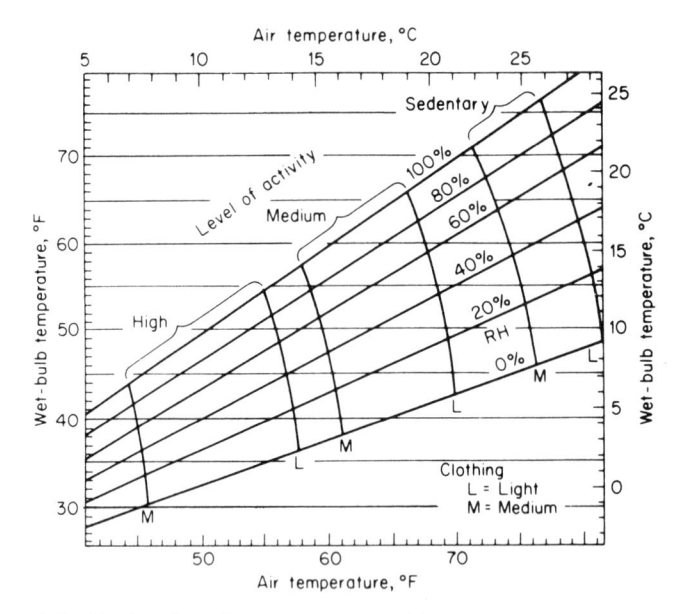

Figure 3.5 Comfort lines for persons engaged in three levels of work activity with light clothing (0.5 clo) and medium (1.0 clo). These data are for a low level of air velocity. (From M. S. Sanders and E. J. McCormack, *Human Factors in Engineering Design*, 6th ed, 1978. Used by permission of McGraw-Hill Book Company.)

reflected (or emitted) by a surface is measured in luminous flux per unit area, such as foot candle (fc) or lux (lx) (1 lumen/ft^2 = 1 fc and 1 lumen/m^2 = 1 lx; 1 fc = 10.76 lx). The Illuminating Engineering Society recommends a range of 2–3 fc for public areas with dark surroundings, but for high-contrast reading or writing tasks ten times more is required. When small or low-contrast visual tasks are performed, 200–500 fc are recommended and microelectronic assembly operations require about 1000 fc. Perhaps of more importance than illumination, however, is glare (either direct or reflected light), for our eyes are not adapted to such brightness. Glare can cause annoyance, discomfort and human error.

Task Analysis

Tasks are the day-to-day, short-term duties which define a job in direct support of the organization's production. Successful task completion is one of the motivators which provide us with job satisfaction, and therefore we should design tasks to minimize human error and ensure task accomplishment, both technically and socially. In fact, we should always conduct a thorough task analysis to ensure that a process is operable and maintainable in a safe, efficient manner.

A task analysis follows certain general procedures, first listing sequentially all steps that make up the task in which the worker plays a part. Then the components which make up each step can be examined in detail: speed, motion, strength, time and timing, feedback and how critical the task is are all major considerations.

Task analyses often mirror time-and-motion analyses used to establish piecerates or improve productivity, with one exception—the goal of task analyses are to identify potential error-producing situations so that these might be eliminated. Let us look at what first appears to be the simple task of operating a chain saw. Before beginning cutting, we should inspect and make decisions about the equipment: What is its condition? When was the chain last sharpened? Does the chain brake work properly? We must also consider the material to be cut: Is it hardwood or softwood? Are there any knots where the cut is to be made? Finally, we should think about the tasks of cutting: We know that pressure should be applied as far away from the tip as possible, but how much pressure do we apply and how does it vary with time? We know how the cut is to be made, but we have to guide the cut (an added time-dependent factor). How do we maintain speed, and lubricate so as not to bind the chain in the cut? These and other questions have been considered by Drury and demands posed as a function of time have been estimated. Remembering that we receive only about 4 bits of information out of the 10^5 bits provided each second, Drury has shown how overload can occur. Figure 3.6 shows that load exceeds capacity after some time, creating a high probability of human error.

State Considerations

The concept of state, in a sense, describes the day-to-day fluctuations in our ability to perform. We all have days when we feel great, but there are also days when we feel we should have stayed in bed. Nothing seems to go right! Much of our

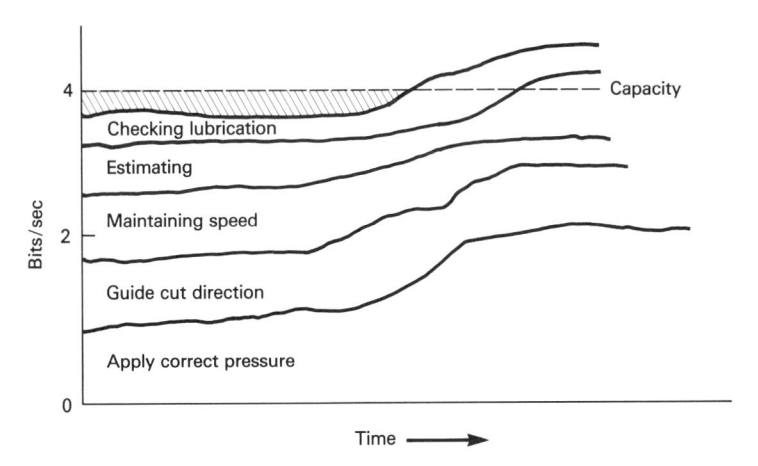

Figure 3.6 Task loads in chain saw cutting. (Courtesy of Dr. C. G. Drury, State University of New York at Buffalo.)

research into state deals with motivational state, a complicated combination of personality factors (both those of the worker and the supervisor), the corporate climate, peer relationships, union participation, and so on. Attitudes and arousal are other states which also must be considered. For example, chemical substance abuse in the workplace is a major concern because of how it can alter the state of a worker and therefore lead to human error.

The role of a company and its supervisors in analyzing the motivational state of its workforce is the subject of Chapter 4. Here we shall look briefly at Argyris's Conflict Theory, which is excellent for describing the difficulties we face. Argyris analyzed the way we behave during different stages of our lives. As children, we are passive and very dependent on our parents; our interests are limited and of short duration. However, as we mature to adulthood, we become increasingly active, self-sufficient, and our interests become both more comprehensive and enduring. Organizations, according to Argyris, have traditionally been structured along principles which conflict with adult behavior. They are vertical organizations, with hierarchies that are based on superior/subordinate relations. Consequently, they create dependence and induce boredom, and because work is usually specialized, interest and self-fulfillment are limited. When state considerations are discussed, the KISS principle should be on the agenda.

Reducing Human Error

Our study of human error is based on the reduction of accidents and thereby injuries. The vast majority of human errors, however, do not result in accidents, but we cannot differentiate between those errors which might or might not result in an accident. In order to develop programs for reducing human errors, though, we need to consider human reliability, i.e. the probability of completing an assigned task successfully within the time specified. The Technique of Human Error Reduction Prediction (THERP) identifies all human operations which affect a

TABLE 3.1 Examples of Human Reliability Estimates from THERP

Description	Reliability
Select wrong control on a panel from an array of similar-appearing controls identified by labels only	0.997
Omit an Instruction when written procedures are available and should be used but are not	0.950
Use a checklist properly	0.500
Error in reading and recording quantitative information from:	
Analog meter	0.997
Digital readout (4 or less digits)	0.999
Chart recorder	0.994

Source: A. Swain and H. Guttman, *Handbook of Human Reliability Analysis with Emphasis on Nuclear Power Plant Applications,* Nuclear Regulatory Commission, Washington, D.C., 1983.

process or event, and then computes the probabilities that specific human errors can produce systems failure based upon associated basic human reliabilities, examples of which appear in Table 3.1. Although THERP has been useful in determining human reliability in nuclear power plants, the difficulty of establishing reliable error rates in conventional workplaces has limited its application.

Other methods for reducing human error are based on systems concepts (see Chapters 4 and 5) or the interrelationships between man, machines and environment—Human Factors Engineering or Ergonomics.

ERGONOMICS OR HUMAN FACTORS IN THE WORKPLACE

Ergonomics is seemingly new and fashionable, receiving a great deal of attention in trade journals and advertising campaigns, and many fine textbooks have appeared over the last 10 years. Ergonomics is not new, however, as it has its roots in prehistoric tool-making. Ergonomics has only been recognized as a separate discipline since the late 1940s, but the military, and then more slowly the pharmaceutical, automobile, consumer and computer industries, have adopted it. Today, work station design, one application of ergonomics, takes up much of the time of computer-based companies.

The Europeans have adopted the term "ergonomics," whereas in the U.S.A. "human factors" has been used instead; however, there is no difference in meaning and we will use them interchangeably. Ergonomics is simply the study of the interrelationships between man, machines and the work environment, with the focus on human beings and how other factors affect them. The elimination of human error is one of the objectives of human factors because work effectiveness and efficiency are improved, but we also find broader goals, e.g. improved safety, increased comfort, increased job satisfaction, and improved quality of life. To accomplish these goals and objectives, we have to understand much more than the causes of human error—we have to know what our capabilities and limitations are as human beings, and how we relate to machines which can perform better than us (see Table 3.2). In many ways, it is the successful marriage of these concepts which results in safe engineering designs and processes.

In order to accomplish these goals, we have to learn more about ourselves, and what characteristics we possess that make a job easier or more difficult. One area of such interest is anthropometry.

Anthropometry

Anthropometry is the study of the physical dimensions of man, such as height, weight, reach, etc. There are two types of measurements to consider: static or fixed position measurements, such as those summarized in Fig. 3.7 and Table 3.3, and functional or dynamic dimensions, taken when the body is engaged in some activity. The practical limit for arm reach, for example, is not simply the arm

TABLE 3.2 Some Activities in which People or Machines Excel

People Excel In	Machines Excel In
Detection of very low levels of light and sound	Performance of routine, repetitive or very precise operations
Sensitivity to an extremely wide variety of stimuli	Ability to respond very quickly to control signals
Perception of patterns and formulation of generalizations about them	Exertion of great force, smoothly and with precision
Ability to store large amounts of information for long periods and to recall relevant facts at appropriate moments	Ability to store and recall large amounts of information in short time periods
Ability to exercise judgment where events cannot be completely defined	Performance of complex and rapid computation with high accuracy
Ability to improvise and adopt flexible procedures	Sensitivity to stimuli beyond the range of human sensitivity (infrared, radio waves, etc.)
Ability to react to unexpected, low-probability events	Ability to do many different things at one time
	Insensitivity to extraneous factors
Ability to apply originality in solving problems, i.e. to find alternate solutions	Ability to repeat operations rapidly, continuously, and precisely the same way over a long period of time
Ability to profit from experience and alter course of action	Ability to operate in environments hostile to people or beyond human tolerance
Ability to perform fine manipulation, especially where misalignment appears unexpectedly	

Source: G. Marshall, *Safety Engineering*, Brooks/Cole, 1982.

length, but is affected by shoulder and back movement as well as trunk rotation and hand movement.

There are limits to our reach and vision, and many other anthropometric measurements, which must be considered when designing equipment and work areas. We call the limit of our reach the work space envelope, a concept which is important when designing tasks, controls and emergency shut-down switches (see Chapter 8). For example, Fig. 3.8 illustrates the functional arm reach for the 5th percentiles of men and women. The three-dimensional work envelope represented by the information can accommodate 95% of the population. When designing a piece of equipment, it is better to use a person of average build as the model, as this will cause less difficulty and discomfort than when designing around specific individuals.

Case History 3.3 *Anthropometry in a Hospital Chemistry Laboratory*[1]

Let us look at how a simple ergonomic analysis of a repetitive task in a hospital laboratory has been effective in reducing worker complaints and in improving worker efficiency. Close to 1000 samples of blood, urine and stool are

[1] Based upon S. Shah and D. A. Colling, *Campus Safety Newsletter*, National Safety Council, July/August 1989.

Ergonomics or Human Factors in the Workplace

(a)

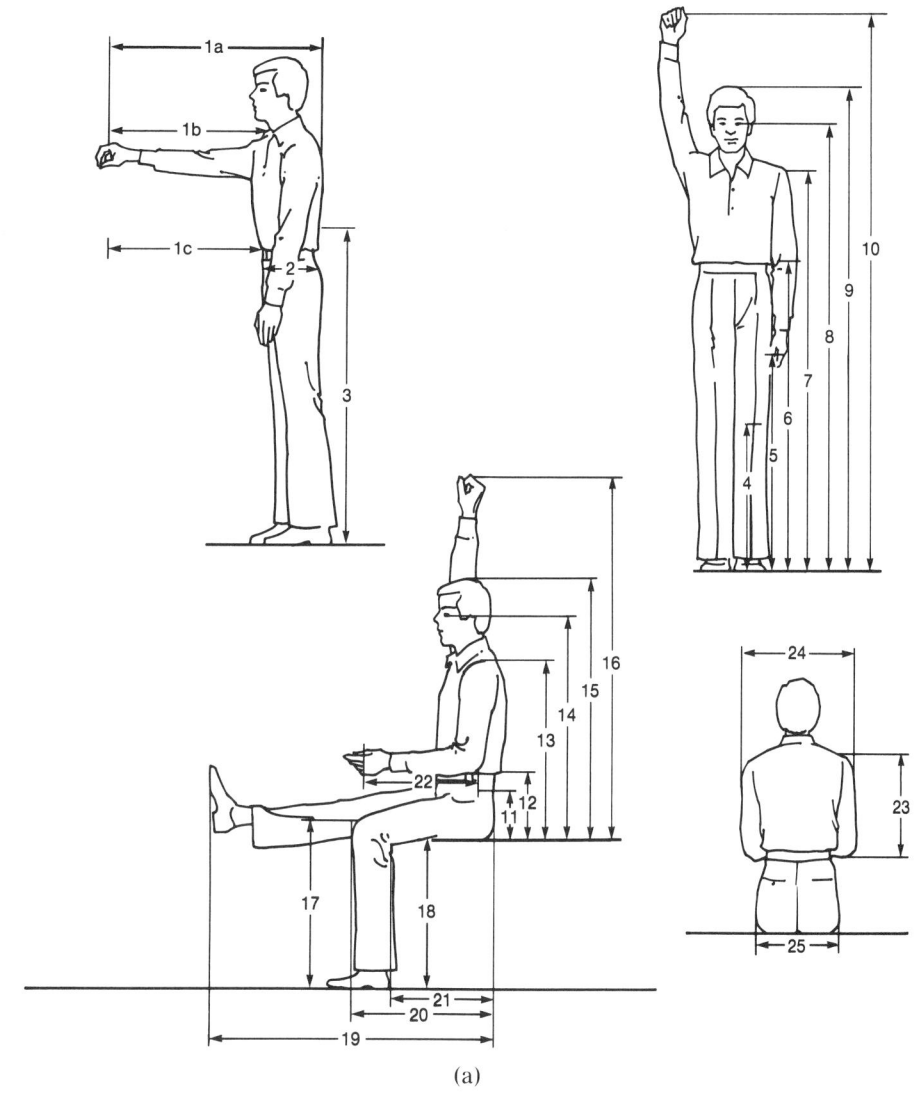

Figure 3.7 Anthropometric dimensions. Each number corresponds to measurements given in Table 3.3. (a) Standing and sitting; (b) dimensions of the hand, face and foot. (Reprinted per permission of Eastman Kodak Co.)

processed within any 24-hour period in this laboratory. The procedures for handling these samples are standardized only with respect to sample identification and test requisition records. Existing laboratory benches are used to store the samples on.

The samples are brought to the laboratory by doctors, nurses, nurses' aides

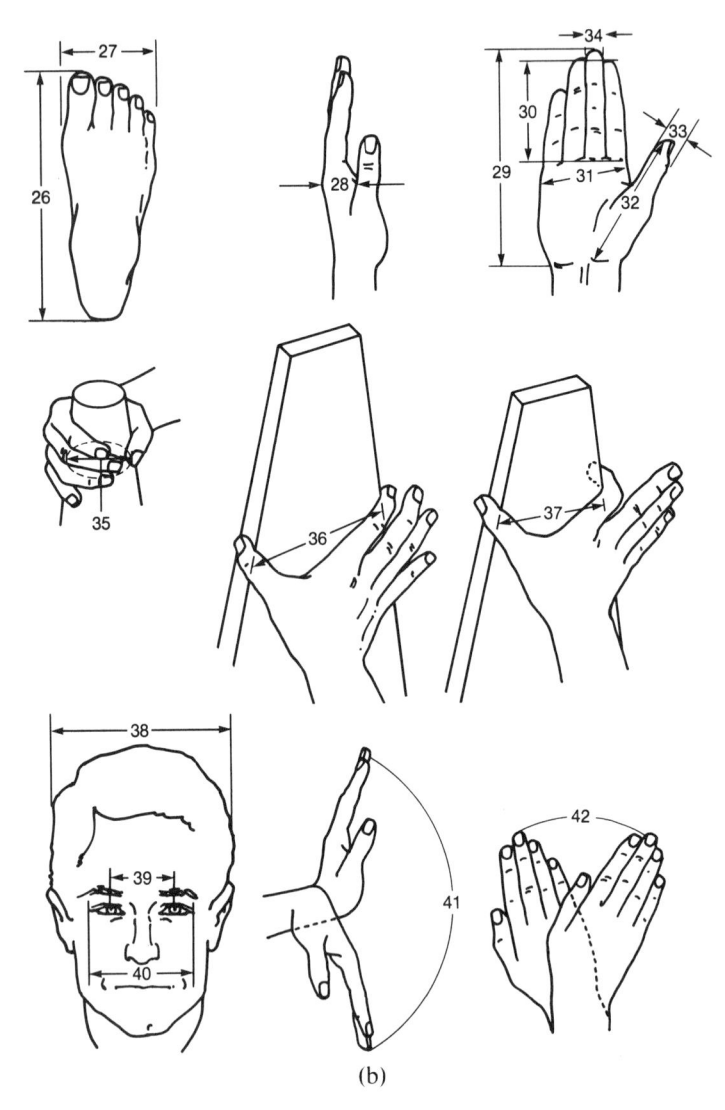

Figure 3.7 (continued)

and by a transport service. Once the sample is received, the procedure is as follows:

1. The requisition slip is time-stamped.
2. The tube and slip receive a sample number.
3. The tube is placed in the centrifuge.
4. The tests, which are ordered on the requisition slip, are entered into the computer.

TABLE 3.3 Anthropometric Data (Inches)

Measurement	Males (50th percentile)	Females (50th percentile)	Population Percentiles, 50/50 Males/Females 5th	50th	95th
Standing					
1. Forward functional reach					
a. Includes body depth at shoulder	32.5	29.2	27.2	30.7	35.0
b. Acromial process to functional pinch	26.9	24.6	22.6	25.6	29.3
c. Abdominal extension to functional pinch	24.4	23.8	19.1	24.1	29.3
2. Abdominal extension depth	9.1	8.2	7.1	8.7	10.2
3. Waist height	41.9	40.0	37.4	40.9	44.7
4. Tibial height	17.9	16.5	15.3	17.2	19.4
5. Knuckle height	29.7	28.0	25.9	28.8	31.9
6. Elbow height	43.5	40.4	38.0	42.0	45.8
7. Shoulder height	56.6	51.9	48.4	54.4	59.7
8. Eye height	64.7	59.6	56.8	62.1	67.8
9. Stature	68.7	63.8	60.8	66.2	72.0
10. Functional overhead reach	82.5	78.4	74.0	80.5	86.9
Seated					
11. Thigh clearance height	5.8	4.9	4.3	5.3	6.5
12. Elbow rest height	9.5	9.1	7.3	9.3	11.4
13. Midshoulder height	24.5	22.8	21.4	23.6	26.1
14. Eye height	31.0	29.0	27.4	29.9	32.8
15. Sitting height, normal	34.1	32.2	32.0	34.6	37.4
16. Functional overhead reach	50.6	47.2	43.6	48.7	54.8
17. Knee height	21.3	20.1	18.7	20.7	22.7
18. Popliteal height	17.2	16.2	15.1	16.6	18.4
19. Leg length	41.4	39.6	37.3	40.5	43.9
20. Upper-leg length	23.4	22.6	21.1	23.0	24.9
21. Buttocks-to-popliteal length	19.2	18.9	17.2	19.1	20.9
22. Elbow-to-fist length	14.2	12.7	12.6	14.5	16.2
23. Upper-arm length	14.5	13.4	12.9	13.8	15.5
24. Shoulder breadth	17.9	15.4	14.3	16.7	18.8
25. Hip breadth	14.0	15.0	12.8	14.5	16.3
Foot					
26. Foot length	10.5	9.5	8.9	10.0	11.2
27. Foot breadth	3.9	3.5	3.2	3.7	4.2
Hand					
28. Hand thickness, metacarpal III	1.3	1.1	1.0	1.2	1.4
29. Hand length	7.5	7.2	6.7	7.4	8.0
30. Digit two length	3.0	2.7	2.3	2.8	3.3
31. Hand breadth	3.4	3.0	2.8	3.2	3.6
32. Digit one length	5.0	4.4	3.8	4.7	5.6
33. Breadth of digit one interphalangeal joint	0.9	0.8	0.7	0.8	1.0

(continued)

TABLE 3.3 (continued)

Measurement	Males (50th percentile)	Females (50th percentile)	Population Percentiles, 50/50 Males/Females		
			5th	50th	95th
Hand					
34. Breadth of digit three interphalangeal joint	0.7	0.6	0.6	0.7	0.8
35. Grip breadth, inside diameter	1.9	1.7	1.5	1.8	2.2
36. Hand spread, digit one to two, 1st phalangeal joint	4.9	3.9	3.0	4.3	6.1
37. Hand spread, digit one to two, 2nd phalangeal joint	4.1	3.2	2.3	3.6	5.0
Head					
38. Head breadth	6.0	5.7	5.4	5.9	6.3
39. Interpupillary breadth	2.4	2.3	2.1	2.4	2.6
40. Biocular breadth	3.6	3.6	3.3	3.6	3.9
Other measurements					
41. Flexion–extension, range of motion of wrist (degrees)	134	141	108	138	166
42. Ulnar-radial range of motion of wrist (degrees)	60	67	41	63	87
43. Weight (lb)	183.4	146.3	105.3	164.1	226.8

Adapted from Eastman-Kodak, *Ergonomic Design for People at Work,* Van Nostrand Reinhold, 1983; reprinted per permission of Eastman Kodak Co. *Note*: Values should be adjusted for clothing and posture.

5. The spin sheet (a sheet detailing the ordered tests) is given to the assistant.
6. The serum is separated from the cells after centrifuging and poured into the test tube(s).
7. The test tube(s) are distributed to the appropriate racks.
8. The technician picks up the tube(s) from the racks for testing.

An ergonomic analysis was undertaken in this laboratory as a result of a number of complaints by the assistants working on Steps 6 and 7. They experienced aches in their shoulders, sides, lower back and even their thighs. Many of them remained standing while working, which led to spills, dropped pipettes and decreased efficiency. Step 6 involves the assistant in taking a test tube and labeling it, then taking a clean pipette, placing a rubber suction bulb on it, pipetting serum from the centrifuged samples and placing it in the correct test tube. Finally, the pipette is discarded and the test tubes are placed in the appropriate test rack. The processing of each sample required that the assistant twist to each side and reach as far away as 40 inches, way beyond the normal unrestrained maximum of 26 inches (see Fig. 3.8), whether standing or sitting. In addition, those test racks which were needed most frequently, were for some reason farthest from reach.

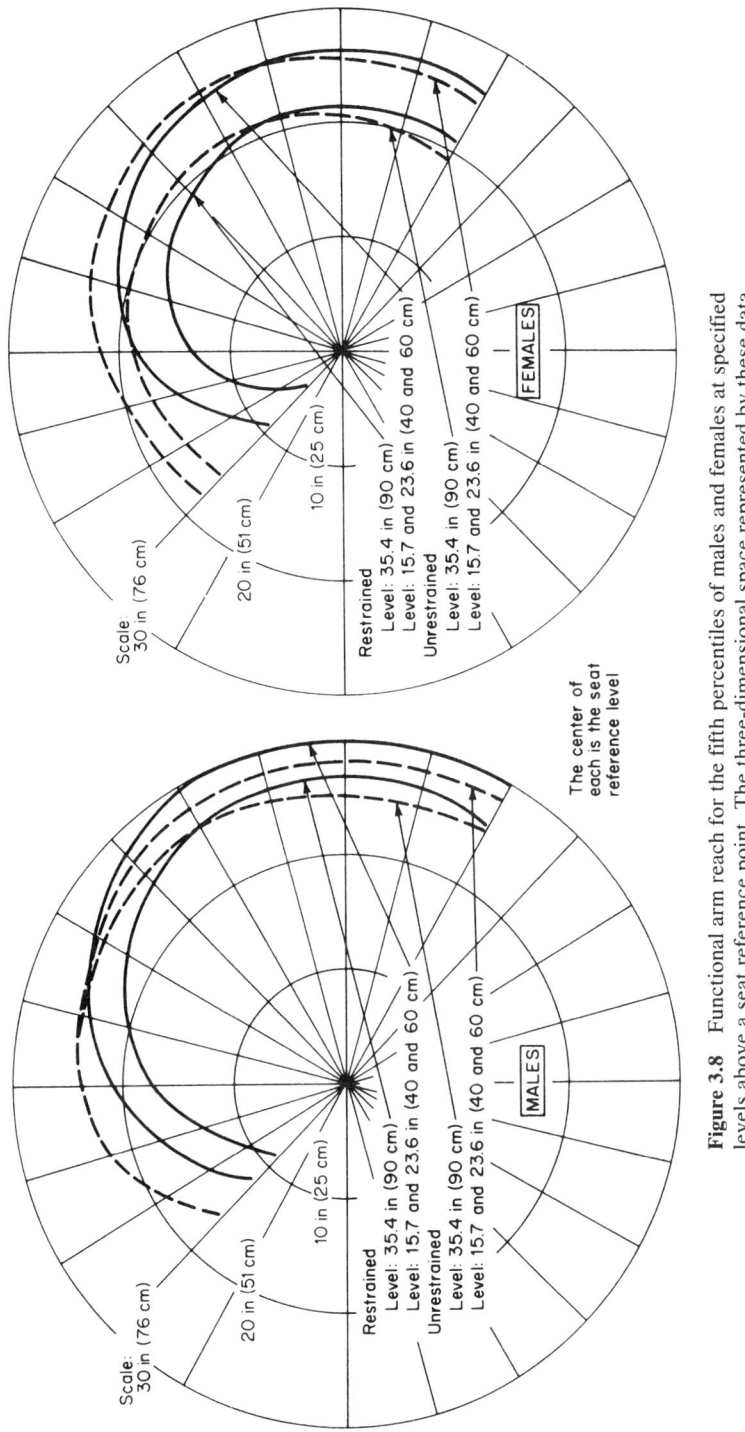

Figure 3.8 Functional arm reach for the fifth percentiles of males and females at specified levels above a seat reference point. The three-dimensional space represented by these data can be considered as forming a work-space envelope that would accommodate 95% of the population. (From Sanders and McCormack; reprinted by permission of McGraw-Hill Book Co.)

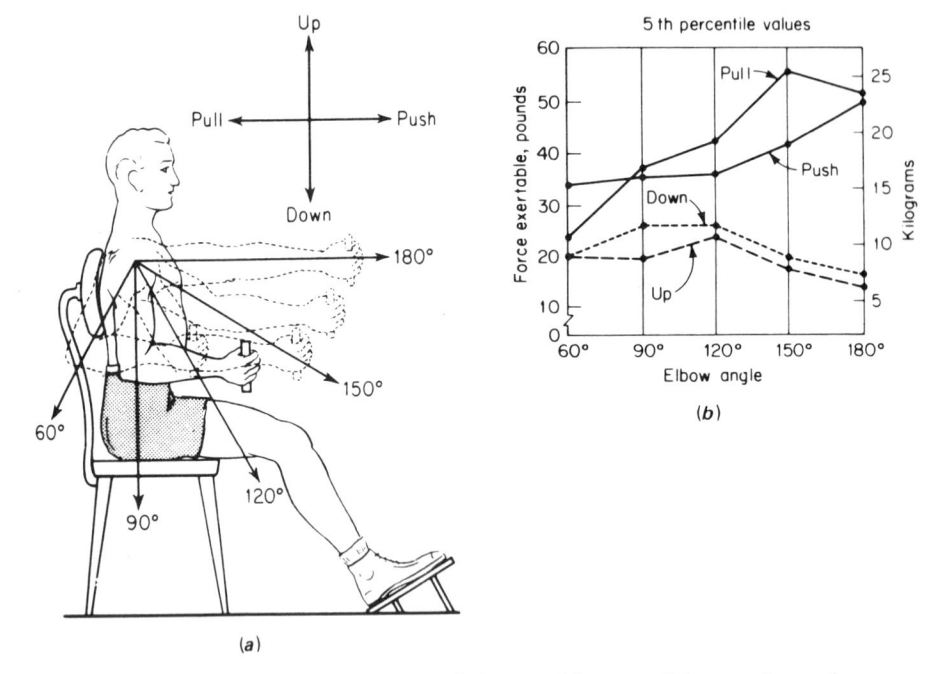

Figure 3.9 (a) Side view of subject's being tested for strength in executing push, pull, up, and down movements at each of five arm positions. (b) Maximum arm strength of fifth percentile of 55 male subjects. (From Sanders and McCormack.)

The analysis showed that most complaints were caused by the poor placement of the laboratory benches. The optimum height for a bench in a seated position with no footrest is 26 inches and for a standing position, 42 inches; however, the actual height of the bench was 34 inches. With the laboratory chair raised to compensate for the bench height, the popliteal height (see Fig. 3.7 and Table 3.3) was about 4 inches higher than it should have been.

After such an ergonomic analysis, it was very easy to redesign the work station. A footstool is now provided to ensure the proper popliteal height, test tubes and pipettes are closer to the assistant and on the same side, thereby eliminating twisting, and those test tube racks which are most frequently used are now well within a 20-inch functional arm reach. The workers' complaints have now ceased and they are working more efficiently. ∎

Biomechanics

Besides anthropometry, we need to deal with the mechanics of human motion, physical strength, and speed of movement to understand the interrelationships between man and machine. This field of study is called biomechanics. The study of human motion in relation to our musculoskeletal system is more specifically

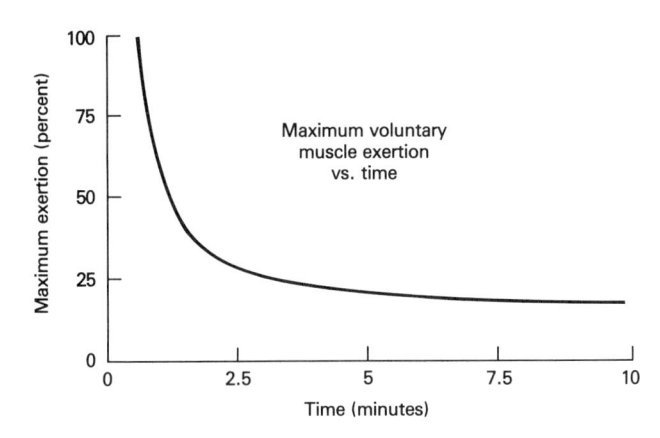

Figure 3.10 The maximum force that can be exerted drops rapidly with time. You can find maximum exertion values for muscles in most workplace design texts. Exertion should never exceed 50% of the maximum. For repeated activities, it should not exceed 30%; for continuous applications, 15%. (From W. Keyserling and T. J. Armstrong, "Ergonomics," in *Maxcy-Rosenan Public Health and Preventive Medicine*, Twelfth Edition, J. M. Last, ed., Appleton-Century-Crofts, Norwalk, Conn., 1986.)

termed kinesiology. Kinesiology views movement as being concerned with a series of levers, fulcrums and muscular forces when bending and twisting.

Strength and endurance are important in many tasks; strength is usually associated with the movement of the arms or legs and can involve either static or dynamic measurements. Although strength is a function of age (where peak strength occurs at about 30 years of age) and sex (where women have about two-thirds the strength of men), it is not absolute (Fig. 3.9 illustrates the effect of different arm motions and arm position on strength). Also, the maximum force we are able to exert drops rapidly with time (see Fig. 3.10). For the continuous application of force we should design tasks which require only about 15% of maximum strength.

Speed of movement is also important for certain applications, such speed being composed of our response or reaction time and movement time. Response time is typically assumed to be 0.5 sec, but this varies because of our need to choose a response, anticipatory factors, intensity, discrimination, and even the frequency with which we encounter the stimulant. Movement time can be affected by the type of movement involved, e.g. we move more slowly from right to left than we do from left to right, and in and out more slowly still.

Ergonomics of Machine Controls

The type of information we obtain from anthropometry, biomechanics and kinesiology is of extreme importance in certain task designs, as well as improving protection against falls (Chapter 9) and materials handling (Chapter 7). One area where we find numerous applications in the workplace is in the design and location of controls. A control is anything which activates, deactivates or changes a machine, and they can be mechanical, electrical, and audible or visual. When designing controls, whether for machines or for processes, we should be aware of many of the goals of human factors. In particular:

- reduced human error;
- increased reliability of the system;
- reduced training requirements;
- increased efficiency;
- increased convenience; and
- increased safety.

Sometimes, designs are superior with respect to increased safety, yet they are unacceptable because of their inconvenience and rejection by the worker. For example, many workers have not accepted two-hand controls because of the inconvenience caused, and they have found ways to defeat the controls, thus negating improved safety. Therefore, ergonomics must include a system of checks and balances when designing controls. Some fundamental principles should be followed whenever possible:

1. *Identification stereotypes:* do not contradict people's expectations.
2. *Control selection:* what characteristics are important to ensure the ease, speed, accuracy and safety of the control?
3. *Location and spacing:* the wrong location or incorrect spacing between controls can interfere with safety and with controlling.

A stereotype is anything which has conventional, almost universal meaning. For example, red stereotypically means stop or emergency off, and a clockwise motion indicates increasing or opening as opposed to decreasing or closing. Workers expect controls to work in certain ways and their expectations should be satisfied wherever possible. This is particularly true in emergency conditions when people revert to their instincts despite training to the contrary. In such cases, we do not have sufficient time to consider all of the information before we need to act. Other examples of stereotypes include:

Color Red—stop, emergency off, alarm, failure in system
Yellow—caution, stand-by, jog
Green—start, on, run
Motion Up—on, start, in, increase, positive, forward
Down—off, stop, out, decrease, slow, reverse
Clockwise—raise, open, move right
Counterclockwise—lower, close, move left
Pull—on
Push—off
Sound Loud, intermittent—urgency, excitement
High pitch, intermittent—vehicle backing
Buzzer—cycle completion

Visual Blinking—emergency, urgency
           Steady—on, system okay
Size   Large—heavy, bulky
           Small—light

We select controls for a number of reasons based upon function, speed and accuracy, identification within a group of similar controls, the force required to activate it, and many others. Many studies have been made of different designs, and the recommended design characteristics for some common switch controls are shown in Fig. 3.11. Even the shape of the control and the surface finish have been considered, yet many examples exist where these and other ergonomic considerations have been absent, causing accidents and injuries.

Case History 3.4 A Simple On/Off Switch Choice

Small high-technology companies are often formed to capitalize on unique ideas which can be turned into useful products. The entrepreneurs behind them find the development and potential growth of these companies both rewarding and challenging. Initially, they employ only a few workers, each of whom must accept multiple roles within the organization. This typical scenario has been the experience of A.B. Lisbon Company, which manufactures custom-printed circuit boards for the electronic toy market. The company employs 30 staff, most of whom work in administration and manufacturing. One area which has received little or no attention is the shipping department. The shipping supervisor, John Maxwell, is also in charge of quality control and inspection for Lisbon.

John has found that cardboard boxes with rigid foam separators are the best packaging for Lisbon products. The foam is purchased and cut to size, most frequently with a table saw. However, this was found not to be effective for cuts 0.25 inches thick or less, which were needed for some separators. A friend suggested he try a slicing machine which he was going to discard because the switch had been removed and his company no longer needed it. John found that he only needed a simple on/off toggle switch, which he bought from a local store. In his basement, he wired the switch in place and mounted it on to the base of the slicer with the off position to the left and on position to the right. The unit was then mounted on a bench in the shipping area where it was used successfully for some time.

As Lisbon prospered, John hired an 18-year-old assistant to train to take over the shipping department. One day, when the assistant was cleaning the area, he removed the guard from the slicing blade and, in doing so, he moved to the right and inadvertently knocked the projecting toggle switch, turning the unguarded blade on and lacerating his left forearm.

The actual toggle switch projected 0.75 inches, displaced 90°, and had an activation resistance of 34 oz, all within the recommended guidelines (see Fig. 3.11). This shows that the ergonomic design of the switch was not enough, and

Toggle Switch Characteristics (Adapted from Department of Defense, 1974.)

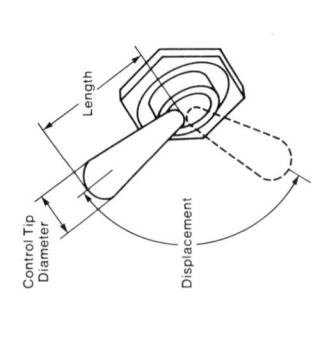

Parameter	Recommended Design Values (Minimum–Maximum)
Control Tip Diameter	3–25 mm 0.12–1.00 in.
Length	
normal	12–50 mm 0.5–2.0 in.
if operator wears gloves	38–50 mm 1.5–2.0 in.
Displacement	
2-position switch	30°–120°
3-position switch	18°–60°
Resistance	
normal	3–11 N 10–40 oz
if control tip is small	3–5 N 10–16 oz

The recommended range of dimensions and resistances for two- and three-way toggle switches is given. The minimum resistance is specified to reduce the potential for accidental activation of the switch. The minimum and maximum displacements of the control are also specified.

Push-Button Characteristics (Adapted from Department of Defense, 1974; Moore, 1975; Murrell, 1965.)

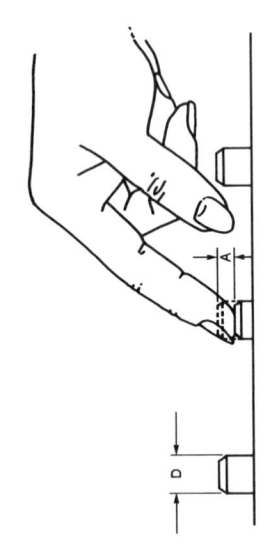

Parameter	Recommended Design Values (Minimum–Maximum)
Diameter (D)	
Fingertip activation	10–19 mm 0.38–0.75 in.
Palm or thumb activation	19–NA mm 0.75–NA in.
Emergency push buttons	not less than 25 mm not less than 1.0 in.
Displacement (A)	
Finger activation	3–6 mm 0.12–0.25 in.
Palm or thumb activation	3–38 mm 0.12–1.50 in.
Resistance	
Finger activation	2.8–11 N 10–40 oz
Thumb activation	2.8–22.7 N 10–80 oz

Note: NA indicates data are not available.

Recommended diameter (D), displacement (A), and resistance ranges for a standard push button are shown. Distinctions are drawn between push buttons operated with the index or middle finger and those activated by the thumb or palm. Maximum diameter is not indicated for the latter condition because it varies with the location of the push button in the workplace.

Recommended Design Criteria for Rotary Selector Switches (Adapted from Department of Defense, 1974.)

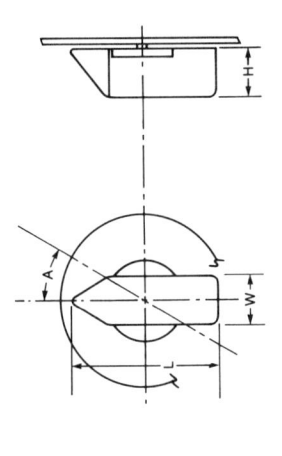

Parameter	Recommended Design Values (Minimum–Maximum)
Dimensions	
Length (L)	25–100 mm 1.0–4.0 in.
Width (W)	NA–25 mm NA–1.0 in.
Depth (H)	16–75 mm 0.6–3.0 in.
Displacement (A)	
Closely grouped controls	15°–40°
Widely separated controls	30°–90°
Resistance:	0.110–0.675 N·m 1–6 lbf·in.

Note: NA indicates data are not available.

Recommended dimensions (L, W, and H), displacement (A), and resistance ranges for bar-type rotary selector switches are given. Minimum widths (W) are not given since this value will vary with the characteristics of the material used to fabricate the switch. The marks at either end of the displacement path represent stops.

Figure 3.11 Design characteristics of common switches. (Reprinted by permission of Eastman Kodak Co.)

that the ergonomic analysis of task performance should also have been applied. Such an analysis would have focused on operator activities while the blade was unprotected. ■

The location and spacing of controls, therefore, can affect safety as well as the control function itself. Where a control is best located depends on the function of the control, how it is activated and how frequently it is used. Of course, those controls which are most frequently used should be the easiest to reach, just like the test tube racks in Case History 3.3. All controls should be located—and guarded—where they cannot be inadvertently activated, as demonstrated by Case History 3.4. However, there are other factors which should be considered when locating controls:

1. Frequently used controls should be placed away from each other and barrier shields should be used to prevent inadvertent activation.
2. In sequential operations, arrange controls so that worker motions are continuous.
3. Use hand controls for precision or dexterity, and foot controls or power-assisted controls when large forces must be applied.
4. For multiple controls of the same function, provide a line of vision between the controls.
5. Distinguish between functional controls and emergency controls, and make sure that the emergency controls are disproportionately large and easy to activate (see Fig. 8.7 for an example of how *not* to do this).
6. Avoid flexible leads for foot controls.
7. Locate emergency controls within reach of either hand of the worker.
8. Whenever possible, test for the rate of human error—the touch telephone key was adopted because it was found to combine speed with lower error rates than other key layouts.

When similar controls are used on a panel, separation is necessary to minimize human error and eliminate confusion. Recommended separation distances depend on the type of control and what it is used for. For example, for random activation, it is desirable to separate push button controls or toggle switches by 2 inches, and foot levers by 10 inches. For sequential operations, however, these separation distances are 1 and 8 inches, respectively.

Case History 3.5 Who Controls?

Metallized plastic balloons have become a popular means of advertising as well as party favors, thereby increasing demand. As a prime producer of the metallized polyethylene terephthalate (PET), Ballardvale Balloons recently installed a semi-continuous metallizing system, the heart of which is a rotating pay-

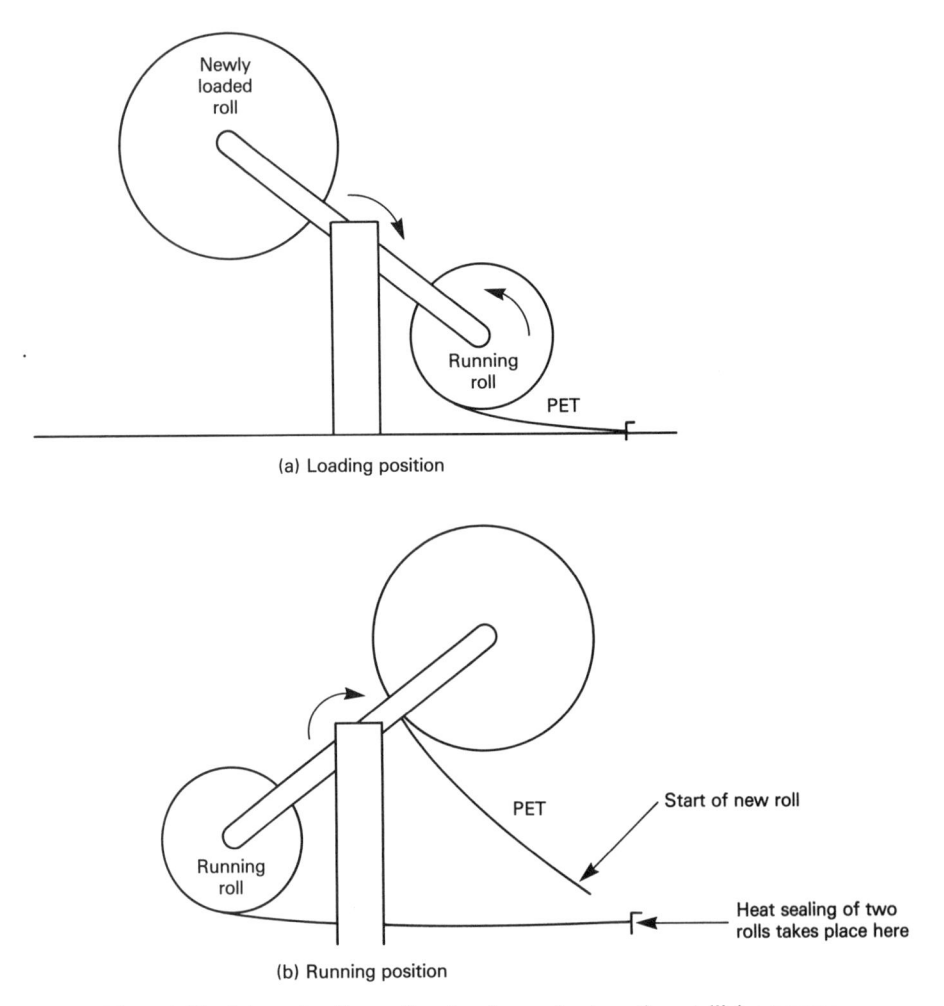

Figure 3.12 Schematic of pay-off system for semi-automatic metallizing process.

off device for handling large, 1000-lb rolls of PET which are drawn through the vacuum metallizing furnace by a take-up drive reel at the finished end (this pay-off device is shown schematically in Fig. 3.12). PET rolls are greater in diameter than they are in length when loaded on to the pay-off device and a special loading fixture was fabricated to lift the new rolls into the loading position with a 1.5-ton electric crane.

When Ballardvale designed this pay-off system, they contracted a machinery builder to construct the equipment, a millwright firm to install it and an electrical contractor to install the controls as well as provide electrical connections. This installation was expected to increase metallizing capacity by 250%, while only requiring one full-time operator. A part-time craneman, however, would be necessary to remove the empty rolls and install new ones with the crane and loading fixture. Indeed, Ballardvale was satisfied with production for the first 4 months, until the day when the craneman was injured while loading a new roll. Two fingers of his right hand were amputated as a result of being crushed between the hub of a new roll and the loading fixture. The machine operator saw the new roll in place and activated the control to rotate the system from the loading to running position, without noticing that the craneman had not removed the roll from the loading fixture. The craneman shouted this to his colleague who reacted by hitting the reverse button, which caused the crushing injury.

This was a violation of the ergonomic principle that an operator should be able to view any operation he or she is controlling. This was easily remedied, however, by relocating the rotational control by the craneman. This would have been specified by correct ergonomic system design. ■

Ergonomics and Information Display

The display of information is necessary to provide us with the status or condition of a process or piece of equipment. The communication of information to the operator or user is of prime importance. We should consider the perception of the user, his ability to discern the message from among all other information, and the importance of the message conveyed. In order to improve efficiency and reduce errors by users of displays, we should consider first the message, then how to display it and, finally, how to ensure detection and communication of the message.

The message which we want to convey can be categorized on a need-to-know basis. When an operator is required to act, we do not want to obscure the message with less important or trivial information. Therefore, we should exercise good judgment in determining what is and what is not necessary. Messages can be presented by visual or auditory means, or a combination of the two. Table 3.4 summarizes research on which form of presentation to use.

It is also important that the displayed information is detected. For example, detection is more likely if both audio and visual displays are presented simultaneously. Artificial signals which an operator must respond to increase the probability of detection; however, if a response is not necessary, the signal is likely to go undetected. Also, we must neither use too many nor too few signals to ensure that the display is detected. Table 3.5 summarizes the preferred displays for the detection and comprehension of task-oriented information.

Having the message understood is also a matter of ergonomics. For example, in viewing a display we are most comfortable when looking straight ahead and down about 15° from the horizontal, and when the viewing distance is less than 24 inches (and more than about 7 inches). This is the type of anthropometric information that can be used for effective work station design. There are also useful guidelines which have evolved from perception research, for the presentation of symbols. The typographic design of characters, such as their width-to-

TABLE 3.4 When to Use the Auditory or Visual Form of Presentation

Use Auditory Presentation If:	Use Visual Presentation If:
1. The message is simple	1. The message is complex
2. The message is short	2. The message is long
3. The message will not be referred to later	3. The message will be referred to later
4. The message deals with events in time	4. The message deals with location in space
5. The message calls for immediate action	5. The message does not call for immediate action
6. The visual system of the person is overburdened	6. The auditory system of the person is overburdened
7. The receiving location is too bright or dark-adaptation integrity is necessary	7. The receiving location is too noisy
8. The person's job requires moving about continually	8. The person's job allows him or her to remain in one position

Source: VanCott and Kinkade, 1972, p. 124, table 4-1.

height ratio and their overall size, can affect their legibility and visibility under different viewing conditions. For example, with reasonably good illumination, a width-to-height ratio of 1:6 to 1:8 is preferred, whereas 1:5 or less is recommended for low lighting. The basic perceptual principles relevant to the design of visual code systems are often ignored (see Fig. 3.13 for examples which relate to machine displays).

TABLE 3.5 Recommended Displays for Information

Preferred Display	Type of Information	How Display Is Used
Auditory		
Siren	Alert	Immediate, short-term attention
Whistle	Attention	Frequently required attention—getting in low level noise
Bell	Attention	Infrequently required attention—getting in low level noise
Visual		
Digital read-out	Quantitative	Minimum reading time and error rate
Moving dial	Quantitative or check	Trends and deviation from normal, easy to detect
Moving dial or digital read-out	Adjustment	Dial movement and control change must be coordinated
Lights	Status reading	On/off, color coding
Annunciator lights	Instructions	Blinking, gives instruction on light
Video display terminals, charts	Graphic	Continuous surveillance

Adapted from Burgess, and Grether and Baker.

Ergonomics or Human Factors in the Workplace 71

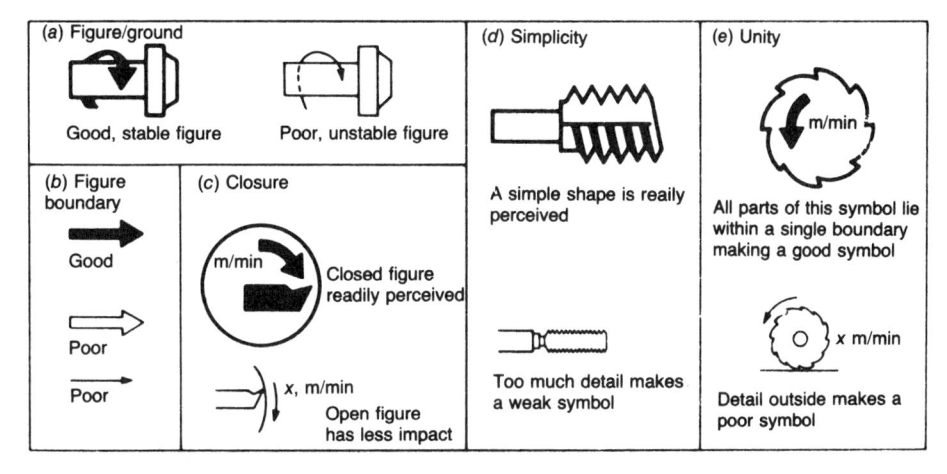

Figure 3.13 Examples of certain perceptual principles relevant to the design of visual code symbols. These particular examples relate to codes with machines. (From Sanders and McCormack.)

Ergonomics, Cumulative Trauma Disorders and Tool Design

Within the workplace, ergonomics has been applied to cumulative trauma disorders, caused by the hand and body motions required for many types of repetitive work. Frequent or continuous use of the same muscles is required in certain work activities, e.g. electronic or mechanical assembly procedures, meat processing, textile manufacturing and packaging. The most common manifestations of such repetitive work are inflammations and injuries to the hand, wrist, arms and shoulders, and the best known are carpal tunnel syndrome, tenosynevitis, and white finger. It is only recently that we have begun to recognize how extensive these disorders are, how they occur and what we can do about them.

The human hand is very complex, and is composed of bones, arteries, nerves, ligaments and tendons. By means of the tendons which connect them, the muscles in the forearm control the flexion of our fingers. These tendons and the median nerve pass through the carpal tunnel, a channel in the wrist formed by the bones of the back of the hand and the transverse ligament which forms the heel of the palm. The configuration of the wrist joint allows the hand to move in only two planes: up and down (flexion) or side-to-side (deviation). When the wrist and forearm are lined up, we do not experience any problems; however, when we bend the wrist down (palmar flexion) or sideways away from the thumb (ulnar deviation), the tendons bend and interfere with each other in the carpal tunnel. Tenosynevitis, an inflammation of the tendons and their sheaths, is caused by continued bending. This continued interference and bending of the tendons compresses the median nerve, causing numbness, loss of grip and even muscle atrophy in the forearm—the symptoms of Carpal Tunnel Syndrome. White finger is a circulavascular disorder prevalent among workers who use vibratory tools, such as jackhammers. It is identical to Raynaud's Disease where the hands or fingers turn white because of a reduction in the flow of blood to those areas, manifested by tingling, numbness and even pain.

Ergonomics can be used to examine the types of motion involved in worker tasks, e.g., painting involves palmar flexion, and gripping a conventional pair of pliers results in ulnar deviation. Many tasks, however, have not been analyzed and new tasks appear constantly. In many of these cases, videotaping has proved invaluable in determining the frequency, extent and type of motion.

Although we have much to learn about the ergonomics of repetitive tasks, many advances have already been introduced, principally in the redesign of hand tools, e.g. curved pliers have eliminated ulnar deviation when gripping them.

Five factors have been identified for hand tools that affect the health and performance of their users:

- static loading of arm and shoulder;
- awkward hand position;
- pressure on palm or fingers;
- vibration and noise (in particular power tools); and
- pinch points on double-handled tools.

Redesigned hand tools that have taken these factors into consideration are now available, though further research is required to reduce the incidence of cumulative trauma disorders further.

FURTHER READING

D. C. ALEXANDER, *The Practice and Management of Industrial Ergonomics,* Prentice-Hall, Englewood Cliffs, N.J., 1986.

C. ARGYRIS, *Personality and Organization,* Garland STPM Press, 1987.

B. J. BELL, *Evaluating the Contribution of Human Errors to Accidents,*

J. G. BURGESS, *Designing for Humans: The Human Factor in Engineering,* Petrocelli Books, 1986.

R. S. EASTERBY, "The Perception of Symbols for Machine Displays," *Ergonomics,* **13,** 149, 1970.

EASTMAN KODAK, *Ergonomic Design for People at Work,* 2 Vols, Van Nostrand Reinhold, 1983.

F. HERZBERG, B. MANSNER and B. BLOCH SYNDERMAN, *The Motivation to Work,* John Wiley, New York, 1967.

J. B. REVELLE and L. BOULTON, "Worker Attitudes and Perceptions of Safety," *Professional Safety,* **26,** 28, December 1981 (Part I); **27,** 20, January 1982 (Part II).

M. S. SANDERS and E. J. MCCORMACK, *Human Factors in Engineering and Design*, 6th Edition, McGraw-Hill, New York, 1987.

A. SWAIN and H. GUTTMAN, *Handbook of Human Reliability Analysis with Emphasis on Nuclear Power Plant Applications,* Final Report (NUREG/CR-1278), Nuclear Regulatory Commission, Washington, D.C., 1983.

H. P. VANCOTT and R. KINKADE, *Human Engineering Guide to Equipment Design*, Rev. Ed., Washington, DC, American Inst. of Research (B. H. Deatherage, pp. 123–160, W. F. Grether & G. A. Baker, pp. 41–121), 1972.

QUESTIONS

3.1. List the top three motivators in Figure 3.1. Explain how you, as a supervisor, can use these motivators to improve the safety of your workers.

3.2. How can we apply the information in Figure 3.2 in getting the message of safety across?

3.3. Explain why witnesses to an accident, including the injured worker, have different descriptions of what happened. Does it matter whether statements are obtained before the end of work or the next day?

3.4. In your own words, explain what human error is. Did your understanding of human error change while reading this chapter?

3.5. What criteria would you follow in selecting employees for a manual labor position? Explain.

3.6. List three situations in which you have been restricted by anthropometry.

3.7. Explain in your own words what the work envelope is and how it affects work station design.

3.8. Choose your own example of a machine, and then explain how we can achieve some of the goals of ergonomics in designing its controls.

4

Systems Safety: Management

Those safety programs in effect today have evolved from the combined endeavors of safety engineers, concerned legislators and sound management. Anyone who follows the business world recognizes, however, that drastic changes have taken place in management in the past 20 years and that new books on management are often found at the top of the bestseller lists. Somehow, these changes have not been felt in safety management. This chapter examines the traditional role of management in safety, and the impact of some of the new management theories which are built on trust, subtlety and interpersonal relations.

Industrial safety continues to progress, largely through the continued application of knowledge and techniques acquired over many years. Though industrial safety measures are now well-known, they are only effective if they are executed properly. This is why management is so important—in essence, the function of management is to optimize performance, whether it is in safety, in business, or in an institutional organization.

An organization is composed of people working together to achieve a common goal. Organizations include business, religious, fraternal, educational and governmental agencies, formed to overcome the limitations of individuals. In the natural evolution of the hierarchy of organizations, a leader directs the activities of the group. He is a *manager* who must act, using what little might be known, if the institution is to flourish. He practices *management*. Management is multidisciplinary: it denotes the function, social position and rank of those who practice it; it is a field of study; and it is a profession. Management is the effective heart of any organization, and provides the needs which give life to it.

All organizations have three things in common: management functions, management tasks, and management work. Traditionally, management approaches have fallen into three categories:

1. *Philanthropic:* the desire to look after the needs, housing and welfare of those less fortunate.
2. *Procedural:* the orderly administration of employees and their work.
3. *Prevention:* solving problems, particularly those involving people.

These traditional approaches have led to management by objectives, made popular by Peter F. Drucker. Management, in its leadership role, dictates the direction in which an organization will go, and this is only successful if it knows the goals of the organization, if it sets objectives, if it organizes and, above all, if it performs.

Safety management is a relatively new discipline, and therefore there are insufficient data necessary for valid evaluation. Management first emerged in the business world, and it is only business management which is capable of providing us with sufficient information on the allocation of resources and decision-making. Nevertheless, we shall attempt to examine management concepts with safety in mind whenever possible.

THE TASKS OF MANAGEMENT

Management is the means by which an organization flourishes, and it must never become an end unto itself. Tasks have to be performed for management to fulfill its role. The three tasks which management must perform are as follows:

- carry out the purpose and mission of the organization;
- increase productivity; and
- manage social impacts and social responsibilities.

An organization exists for a specific purpose and mission, and economic performance must always come first in every business decision and action. However, at the same time, organizations have become the means by which we find our livelihood, access to social status, individual achievement and satisfaction. Keeping workers both productive and achieving has become more and more important in measuring the effectiveness of an organization. Safety is a part of the third task of management, i.e. meeting their social responsibilities and managing social impacts. No organization can escape such responsibilities, simply because they address our values and needs which remain after our economic needs are fulfilled.

It should be realized, however, that these management tasks are further complicated by another element, i.e. time. Perhaps we are more cognizant of this than ever before in our history, when we hear daily of environmental issues, such as acid rain, nuclear waste and pollution. Every short-term management decision must consider its long-term impact. Management must live simultaneously in the present and in the future where the only true forecast is that of change.

Administration or procedural management is an important, unavoidable management task. A manager has to manage and improve that which already exists and redirect resources to improve future performance. The primary concern of a manager is to optimize the effectiveness of those activities which yield optimum performance, and strive to fulfill the organization's objectives.

Each management task has its own requirements and skills, but they must all be integrated, and each must focus at all times on what the organization is, and what are its mission, objectives and strategies. To accomplish all of this successfully, we must understand what the roles of management are and what management skills are needed.

MANAGERIAL ROLES AND SKILLS

Henry Mintzberg systematically studied the activities of business managers, and found that they typically face long working hours, intense activity with few, if any, breaks, and that they are plagued by interruptions; little time can be spent

TABLE 4.1 The Roles of Management

Interpersonal	
Figurehead	To attend ceremonies and represent the organization to external constituencies
Leader	To motivate subordinates and integrate the needs of subordinates and the needs of the organization
Liaison	To develop and maintain contacts with outsiders
Informational	
Monitor	To seek and receive information (mail, reports, etc.) of relevance to the organization
Disseminator	To transmit to members of the organization information relevant to the organization
Spokesperson	To transmit to outsiders information relevant to the organization
Decision-making	
Entrepreneur	To seek problems and opportunities and to initiate action in the best interests of the organization
Disturbance handler	To resolve conflicts between persons within the organization or between the organization and external parties
Resource allocator	To make choices allocating resources within the organization
Negotiator	To represent the organization in formal negotiations with third parties such as union officials or government regulators

Source: Schermerhorn, Hunt and Osborn, *Managing Organizational Behavior*, 1985. Reprinted by permission of John Wiley & Sons, Inc.

concentrating on a single subject. Mintzberg was able to identify ten separate managerial roles, separable into three categories: interpersonal, informational and decision-making (see Table 4.1).

In order to fulfill these roles, managers must possess competent managerial skills, those skills which will promote the desired performance. Skills have been classified into three categories:

- *technical:* acquired through formal education;
- *human:* involving interpersonal relations and the ability to work well with others;
- *conceptual:* involving the ability to identify discrete problems and integrate their solutions into acceptable solutions which serve the organizational objectives.

Technical skills are relatively more important at lower management levels, whereas conceptual skills gain in importance at higher levels. Human skills are consistent in their importance at all levels of management. Of course, this is true for safety management skills as well as business management skills.

MANAGEMENT BY OBJECTIVES

Management, remember, provides direction, has to know the mission of the organization, has to set objectives, has to organize and, above all, has to perform in order to be effective. Five managerial functions have been traditionally recognized: planning, organizing, staffing or coordinating, directing or leading, and controlling. This tradition has also been developed for safety management by such authors as Firenze, Ferry, Petersen and Grimaldi.

We recognize that management is faced with multiple tasks—tasks associated with productivity and social responsibility—and the same is true for safety management. These multiple tasks affect our five managerial functions. For example, the purpose of an organization includes profitability and the creation of customers (marketing). This means that we must include safety costs and define who our customer is in safety planning. Social impact and responsibility tasks certainly encompass safety, probably a major reason for the existence of safety management. And, finally, productivity depends on the worker remaining healthy and free of injury.

Planning

The planning function consists of four distinct areas: forecasting, goal and objective setting, policy making, and budgeting. Planning also has two essential characteristics: first, a plan involves the future; and, secondly, it involves action.

Forecasting for an organization is usually performed at the highest levels of management; only in matters of legislation impact, such as the enactment of the 1970 Occupational Safety and Health Act, does safety management become involved.

Setting goals and objectives, however, is very much a part of safety planning. These goals and objectives, of course, must be compatible with the corporate goals and objectives. This means that we have to address the effect of our safety program on corporative profitability, market standing and productivity. Traditionally, we have tried selling safety programs based on the impact of accident costs on profitability or on how adverse products liability litigation might affect market standing. Such tactics have limited appeal, however, for the same reasons that no one thanks the safety officer when he goes home a whole person. However, they are also dangerous for another reason—accidents are rare events, and therefore it is conceivable that complete abandonment of the safety program might have no immediate impact, leading to the erroneous conclusion that it was unnecessary. Safety costs are more appropriately addressed in budget planning than in profitability planning. Selling safety is something we can do better on other grounds, such as social impact and productivity.

The National Safety Council suggests that safety program objectives include:

1. Gaining and maintaining support for the safety program at all organizational levels.
2. Motivating, educating and training program participants to recognize and correct or report hazards.
3. Engineering hazard control into the design of machines, tools, processes and facilities.
4. Providing a program for inspection of machines, tools, processes and facilities.
5. Complying with established safety and health standards.

Once goals and objectives have been set, policy making should be adopted. Policy is not specific like objectives, but is general in nature, giving notice that the program will be carried out. A written policy statement, signed by the Chief Executive Officer (CEO), should be available to each employee. It is important that this statement be written to demonstrate the support of top management for safety and the corporate commitment to safety. The wording of such a statement is almost secondary, but it should describe the purpose behind the program and solicit the active participation of all employees. The National Safety Council recommends the policy statement also reflect:

1. The importance that management places on the health and well-being of employees.
2. Management's commitment to occupational safety and health.

3. The emphasis the company places on efficient operations, with a minimum of accidents and losses.
4. The intention of integrating hazard control into all operations, including compliance with applicable standards.
5. The necessity for active leadership, direct participation, and enthusiastic support of the entire organization.

Budgeting is the last aspect of planning. When it comes to budgeting for safety, we have one outstanding measure; it is also the one cost which we can use validly to sell the safety program to upper management, i.e. the cost of workmen's compensation insurance. There are many other cost factors which enter into safety budgeting, some of which are apparent, such as staffing salaries, but many of which are less apparent and hard to identify; for example, the time spent by a line supervisor doing research on the best respiratory equipment for a new task in his/her department.

Budgeting for safety differs only in concept from traditional considerations of the cost of an accident. In considering the cost of an accident, there are two major classes: insured and uninsured costs. Insured costs include compensation payments, medical expenses, repair or replacement and legal fees. Uninsured costs of an accident are harder to measure and are reflected in the increased cost of doing business: these include wages paid to workers who were not injured, but whose time was spent assisting the injured worker or whose time was spent in accident investigation; wages paid to workers who suffered minor injuries and spent time away from their jobs; the extra costs of overtime; the costs of lost productivity; the costs of training new workers; and costs attributable to the loss of customer confidence. In considering a safety budget, insurance premiums represent the largest simple direct cost. Staff salaries are also direct costs and should include partial values for those who have additional responsibilities beyond safety and health and for those who serve on safety committees (or safety circles) as well as salaries for full-time safety personnel (although there is no absolute rule, only companies with more than 500 employees will usually have full-time safety personnel). Other direct costs include training and personal protective equipment supplied by the company. Indirect budget costs include salaries of line staff when performing safety tasks, public relations, purchasing and liaison—both internal and external to the organization.

Organizing

Organization structure is the oldest and most thoroughly studied area of management. Organizing essentially provides for the distribution of the work of management and establishes responsibility and authority. The structure, however, does not just evolve, it requires thinking, analysis and a systematic approach. For example, surveys have shown that the safety experience in industry is appreciably better when safety personnel report to senior management, yet more than half of safety personnel report to plant, personnel or security managers. We would stand a better chance of success with our safety program if we consider this fact during the organizing stages.

In studying business organization, we have learned that the first step is *not* to design the structure; indeed, it should be the final step. The first step is to identify and organize the building blocks of the organization. Building blocks carry the load and are part of the final structure; they are determined by the contribution they make. In designing the building blocks of our safety organization, we have to address four questions:

1. What should the units of organization be?
2. What components should interface with each other and which should be kept apart?
3. What size and shape should the components have?
4. What is the appropriate placement and relationship of different components?

In order to identify the basic units of organization, it has been traditional to analyze all the activities needed: record-keeping, safety committees, inspections and conformance are typical activities which we might consider. But this approach has not been successful. We have to begin organizing with desired results, then identify the *key activities*. We have to answer these questions:

- In what safety area is excellence required to achieve the corporate objectives?
- In what areas would lack of performance harm the safety program?

We have to realize that in searching for the answers to these questions, many arguments concerning task-focus and person-focus in job safety and organizational structure have been nonproductive. This is not an either/or proposition, but more correctly both—in varying proportions. How different our safety organization might be if hazard identification were the key activity rather than conformance.

Staffing or Coordinating

What activities belong together and which should be treated separately has long been controversial, and has led to organizations which define "line" and "staff" responsibilities. Drucker has pointed out that the distinction has merit, but a more searching analysis is needed which organizes activities by the kind of contribution they make. He defined four major groups, based upon their contribution:

- Those which result in producing activities, i.e. those which produce measurable results.
- Support activities which do not produce results, but feed the result-producing activities.
- Housekeeping activities which have no direct or indirect relationship to results.
- The activities of senior management.

Identifying key activities and analyzing their contributions help to define the building blocks of our safety organization, but how they fit together requires decision analysis and the analysis of relations. Decision analysis deals with the issues of responsibility and authority. We first have to consider the four characteristics of a decision (i.e. future effects, impact effects, qualitative human factors and frequency), and then make it from the lowest possible level nearest to the activity consistent with ensuring that all activities and objectives affected are fully considered in the decision.

The final step in designing the building blocks of our safety organization is an analysis of relations; it tells us where a specific component belongs, with whom we work and what influence we have on each other is significant. The basic rule in placing an activity within an organization structure is to have the smallest possible number of relationships—*keep relationships to a minimum but make each count*. The relationship on which the success of the unit depends, however, should be easy, accessible and central to the unit.

Drucker acknowledges that there is conflict between placement according to decision analysis or relation analysis, and suggests we follow the logic of relations as far as we can. The four analyses—of key activities, of contributions, of decisions and of relations—should be kept as simple as possible, yet never be overlooked. When we consider all of Drucker's ideas for safety, two organizational problems emerge—placement of safety within the business organization and placement of the individual building blocks within safety itself. It would be remiss not to point out again that safety experience is appreciably improved when safety personnel report to senior management. Similarly, we should also apply relations analysis to collateral personnel, i.e. those who are not full-time safety personnel.

Directing or Leading

Staffing and coordinating functions are directed toward performance by linking the actions of the building blocks into a consistent pattern. Various methods are employed and, again, both task-focus and person-focus methods should be used, consistent with the values of the organization. However, coordination involves communication and human psychology almost demands that person-focus methods are emphasized.

Leadership is also directed toward performance; we define leadership as the process of using power to obtain influence. Within organizations, we encounter both formal leadership, derived from designated positions of responsibility and authority, and informal leadership, based upon the special skills or resources that certain individuals possess. Modern views of leadership emphasize contingency approaches which focus on the nature of the task, on the subordinate and on the group. In other words, leadership is also task-focused or human-focused. Task-motivated leaders like clear guidelines and standardized procedures; they function best both in high-control situations where they are comfortable with the structure, *and* in low-control situations where the objectives are consistent with their leadership qualities.

On the other hand, human-focused leaders are concerned with interpersonal relations and are best in situations where different viewpoints are encountered and when creative and resourceful thinking is needed to solve complex problems. Such leaders are best suited in moderate-control situations, where concern with interpersonal relations is most appropriate in problem solving.

Leadership and coordination both seek to optimize performance. Any organization must focus on performance and establish high performance standards, both for groups and for each individual. Simply stated, performance depends upon ability and motivation, the combination of which can enable the average person to achieve extraordinary goals. Performance does not mean that we have to succeed each and every time, but it does mean that we should not become complacent or accept inferior solutions. The focus on the ability of people should be positive, based upon what we can do rather than what we cannot do. Performance depends upon us, upon our values and morality. Because of this, management decisions that affect people, particularly the designation of roles, are important because they reflect the true values and beliefs of the organization. In people decisions, integrity is the one absolute requirement of the management and the manager.

These aspects of performance emphasize the ability of people, but what about motivation? In Chapter 3 we examined people perspectives; here, we look at management perspectives which must focus on communication. Without communication, our safety programs are doomed to failure. Although there is an information "explosion", and the study of communication has been extensive, it is clear that less and less communicating occurs in light of all the new information.

Peter Drucker has pointed out four fundamental aspects of managerial communications which we have learned, mostly by trial and error, and which we can use to our advantage for motivation and performance:

1. Communication is perception.
2. Communication is expectation.
3. Communication makes demands.
4. Communication and information are different—yet interdependent.

Communication requires both a sender and receiver, but we know that a concept cannot be communicated unless the receiver can perceive it. This means we must work out our own concepts before we communicate, then communicate in terms that the receiver can understand. This means that we must know the receiver, his beliefs, motives and aspirations. Before we can communicate downward in an organization, we have to have upward communication.

Let us look at ourselves as receivers for a moment. What do we perceive from a message? As a rule, we perceive what we want or expect to perceive, hear what we want to hear or see what we want to see. We all tend to fantasize to a certain extent, fitting impressions into what our expectations are, defending rigorously against any change. Therefore, when reversing our role to that of sender, we must be aware of what is expected. Our communication must utilize that expectation, or shock techniques designed to break through those expectations must first be employed.

Motivation implies perception of the receiver which is focused on the values of the receiver, making demands upon the receiver to do something, believe something or become somebody. We face dangers in the safety profession, however, because too much communication (or too many laws) can be perceived as propaganda, and then all safety communications become suspect. It is easy to confuse presentation of information with communication, yet the requirements of information are the opposite to those of communication. Thus, the information contained in safety regulations is specific, but effective communication skills are required if it is to be understood by the receiver.

Controlling

In management, controlling provides direction in accordance with the organization's objectives. Controls, on the other hand, are the means by which management maintains such direction is achieved. Drucker has pointed out that the true control of an organization which wants performance is dependent on "people" decisions, e.g. placement, pay, promotion, demotion and terminating employment. These decisions model and mold behavior far more than any other decisions. Nothing could be more true in safety management: it is the people we select for the jobs in safety—whether for safety committees, safety training and motivation, or for supervision—that will establish the perception of our safety programs.

Controls are needed to establish this perception, and therefore they must be subjective, involving both the observer and the event being measured. Controls must also be focused on results, which suggests that inspections and recordkeeping are critical safety management functions. These two factors—measurement and information—are indeed what make up controls. To give the safety manager control, we must establish controls which have certain specifications, many of which we can derive from legislated safety standards:

1. They must be economical, with better controls requiring little effort. For

example, computer applications in safety have proved economical tools for record-keeping.
2. They must be meaningful and related to key activities. For example, safety standards should be incorporated into task designs at an early stage.
3. They must be appropriate, i.e. priorities must be correctly established. Past safety programs were severely limited because they focused on accidents which we know to be rare events.
4. They must be congruent, i.e. not misleading. One of the problems which we face in safety is false complacency because of inadequate hazard identification (see Chapter 5).
5. They must be timely. In safety, this means we must have controls which are both stable and maintainable over time.
6. They must be simple and understandable. In today's language, this means that controls must be "user-friendly."
7. They must be operational, i.e. focused on action. Safety controls require analysis which provides feedback or communication to improve safety management control.

Case History 4.1   High-tech Industrial Safety Management

High-tech industries are a recent phenomenon, spawned by spin-off companies formed from academic and other basic research efforts in electronics, physics, chemistry, computer science and other fields with rapidly developing technologies. Many Fortune 500 companies (e.g. IBM, 3M and Xerox) epitomize the meaning of high-tech industries. In many respects, these companies have been considered to be clean and safe places to work; this image is not entirely correct, although accident records within the industry are far better than for construction or heavy manufacturing industries.

Comput-all is a typical high-tech company, formed in the 1960s when microelectronics began the trend toward newer and smaller electronic products. Comput-all began with the idea of miniature calculators, but began producing computers in the early 1970s and have continued to expand, and today it has a workforce of approximately 30,000. Safety within Comput-all was haphazard with little or no emphasis and no safety management during the formative years. As a result, the frequency of accidents which involved workmen's compensation escalated to about 65% above the computer industry average, a fact brought to the attention of Comput-all by their insurance company.

Comput-all responded by hiring a safety management consultant who provided immediate relief while seeking to hire and establish meaningful safety management for the company. The first step he took was to emphasize the importance of safety. A policy statement was posted and *sent to the homes* of all employees with a letter from the President of Comput-all. The policy statement (which has been retained to the present) says:

Management by Objectives

Comput-all intends to provide a safe working place for all of its employees and will not knowingly subject them to unsafe conditions. Comput-all will inform employees of all known hazards in their working environment and what precautions have been taken to minimize risks from those hazards. Comput-all expects that employees will cooperate by following all safety procedures and take reasonable care to protect themselves in the performance of their jobs. Accident prevention, maintenance of a safe working environment and property protection are the responsibility of each and every one of us at Comput-all.

Working with the senior management of Comput-all, the consultant established simple but effective goals and objectives for their safety program. Each plant manager was assigned responsibility for safety and loss prevention at his facility, with the corporate safety program being based upon control of the work environment *and* behavior in the workplace. The size of the organization, as well as its activities, was taken into account. Comput-all established the position of the corporate safety and loss prevention manager who reported directly to the vice president of administration. Beneath the corporate manager were a staff of safety professionals, including risk managers, safety engineers and safety technicians. The total number of safety personnel assigned was 1 per 1000 production workers and 1 per 2500 administrative staff. A safety committee and emergency preparedness organization was also established for each facility. The activities of these safety personnel were expected to include:

- job observation and analysis,
- hiring and placement,

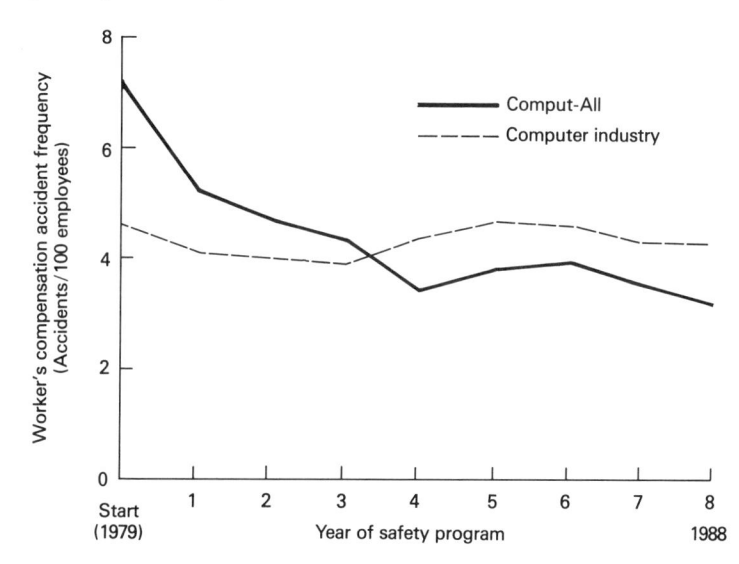

Figure 4.1 Effect of safety management on Comput-All's safety performance in initial years.

- training,
- establishment of safety rules and regulations,
- work safety standards,
- maintenance safety standards,
- purchasing safety standards,
- record-keeping.

Under the leadership of the corporate safety and loss prevention manager, Comput-all's safety record has steadily improved while that of the computer industry has been relatively unchanged. In 1989, Comput-all's safety record improved to the point where the frequency of accidents requiring workmen's compensation was 25% less than the computer industry as a whole. The steady improvement directly traceable to the improved management of Comput-all's safety program is shown in Fig. 4.1. ∎

TRANSITION IN MANAGEMENT THEORY

Management by objectives represents what Petersen calls the Contingency School of Management. The assumption behind the theory is simply that "everybody is different," and that management style and how we deal with workers must be contingent upon the situation, the worker and his needs. Petersen also refers to earlier management efforts as the Classical School, (based upon Theory X management) and the Human Relations School (based upon Theory Y management), and is righteously critical that too often our safety programs have remained firmly rooted in the classical school of thought, depending on standards of performance, corporate safety manuals, and rules and regulations to get the safety job done. What we need to do is to build our safety program so that it will be compatible with our style of management.

If we extend the contingency method of management (or management by objectives) to current thinking, the discord between safety program thinking and management thinking is even more noticeable. Peters and Waterman, in their popular book *In Search of Excellence,* examined the management of many companies and described the management of the most successful companies as being brilliant on the basics, always intense, customer-oriented, but also willing to listen and treat people well. Their research showed eight attributes of good companies:

1. Bias for action, i.e., a preference for doing something—anything—rather than subjecting a question to exhaustive analyses and committee evaluation.
2. Staying close to the customer, i.e. real customer service.
3. Autonomy and entrepreneurship, i.e. breaking management into small units and encouraging independence and competition.
4. Productivity through people, i.e. creating in *all* employees the awareness that their best efforts are essential.

Transition in Management Theory

5. Hands-on, value-driven, i.e. insisting that executives keep in touch with the essential business of the firm.
6. Stick to the basics, i.e. remaining with the business the company knows best.
7. Simple form, lean staff.
8. Simultaneous loose–tight properties—dedication to central values with tolerance for all who accept the values.

We can redefine these attributes in order to relate to good SAFETY management:

1. Bias for action—be ready to innovate, secure enough to try and fail, but move toward success.
2. Close to customer—the workers are our safety customers. Learn their preferences and indulge them.
3. Autonomy and entrepreneurship—how effective are supervisors in the safety effort?
4. Produce safety through people.
5. Hands-on, value-driven—involve all parties in safety.
6. Stick to basics—the KISS principle.
7. A lean staff promotes effective communication.
8. Simultaneous loose–tight properties promote autonomy while adhering to the real safety goals and objectives.

Innovation is crucial to the success of safety programs because all the elements of safety technology can be present, but our own social innovations are needed to make the technology work. In many respects, the other attributes Peters and Waterman have emphasized differ little from similar observations made by Likert in 1967. Likert found corporate success depended strongly on the amount of confidence and trust between workers and management, that teaching workers how to solve their own problems was better than giving them the answers, that information sharing was more important than only "need to know," that a better climate resulted from actively soliciting ideas, and that recognizing accomplishments also created a successful climate. In many respects, innovation is the ingredient which translates what should be done—the attributes of success—into true success.

If we look at the ideas emphasized by other popular management authors, such as Ouchi (Theory Z), Blanchard and Johnson (One Minute Manager), and more recent books or articles by Drucker, Petersen, Peters and others, we find some attributes to sound management which may not differ a lot from those of Peters and Waterman, but which have a different emphasis:

1. Commitment, relentless dedication, pride in work.

2. Involvement, a key to increased productivity.
3. Communication, particularly feedback and perception.
4. Trust, self-confidence.

Peters has described the results produced by commitment, using examples of the persistent champion whose best characteristic is the ability to select and nurture other champions. A main benefit of the resulting interpersonal relationship is the ability to focus on problems without being judgmental. Drucker has pointed out that the most important management decision is employee selection, because those people determine the performance capacity of the organization. In addition, Drucker points out that workers have the right to competent command.

Ouchi, in his description of Theory Z management, approaches commitment through trust. People committed to long-term relationships have strong commitments to behave responsibly and equitably toward one another. Ouchi also points out the subtlety of relationships, their complexity and dynamic nature, and that intimacy is an essential ingredient in a healthy society. Marineau[1] focuses on self-confidence, which he defines as positive regard, both self-regard and regard for others. Positive self-regard permits the recognition of strengths, compensation for weaknesses, allowance for failure and continued growth or development. Positive regard for others includes acceptance of people as they are—with their strengths and weaknesses—dealing with people in the present not the past, and dealing with people in a courteous manner, promoting their self-esteem.

The "One Minute Manager" is also built on the simple principle of self-confidence, i.e. that people who feel good about themselves produce good results. The One Minute Manager encourages goal setting, then matches performance to goals, with problems only arising if there is a difference between what is happening and what is supposed to happen. The premise to this concept is that feedback (communication) is the prime motivator and that behavior can be managed through one-minute praisings and one-minute reprimands.

The common ingredient of all these attributes and keys to successful management is *people* (Fig. 4.2). When we now try to integrate these ideas into safety management, we must focus on the people and their behaviors and attitudes. Although we know that there is no single person or plan which can make safety work in the workplace, safety must be shared and recognized as a basic part of the corporate culture.

Case History 4.2 The Alcoa Experience[2]

Alcoa has had a traditional safety program, consisting of posters, programs, some engineering and a lot of hopeful talk with mixed results. There was clear room for improvement, particularly in viewing safety as being as basic a part of

---

[1] Based on presentation to ASSE Professional Development Conference, Baltimore, June 1987.

[2] Based on the remarks of C. F. Fetterolf, President, Aluminum Corporation of America to the American Society of Safety Engineers, 1984.

Transition in Management Theory

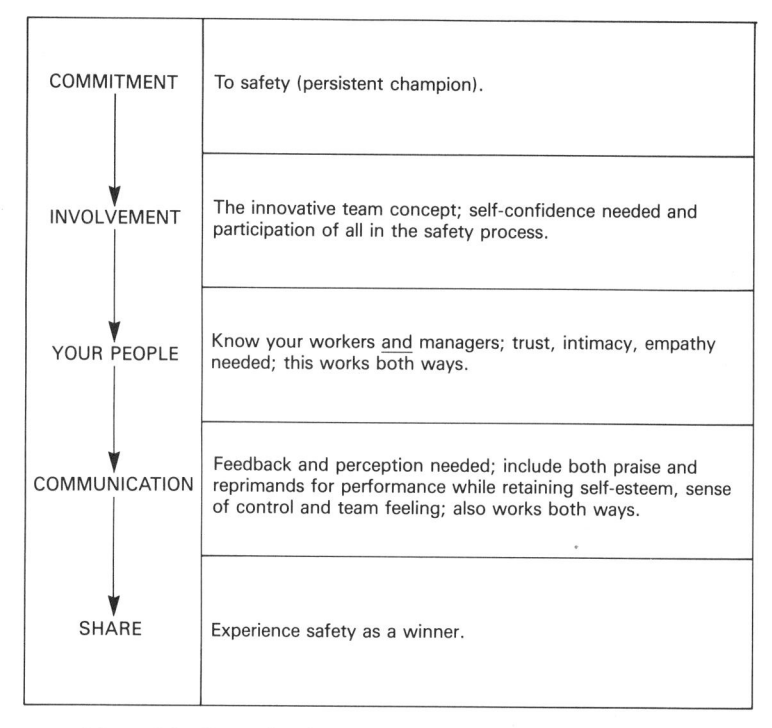

COMMITMENT	To safety (persistent champion).
↓ INVOLVEMENT	The innovative team concept; self-confidence needed and participation of all in the safety process.
↓ YOUR PEOPLE	Know your workers and managers; trust, intimacy, empathy needed; this works both ways.
↓ COMMUNICATION	Feedback and perception needed; include both praise and reprimands for performance while retaining self-esteem, sense of control and team feeling; also works both ways.
↓ SHARE	Experience safety as a winner.

Figure 4.2 Key safety ideas from current management theory.

the job as production, quality, delivery and cost control. Alcoa management viewed safety, in the final analysis, as an attitude. In this view, the main job of the safety professional must be to promote and manage that attitude. Alcoa looked at the "new" management concepts and the safety professional in light of these concepts. The four parts of the safety professional were:

1. Basics: the nuts and bolts of safety, accident investigations, record-keeping, employee contacts and job safety analyses; these are the time-tested, measurable fundamentals.
2. Cost-effectiveness: although it seems callous to put a price tag on safety, nothing is without risk and we must make choices.
3. New breed: his main concern is people and uses the latest techniques in managing safety problems in the workplace.
4. Corporate executive: looks at safety from the broadest vantage point, both in terms of people and cost to the company.

The key to the safety professional, however, is the view toward today's issues as opportunities in safety.

Alcoa found that managing an attitude is not easy, but it can be done. The turnaround came about because of a fatality in one of the plants. Though it was not caused through negligence—there was nobody to blame—the loss of the man's life in a plant Alcoa was responsible for had an impact. Alcoa had experienced fatalities before, only a small number, but enough to make people think of such incidents as just part of the cost of doing business. It was *this* attitude that had to be changed. The task was to find a way to make safety basic to a job at Alcoa. Alcoa's first step was to issue a policy statement stating just that, and then develop joint safety committees and ensure the ongoing commitment of senior management. The tough part, however, was to involve the workers.

Alcoa makes aluminum, a basic industry. They were steeped in the traditional labor relations environment of basic industries, a classic adversarial situation. After 50 years of looking at one another from across a bargaining table, it was not easy to sit down one-on-one with a union member and ask him for his cooperation. But that is exactly what the Alcoa safety professionals did, and it paid off.

Alcoa made it part of their safety program to involve everyone. They brought hourly and salaried people together into problem-solving situations and gave both authority to work out solutions. The results were slow at first, and there was a lot of harsh experience to overcome, but Alcoa kept at it. In the end, injury rates were cut by 50% within 5 years—half the first-aid cases, half the medical cases, half the restricted work schedules, and, most dramatically, half the lost workdays. Although Alcoa was concerned with simply doing what was right and did not hold a cost yardstick to safety, the reduction in the number of accidents and injuries has saved the company $25 million a year. The real success of the Alcoa experience, though, is the commitment of the Alcoa employees, something Alcoa believes could not be bought with all the money in the world. ■

The Alcoa experience represents the innovative management concepts this chapter has emphasized, but it is unique only because of the changes within the organization which had to occur. These changes had to occur in upper management in order for the success to occur. Management's reaction to change therefore determines the success that change can bring and the innovation process is secondary to the reaction. With Alcoa's change and new commitment to safety, we see a behavioral system at work. Behavioral systems are not new, but the innovation and commitment necessary for the change is. The key features of the Alcoa experience are the change in management, the involvement of workers in safety, the recognition of risk management, and the training and patience necessary for workers to change their attitudes. Let us now look at each in turn.

BEHAVIOR MANAGEMENT

Positive reinforcement, like many other concepts, was talked about long before it was tried. The concepts are simple and straightforward, but their practical and successful application in an industrial environment is not easy. The central prin-

Behavior Management

Figure 4.3 Behavior feedback loop. (From T. Krause, J. Hidley, and W. Lareau, "Behavioral Science Applied to Accident Prevention," *Professional Safety,* July 1984, p. 21; reprinted by permission of the authors.)

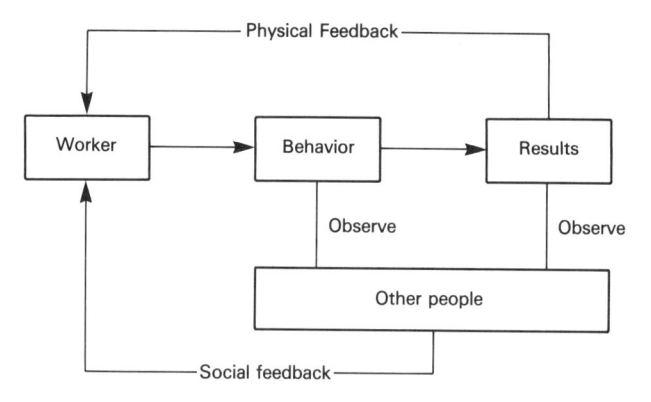

- Positive reinforcement feedback for more behavior
- Punishment for less behavior

ciple of behavior management is performance-related feedback. The most effective way we can modify or change performance is to provide the performer with feedback as soon *and as often* as possible. Figure 4.3 demonstrates how the feedback can affect behavior which is operationally defined, such as using personal protective equipment. We have known for a long time, however, that failure to exercise controls simply by overlooking or enabling improper behavior is a common management error that creates problems for workers. Thus, occasional punishment of unsafe behavior cannot help very much in reshaping behavior. Certain critical problems arise when adopting behaviorally oriented programs: the use of positive reinforcement rather than punishment; the difficulty of defining safe behavior in many instances; and the difficulty of eliminating the social acceptance of unsafe behavior. Petersen has shown that various measures used to evaluate behavioral programs (e.g. how often personal protective equipment is used, the frequency of safe behavior, accident frequency and accident severity) have all shown significant improvements, no matter what rewards were used in behavioral management program studies.

Behaviorally oriented solutions have been used by Krause *et al.* in an equipment manufacturing plant which had a lost-time accident rate of 19.6 per 100 manufacturing employees, despite rigorous traditional safety efforts. The program consisted of two phases—Investigation and Intervention (see Fig. 4.4). Statistical analysis focused on the incidence and severity of injuries per man-hours worked, which required modification of the accident data system. Pareto analysis demonstrated that maximum benefit could be realized by concentrating on eight occupations which accounted for 76% of all accidents and 87% of severe injuries; most lost time was attributable to back injuries.

The most startling revelation during the investigative phase came from interviews which were conducted to assess the level of safety knowledge and to obtain information about behavioral patterns in the work area. Workers who had

Phase I. Investigation

```
Statistical analysis ─────────► Review safety program
        │                              │
        ▼                              ▼
  Modify accident                  Interviews
   data system                         │
                                       ▼
                              Behavioral observations
                                       │
                                       ▼
                               Develop critical
                               behavior index
                                    (CBI)
```

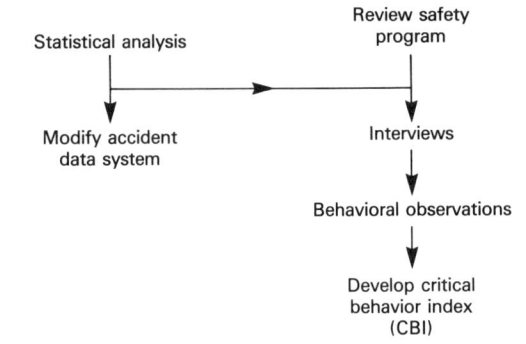

Phase II. Intervention

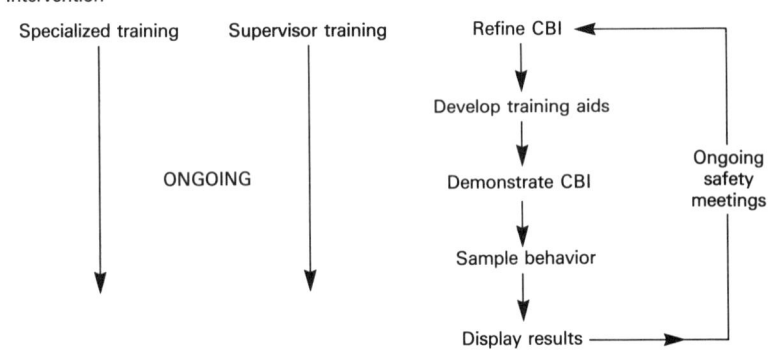

```
Specialized training    Supervisor training    Refine CBI ◄──────┐
        │                      │                    │            │
        │                      │                    ▼            │
        │                      │            Develop training aids│
        │                  ONGOING                  │        Ongoing
        │                      │                    ▼         safety
        │                      │            Demonstrate CBI  meetings
        │                      │                    │            │
        │                      │                    ▼            │
        │                      │             Sample behavior     │
        ▼                      ▼                    │            │
                                                    ▼            │
                                            Display results ─────┘
```

Figure 4.4 Flow chart of behavioral safety program. (From Krause, Hidley, and Lareau.)

good safety records were easily distinguished from those who did not by their attitudes during the interviews—they were more cooperative and less defensive. Both safe and unsafe workers, however, demonstrated a lack of knowledge regarding existing safety programs. Most supervisors were defensive and poorly informed and found it difficult to enumerate unsafe behaviors or behavioral factors important to accident prevention. Instead, they incorrectly tended to blame equipment, poor training and inexperience for accidents.

The statistical analysis and interviews were examined and a strategy was developed to conduct behavioral observations. The objective, of course, was to identify observable, accident-related behaviors and to measure their frequency. Both expected and unexpected unsafe behaviors were discovered; additional interviews and meetings with foremen were necessary to define these unsafe behaviors in objective, observable terms to develop the critical behavior index (CBI). The development of the CBI and the introduction of these meetings were judged important simply because the foremen became involved—they began to view safe behavior as an achievable goal.

Behavior Management 93

Intervention must address the problems of behavioral feedback; in this case, positive reinforcement methods were attempted. Positive reinforcement, however, was not practical if it depended on the supervisor to observe and feed back positive reinforcement. It was unlikely that any supervisor could notice a large enough number of safe behaviors to establish any effective reinforcement schedule. Yet it was recognized that ongoing feedback is needed so that each and every safe behavior is reinforced. This was done by establishing training aids using staged unsafe behaviors based upon the CBI and demonstration of alternative safe behaviors. These training aids were presented to all workers in groups led by manufacturing management or foremen who were coached extensively. Each group was given a chart to show the unsafe behaviors observed in their work area, and was asked to set a goal for reducing such behaviors.

Two other programs in the intervention phase were conducted, one planned and one spontaneous. An extensive back school program was instituted because of the nature of the most frequent and most severe injuries. The foremen's meetings became a year-long course which began with an overview of the behavioral intervention program, but was extended to include resistance to change and how to cope with it, accident investigation, hazard identification, and even how to present behavioral data to the safety committee.

The results of the intervention program were both direct and indirect. There

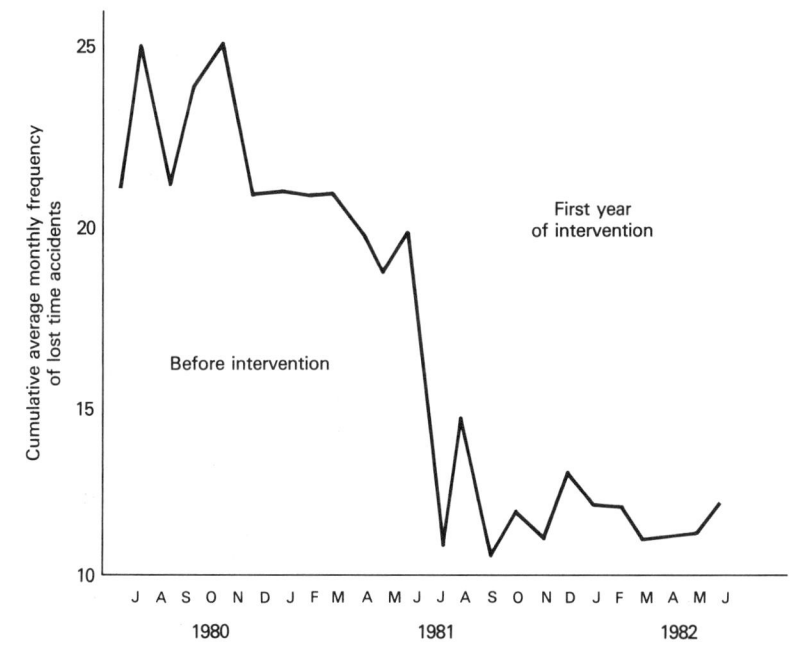

Figure 4.5 Lost-time accident rate before and during the first year of intervention. (From Krause, Hidley, and Lareau.)

were dramatic reductions in the number of observed unsafe acts and, consequently, the number of accidents was reduced (see Fig. 4.5). Injury severity was also reduced by 39% during the year of intervention. The indirect results of the program included improved attitudes on the part of both supervisors and workers, their ability to solve safety problems and their enthusiasm or general involvement with safety.

Behavioral management concepts have been extended beyond reduction of unsafe behavior to other safety-related management areas. For example, many innovative large companies have begun employee assistance programs, which apply intervention techniques to chemically dependent employees or employees experiencing marital, emotional or even financial problems, recognizing that these problems affect safe performance at work.

TRAINING

Employing the right people is the most important aspect of management because those people determine the performance capacity of an organization. In doing so, the job must be understood, and therefore accurate job descriptions are critical when deciding job/worker compatibility issues. Many factors intervene in decision-making accuracy when we consider the selection process for entry-level work. There are fewer applicants for the less desirable positions, they possess fewer qualifications, and affirmative action or equal opportunity programs can influence the selection process. When we also consider that job descriptions endure, but that assignments change continually—often unpredictably—there is a wide gap between the actual work environment and a safe work environment. Therefore, training is necessary to achieve safe working conditions.

The Occupational Safety and Health Act places responsibilities for safety on employers *and* employees. Among these responsibilities are implementing training programs for the safe operation of powered industrial trucks and power presses, as well as right-to-know training for hazardous materials encountered in the workplace. Therefore, training is also necessary because it is mandated.

What is training? Perhaps the best definition is that of Firenze which states that training provides the difference between where we are and where we should be. This definition suggests that training can be flexible and need not be confined to classroom didactics. Training is an extension of management and therefore should follow all of the qualities we have embraced for management—innovation, commitment, bias for action, involvement, etc.

Before we focus on individual job safety, however, all new employees should have safety and health training on the *first* day with the following general information:

- Rights and responsibilities of employers and employees under OSHAct.

- First aid and emergency procedures, including what, how and when questions.
- Corporate safety and health rules and regulations.

The presentation of such information is most easily accomplished by the personnel department at the same time as other employment information is first recorded. Of great importance in this training is the proper communication of the safety message, so that proper perception of the importance of safety within the company is established at the outset. For example, it is commonplace to employ videocassettes for such training, but if not properly integrated with interest, questions and discussion, the new employee can easily miss the message.

It is with respect to the specific hazards of an employee's job where training innovation is needed. Too often we look at the procedures, precautions, safeguards and the personal protective equipment available and assume a new (or even experienced) employee automatically understands them. As we will see in Chapter 5, and as we have already seen in Chapter 3, hazard identification is frequently inaccurate and human error does occur. Training in the safe operation of equipment must consider these facts. We must conduct meaningful hazard identification programs to examine the hazards of a particular job, and then examine the remedies available to reduce or eliminate the hazards. Training has to convey this information to the employee so that it is understood. Then, and only then, will the employee accept and use personal protective equipment if that is what is required.

The methods used for training for a particular job need not be formal. They may include videos, both purchased and specifically prepared, open discussions with foremen (including behavioral modification programs) and other workers. The methods used are not important, only the understanding of the message communicated is.

Case History 4.3 Training for Supervisors: The Federal Hazard Communication Law (OSHAct 1910.1200)

Lowell Associates is an electronics firm employing about 6000 people in three facilities located in the greater Lowell area. Primarily a government contractor, Lowell Associates has always been committed to work safety, with all workers and management being involved. When OSHA issued the national hazard communication standard in November 1983 requiring that manufacturing workers be informed about chemical hazards and trained in safety procedures, Lowell Associates began developing a supervisory training program which could be incorporated *before* the law became effective in mid-1986.

Since OSHA did not define how employees are to be trained, Lowell Associates adopted the following approach:

- Prepare a comprehensive training program for supervisory personnel who have employees that work with hazardous materials.

- Supervisors then give site-specific training to their employees who work with hazardous materials.
- Require all new employees to attend industrial substance awareness presentations (all employees that handle hazardous materials had previously been trained in compliance with local regulations).
- Monitor attendance and participation at all training programs and document in personnel files.

Building the supervisory training program required a solid commitment from management, and time and hard work on the part of many people from various disciplines. Fortunately, they had had the experience of developing the industrial substance awareness program, and therefore could assign the responsibility for training development to the proper personnel—the industrial safety officer and a senior principal chemist. Support personnel included additional safety and medical specialists, a second chemist, media personnel and corporate counsel. These people built upon their previous experience, attending Safety Council presentations, studying requirements of the new law and discussing methods for how best to present the new information so that it would be understood. A decision was made to develop in-house slides with an accompanying text written in collaboration with a corporate technical writer. The company photographer and graphic arts personnel were charged with developing appropriate slides. When the slides and text were combined with narration, the program was videotaped and reviewed by senior management before formal "live" presentation to the first group of supervisors.

The actual training session was condensed to 4 hours. The final program outline was as follows:

- Introduction to the OSHA Right-to-Know Law.
- Copies of the hazard communication plan were distributed.
- Material safety data sheets were explained and the location of MSDS books described.
- Hazardous materials labeling system.
- General chemical hazard information—mode of entry, form, symptoms, etc.
- Substance review.
- Responsibilities of supervisors.

Lowell Associates expended extensive time, work and money to develop this program. Nevertheless, they learned a great deal from the experience. For example, they were surprised to learn that the number of hazardous materials, mostly chemicals, now approaches 2000, and that the storage and tracking of information required computerization. Although compliance to the law could have been accomplished through the purchase of generalized video-cassette programs,

Lowell Associates are comfortable that they now have an effective team that knows the company resources and is able to handle training projects more effectively than would have been possible with any generalized training program. ■

SAFETY COMMITTEES AND SAFETY CIRCLES

Safety committees came into being shortly after the first Workmen's Compensation Laws and the establishment of the National Safety Council in the 1920s. They have been both blessed and cursed as a marriage of industrial safety to the committee approach. Even the National Safety Council states that a safety and health committee is invaluable for hazard control, but they can also be neither productive nor effective. The difference between success and failure lies with the original purpose of the committee, its staffing and structure, and the support it receives from management.

Unfortunately, the traditional workplace safety committee has failed to improve safety and health conditions significantly. There are a variety of reasons for this: they have little real power; they are often established without any thought being given as to their function; and management wants to offer something to employees without relinquishing any real control. In many cases, committees were made up of people who were available, regardless of their disposition. Members often remained untrained in safety, ergonomics and hazard identification.

The result is that most safety committees became merely reactive and not a driving force. Their meetings very often became forms for re-hashing recent injuries or inspections or to discuss new poster acquisitions. In other words, safety committee activities have all too frequently demonstrated a lack of all of the attributes of successful management expounded in this chapter.

Safety circles are more recent developments, originating from the concept of quality circles introduced in Japan in the 1960s and in the U.S.A. in 1974. Safety circles are small groups of employees who undertake similar work and report to the same supervisor, and are therefore different than safety committees. Circles meet voluntarily and regularly to identify, analyze and solve work-related safety problems. They are composed of voluntary members, supervisors of those members, and a facilitator who coordinates activities and is responsible for training members.

To date, safety circles have met with limited success. Those which have been successful have had the full support of participative management whose goal was to develop people who could contribute ways to prevent accidents. Other elements which have been identified with successful safety circles include: a thorough understanding at all levels of the organization of the philosophy and objectives of the circle; selection of an enthusiastic facilitator who understands his role and is neutral; selection of a team of volunteers who have specialized knowledge and experience in key areas; thorough training in interpersonal relations; hazard

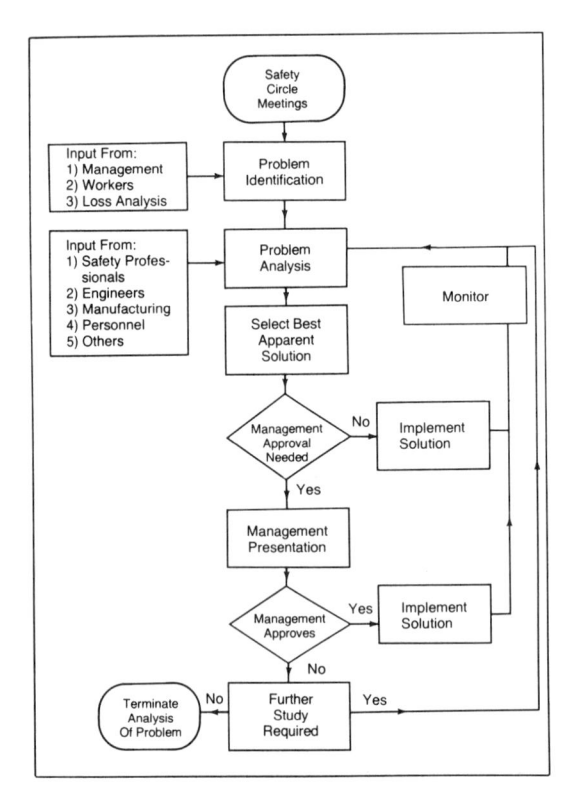

Figure 4.6 Safety circle operations. (From P. Salim, "Safety Circles," *Professional Safety,* April 1982, p. 18.)

identification and problem solving for those volunteers; and a controlled, gradual development of the circle without expectation of immediate benefits.

Safety circles meet weekly during the working day, usually for an hour. The circle membership as a whole decides in which order topics on the agenda will be discussed. That problem is then analyzed by the circle, and some members are assigned to research with outside specialists in between meetings when necessary. The discussion of their findings then leads to recommendations which are actually presented to management. These operations are summarized in Fig. 4.6.

Although safety circles can and do fail, they tend to be more successful than safety committees. Why is this? First, and probably of most importance, is the different management styles of the two. Secondly, the employees are more involved. The main benefits to members are recognition and self-esteem; much is due to extensive training and providing solutions to safety problems. We can argue that safety committees can survive if we instill employee involvement, creativity and motivation into the traditional safety committee, but such arguments only emphasize management attributes. When all is said and done, corporate management needs a safety team which has specific goals and objectives, is trained in safety and health matters, and can perform. The name is less important—committee or circle—than the degree of success achieved.

RISK MANAGEMENT

The Alcoa experience required a change in management thinking, where originally fatalities were viewed as part of the cost of business. However, it remains that a business exists for a specific purpose and mission and that economic performance must always be first in every management action. This concern for economic survival in the face of all adversity led to the concept of risk management—for many years a concept almost synonymous with insurance. The main objective of risk management is to prevent a serious impact on the company's financial structure from unplanned or uncontrolled losses. Risk management consists of:

- identifying and analyzing loss exposures,
- evaluating risk management measures,
- implementation of chosen measures.

A risk management program can be directed at risk control or risk financing. Risk control techniques include risk avoidance, loss prevention, loss reduction, and even segregation of exposure. Risk financing techniques are sources of funds such as insurance to restore losses which cannot be prevented or further reduced.

Today's corporations are far more sophisticated in managing risks, just as they are in managing businesses. Risk managers have gained recognition and influence outside of insurance. In many cases, safety and security personnel have necessarily become involved in risk management. Fig. 4.7 illustrates how risk management and safety management have overlapping responsibilities despite fundamental differences in their goals. In Fig. 4.7, the term static loss is used to differentiate between dynamic and entrepreneurial loss, e.g. the public's rejection of a new product. Entrepreneurial losses once offered the hope of increased business gains, something that static losses never did.

There is another factor, however, which affects the role of risk management besides management changes, i.e. the regulations which have emerged in light of true industrial disasters like that in Bhopal in 1984 and in Chernobyl in 1986. Risk management and prevention programs must be established and, in some cases, be filed with the administering agency. Corporations need to consider trends in insurance, legislation and even litigation when planning a strategy. This information is necessary when "fine-tuning" the balance of risk avoidance, risk reduction, risk retention or risk transfer.

In order to re-evaluate corporate risks, risk screening and risk assessment must be conducted. Risk assessment is explained in Chapter 5, but for our present purposes it is sufficient to note that the type of information needed is most likely available through safety offices. What the consequences of an event are is how risk management differs from safety management—risk management focuses on the economic impact only, estimating lost production costs, for example.

RISK MANAGEMENT
Mission: Funding for static loss

Finance
Law
Insurance

SAFETY MANAGEMENT
Mission: preventing static loss

Hazard recognition

Safety and health controls

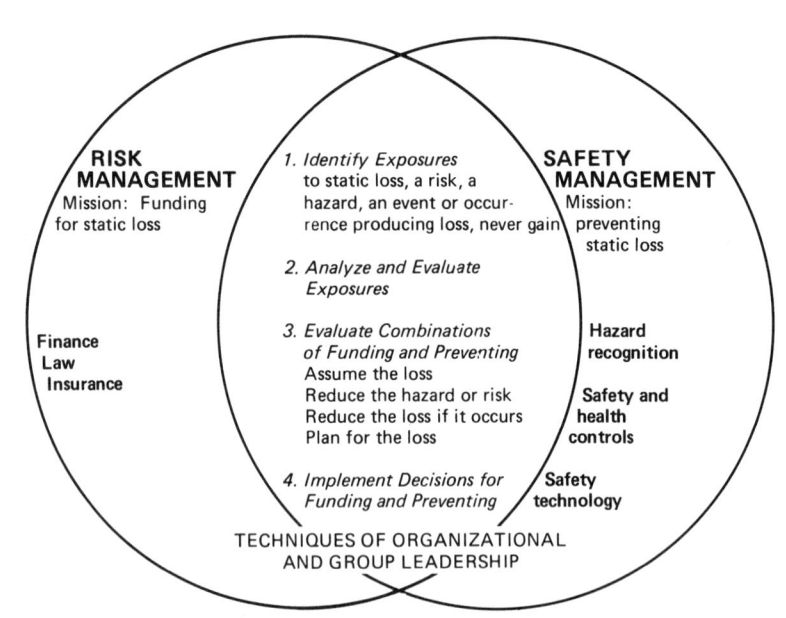

Safety technology

1. Identify Exposures to static loss, a risk, a hazard, an event or occurrence producing loss, never gain
2. Analyze and Evaluate Exposures
3. Evaluate Combinations of Funding and Preventing
   Assume the loss
   Reduce the hazard or risk
   Reduce the loss if it occurs
   Plan for the loss
4. Implement Decisions for Funding and Preventing

TECHNIQUES OF ORGANIZATIONAL AND GROUP LEADERSHIP

INTERFACE: Both Risk Management and Safety Management deal with static loss. Both interface with every function and operation in the organization. They interface vigorously with each other with much merging and overlapping in the range of strategies and methods available to them. They should be in fundamental communication with each other.

Figure 4.7 (From H. W. Heinrich, D. Petersen and N. Roos, *Industrial Accident Prevention*, 5th ed., 1980. Reprinted by permission of McGraw-Hill Book Company.)

In the process of performing the risk assessment and evaluation, we will usually uncover a number of areas for reduction of risk, some clear-cut and easily implemented, but others less clear which require careful scrutiny in the context of overall risk management planning. Recommendations which typically accrue include:

- reduction in inventory quantities,
- process design modifications,
- handling procedure change,
- improved monitoring,
- addition of regular risk management audits.

The risk assessment and evaluation provides the necessary insight to determine the feasibility and desirability of the respective risk management options. If the risk reduction measures cannot realistically be accomplished, cessation of activities (risk avoidance) might be viewed as desirable, but insurance availability is more likely to be considered.

Risk retention and risk transfer techniques both involve insurance. Retention is essentially self-insurance, and is perhaps best illustrated by the deductible portion of any insurance policy. Risk transfer describes the underwriting of a third party—the insurance company—for the financial consequences of a loss in exchange for the premium policy payment. The complexities of insurance policies that are available are beyond the scope, and not a major concern, of this text.

FURTHER READING

Books

K. Blanchard and S. Johnson, *The One Minute Manager,* William Morrow, 1982.

P. Drucker, *Management—Tasks, Responsibilities, Practices,* Harper and Row, New York, 1973.

P. Drucker, *The Frontiers of Management,* Truman Tally Books, 1986.

T. Ferry, *Safety Management Planning Manual,* Merrill, 1982.

J. Grimaldi and J. R. Simonds, *Safety Management,* 4th Edition, Richard D. Irwin, Inc., 1984.

H. W. Heinrich, D. Petersen and N. Roos, *Industrial Accident Prevention,* McGraw-Hill, New York, 1980.

F. McElroy, ed., *Accident Prevention Manual for Industrial Operations: Administration and Programs,* 8th Edition, National Safety Council, Chicago, Ill., 1984.

W. Ouchi, *Theory Z,* Addison-Wesley, New York, 1981.

T. Peters and R. Waterman, Jr., *In Search of Excellence,* Harper and Row, New York, 1982.

D. Petersen, *Human-Error Reduction and Safety Management,* Garland STPM Press, 1982.

J. ReVelle, *Safety Training Methods,* John Wiley, New York, 1980.

J. Schermerhorn, J. Hunt and R. Osborn, in *Managing Organizational Behavior,* John Wiley, New York, 1985.

Articles

H. Cohen, "Employee Involvement: Its Implications for Improved Safety Management," *New Directions in Safety* (Ed., T. Ferry), 1985.

M. Cole, "Quality Circles and Safety Committees," *Professional Safety, 29* (4), 33, 1984.

D. Dailey, "Creativity and Motivation: Key to Effective Safety Committees," *Professional Safety, 31* (12), 17, 1986.

P. Salim, "Safety Circles," *Professional Safety, 27* (4), 18, 1982.

QUESTIONS

4.1. The National Safety Council suggests that there should be five objectives included in a safety program. What are they?

4.2. What are the three tasks of management and under which task does safety come?

4.3. Identify and briefly explain the five traditionally recognized managerial functions with respect to safety.

4.4. There are ten roles of a manager. Select two of these roles and explain how they might be affected by the *skills* of the manager.

4.5. Why is involvement at all levels of management so critical to the success of your safety program?

4.6. Should employees' opinions be considered by management? If so, is this input more valuable for operations management or safety management? Explain.

4.7. Communication involves both a sender and a perceiver. Explain why the sender must clearly understand both roles.

4.8. Training is traditionally thought of as a classroom procedure. Discuss what other training techniques can be used to improve safety.

4.9. Should employees be retrained on a regular basis? Explain.

4.10. Management styles can vary dramatically from company to company. Explain how management styles can affect the success or failure of safety circles *and* safety committees.

4.11. Explain why risk management must be carefully approached by safety management.

5

Systems Safety: Hazard Identification

Nowhere do we need those principles espoused for successful safety management—competency at the basics, always intense, commitment to worker safety, and listening to and treating people decently—more than in the identification of hazards. We fail miserably in our ability to identify hazards before an accident occurs and it need not be so. This chapter not only explores the traditional methods for identifying hazards, but demonstrates how the management attributes can be useful and how to use some new means for technical assistance.

Whenever we examine incidents which cause injuries in the workplace, whether they be people-related or machine-related, we find that many hazards, even commonplace ones, go undetected and/or uncorrected. If they go undetected, we have a major problem which deserves all of our attention. If they are detected, but uncorrected, then we have a major management problem to resolve. Even in the latter case, though, we have to educate management, and therefore still face the problem of hazard identification methods, albeit having a different goal in mind.

But what is a hazard? We can define it as *a workplace condition which exists or can be caused in combination with other variables, which has the potential for accidents, serious injuries, disease and/or property damage.*

Research has shown that hazard identification has been a significant factor in incidents involving machinery safeguarding, industrial fires and even falls; many of these investigations appear as Case Histories in Chapters 8, 9 and 10. So the findings cry out to us—we are not competent at the basics, we do not innovate and we do not fully understand our machines, processes, people and materials. Only by implementing and fostering careful, thorough hazard identification procedures into our safety management programs, can we expect to improve our safety programs and records.

Where should we begin? As safety engineers, we cannot be expected to participate in the evaluation of hazards for each piece of equipment, each process or for each operator within our company. However, we must lead those efforts vigorously, be familiar with the basic procedures for hazard identification and educate others in these procedures. There are many names which describe hazard identification procedures, but there are only three common elements to these methods: *experience, testing* and *analysis*.

Experience applies not only to our own individual experience, which is unique and always expanding, but also to corporate experiences and the experience of others. Corporate experience is perhaps the best beginning point for hazard evaluation; examination of accident and injury records often points the way to where our efforts should be focused. If company records are incomplete or do not provide direction, then industry hazard information can be sought from insurance companies, from standard-making organizations such as OSHA or ANSI, or from information clearing houses such as the National Safety Council.

Such experience provides only general information which we must relate to specific equipment, processes or people. We need and must solicit input from others who are more familiar with all of these factors, e.g., engineers, designers, supervisors and operators. Often these people have lived with all the complexities and problems, and have discovered their hazards and pitfalls. In order to benefit from the experience of others, we have to communicate with them through interviews and discussions. These processes of obtaining information cannot be overemphasized in hazard identification, but must be structured for the best impact.

Interviews must begin by setting the person being interviewed at ease, developing a rapport which is information gathering, not fault-finding. A good beginning is simply to explain why the interview is being conducted. The main part or body of an interview should be where questions are asked and answered, by both interviewer and interviewee. It is important not to ask leading questions nor questions which are answered by a simple "yes" or "no." Open-ended questions such as "What do you think of . . .?" encourage the interviewee to respond at greater length. Of course, we have to be open-minded and listen to the responses.

Case History 5.1   Safety in Research

The properties of metal alloys can be classified as either structure-sensitive or structure-insensitive. The former are affected by processing or by the presence of impurities in the alloy, whereas the latter are not affected by processing and require large composition variations to be noticed. An impurity which is deleterious to steel or other iron-base alloys is sulfur, an element present in iron ore and both difficult and expensive to remove below levels of about 300 ppm (0.03%). Foundry research in the 1960s into methods of sulfur removal during melting included one method with significant safety hazards—inoculation with a calcium alloy which rapidly vaporized, the vapors chemically combining with sulfur as they bubbled through the molten steel. A calcium alloy was used because pure calcium vaporized too fast, explosively spilling molten steel. The reduced sulfur of the cast steel improved the structure-sensitive properties of steel and the technique was adapted for iron–nickel magnetic alloys. It was learned early on that the same calcium alloy could not be used; a special calcium–nickel alloy was prepared and tested. Initial experiments indicated the technique was not only promising, but was much safer with respect to the spattering of the molten steel. During scale-up experiments, however, more calcium was retained in the calcium–nickel alloy, but the alloy disintegrated, i.e. was granular rather than solid. When this was contained in a cloth bag and immersed, the reaction was almost as violent as with pure calcium; fortunately, no one was hurt, but all the molten metal was blown out of the furnace.

What went wrong? The research and production metallurgists agreed that the increased surface area of the granules probably caused a faster, thus more violent, reaction. The safety officer had no experience in this area, but was familiar with hazard identification procedures. He interviewed the melters as well as the metallurgists. Because he asked open-ended questions, he found the melters thought the cloth bag might have caused the violent reaction. No one could agree, but some thought that it might have released the granules too quickly, thus increasing the potential for a violent reaction. As a result of the discussions, the granules for the repeat experiment were wrapped in iron foil *and* increased shielding was used at the time of inoculation.

As a result, no further violent reactions were experienced with calcium–nickel alloy inoculations. This was all the result of disciplined interview techniques, listening to responses and holding open discussion, thus leading to an informed decision. ■

Testing, as it is used in hazard identification, does not refer only to procedures which measure performance criteria; it can include something as simple as observing and recording motions or actions in a process, examination of new equipment, or it can be quantitative, measuring the temperature, humidity or even concentrations of chemicals involved in a process. If testing for safety is planned appropriately, we can obtain valuable data for assuring hazard identification and control.

Test programs should be compatible with safety goals and the significance of unidentified and uncontrolled hazards. For example, a testing program planned for introduction of a new press brake (see Chapter 8) can affect the safety of only a few workers; we would not plan a program as comprehensive or as expensive as that for a process which could release insidious vapors within the plant. We therefore have to balance cost and value of information with the consequences of unidentified hazards. Remember that the test program may be as comprehensive as we need, but also can be limited to only a single task within the whole operation.

What do we test? Well, we can start with what we learned in Chapter 3, looking at worker positions, movements and relationships to the machinery. An effective method wherever motion of workers or machinery is concerned is videotaping. The cost of producing videotapes which provide permanent records for study and for training has become affordable for even the smallest companies. Their value has been demonstrated in numerous ergonomic situations, training programs and in hazard identification. Other valuable test procedures include the examination of blueprints and instructions which are supplied with equipment, and narrative descriptions solicited for processes or tasks. Written narratives in particular are helpful because writing demands a commitment and therefore more careful thought than verbal narratives.

Although these latter examples are testing methods, they overlap somewhat with the last element of hazard identification procedures, i.e. analysis. An analysis is a very careful study of all the components of a work system in order to detect problems, and then to understand the relationship between the system and the problem in order to eliminate the problem and its potential consequences. Although there are many specific analytical methods for hazard identification, there are two reasoning procedures that are actually used. Inductive reasoning is based upon observed phenomena, such as the failure of a machine or the failure of the workers or a combination which could result in an accident, whereas deductive reasoning is based upon an undesired event, then working backwards to identify component actions which might cause that event.

If attempting systematic hazard identification for the first time, one has to determine what order the analysis for each machine or process should take. Perhaps the simplest and most effective way to decide which areas should be examined first is to ask the workers, supervisors and managers to compile a list of the ten most hazardous machines or tasks in the plant. Pareto analysis is then performed on the data, a process named after the nineteenth-century Italian economist who observed that 80% of the world's wealth was controlled by 20% of the people. The analogy is quite general in our case, the results of the survey should point to the small percentage of machines or tasks which most participants regard to be hazardous. Thus we should focus on those 20% for maximum early identification of specific hazards.

Conducting hazard evaluation is not difficult, but organization is absolutely necessary. Page, for example, recommends 4-hour meetings at least once a week until all units are reviewed, then commitment to review every 2–4 years; new projects of course deserve more effort, usually about half-time daily until completed, about 4 weeks. There are many methods for formal hazard evaluation, each having its own contribution for specific tasks and hazard analysis. Although some of these methods bear different names, they overlap sufficiently that there is little meaningful difference. For example, preliminary hazard analysis and "what if" analysis are often described as separate techniques, whereas others have defined preliminary hazard analysis as an approach which uses the "what if" method. The analysis methods we will look at are:

- Preliminary hazard analysis (including the "what if" method),
- Failure mode and effects analysis,
- Hazard and operability reviews, and
- Fault tree analysis.

PRELIMINARY HAZARD ANALYSIS (PHA)

A preliminary hazard analysis, as its name implies, is a hazard review conducted when all information is not necessarily available for a system. This may be applied to a new process, where its main purpose is to recognize hazards early. However, we can also use PHA when any change in our plant occurs, such as the introduction of a new or even replacement machine. We must remember that PHA *is preliminary* and serves as a guide for more in-depth analysis as more information becomes available.

PHA focuses on what is already known about the process, product or change. It often consists of formulating a list of the hazards which might be related to materials (including raw materials as well as intermediate and final process materials), plant equipment, the work environment, maintenance, the safety equipment required and how to interface with other system components. The PHA is not as formal as other analysis methods, but nowhere is this more true than in the "what if" procedure, which is nothing more than the devil's advocate method where we oppose an argument even when we do not necessarily disagree with it, in order to determine its validity. By asking simple questions which begin "What if . . .?", unexpected events which could produce adverse consequences are likely to be considered. How effective it is, of course, depends on the experience and expertise of the staff members involved. A team should be involved, rather than a single person, to avoid biassed errors, and each team member should have a basic understanding of what is intended and to analyze the effect of possible deviations. Usually, questions are focused on different areas of concern, such as electrical safety, personnel safety or long-term hygiene issues. Examples of questions include:

- "What if the operator pushes the wrong control button?"
- "What if the power goes off?"
- "What if the process chemicals are depleted?"

PHA is a good method, but it is limited by its very nature. Within a system, each component of the system must be carefully considered as well as the interface of all of the components, and then the critical questions must be asked. How effective it is then becomes dependent on the commitment of the participants to safety, because time and effort are both required. Other methods are more formal and thus direct the efforts of the participants, usually with more in-depth analysis as a result. Nevertheless, PHA can often serve to provide checklists for formal safety review committees who examine many complex systems on a routine basis, and to provide guidelines for future PHAs.

Case History 5.2   A Model for Analysis

Dalzell Associates is a safety consulting and accident reconstruction firm which has provided much of the information used in the case histories. Last Fall, they were called upon to determine the cause of an incident in an electroplating plant which severely scalded the operator. Like many industrial accidents, there were multiple causation factors, some of which were coincidental, thereby contributing to both the incident and injury. This incident and injury could have been prevented by adequate hazard identification followed by proper control methods. Only the facts are presented here and they will be independently examined for each hazard identification analysis technique.

- *The company.* The incident occurred in the electroplating department, where eight employees oversee anodizing procedures for aluminum parts. This involves handling hazardous chemicals such as sulfuric acid, phosphoric acid and nitric acid solutions.
- *The system.* Anodizing involves chemical treatment to convert aluminum surfaces to aluminum oxide, rinsing and brightening operations. The incident occurred during the rinsing process after a bright dip.
- *The component.* Only the rinse process need be examined. Rinsing is carried out in water at a temperature of 105°F, either by a dip rinse in a 30 × 36 × 30 inch deep open water tank or a spray rinse tank of the same dimensions adjacent to the dip rinse tank.
- *Water temperature control.* Electric heating of the water tank is controlled by a silicone thermocouple. The thermocouple consists of a stainless steel semi-flexible capillary tube filled with silicone, brazed to a stainless steel bellows which expands when the silicon is heated, shutting off the heating control button; contraction upon cooling starts the heating elements. It is the open dip rinse water which is heated. City water is piped through this tank through a valve to the spray rinse tubes of the adjacent tank. Therefore, only the standing water in the pipes within the dip rinse tank is heated for spray rinsing.
- *The component materials.* The tanks were made of stainless steel; the plumbing, including the valve control, was made of PVC plastic piping. (This violated state plumbing codes which limits the use of PVC plastic to drains, waste lines and vents.)
- *The work environment.* There were pungent odors, extensive corrosion and moisture on the floors (slip-resistant open decking was provided).
- *The operator.* The operator had 8 years' experience, but did not have personal protective equipment on.

Failure Mode and Effects Analysis (FMEA)

- *The incident*. The braze joint of the silicone thermocouple failed by a combination of mechanical and corrosive action, permitting silicone to leak out. This permitted uncontrolled heating of the water in the dip rinse tank. When the operator opened the valve for the spray rinse, the PVC plastic joint failed because of the heat and pressure, causing the water to be released.
- *The injury*. The operator was scalded in the groin, thigh and abdomen. ■

Case History 5.3   A Case for Preliminary Hazard Analysis

A preliminary hazard analysis of the anodizing process should have been conducted. The team would probably have been composed of the supervisor, a process engineer and the safety officer. Looking at the process as a system, they would have broken the system down into its components, one of which is the rinsing process. The team would probably start looking at what hazards might be encountered during rinsing. For example, incomplete rinsing might cause unexpected exposure to the hazardous chemicals not rinsed away. The hot water at the right rinse temperature (105°F) would not be a hazard, but *what if* it were somehow hotter?

The team did not identify any other hazards, so it began to look at materials and controls. The engineer immediately raised the question of the use of PVC plastic piping, because he had recently gone over the plans for a new plumbing system for the locker room. He knew its use violated the plumbing codes and insisted it be changed. Also, the supervisor found that the silicone thermocouples were all corroded, and the distributor pointed out that silicone leakage can result from corrosion and bending at the braze. It was felt that inspections should be carried out more often so as to prevent continuous heating. Finally, the team looked at the operator who, because of the hazards of chemicals and hot water, should have always worn protective clothing. ■

If this PHA had really been conducted, the incident and injury would have been prevented. With no plastic plumbing, there would not have been a burst pipe. Also, if there had been more inspections and better indicators, the water would not have overheated, and even the plastic pipe would not have burst. Finally, if the operator had been wearing his protective clothing, there probably would not have been an injury.

FAILURE MODE AND EFFECTS ANALYSIS (FMEA)

FMEAs have been adapted for safety analysis from reliability analysis of complex engineering systems. The analysis consists of a critical review of the system, a breakdown of the system into all its components and systematic evaluation of how components might fail and what the effects of these failures might be. Along with the effects of failures, the FMEA usually requires a criticality rating for each failure mode which requires the analyst(s) to assess the degree of hazard posed by the failure. It is standard practice to use the hazard degree rating system for hazardous materials on a scale of 0–4 as follows:

0 – none
1 – slight
2 – moderate
3 – extreme
4 – severe

One or two analysts usually conduct an FMEA, systematically examining each component of the system and how it might fail, then evaluating the effects of failure, the degree of hazard posed by the failure and how the failure might be prevented or its consequences minimized. All analysts should be familiar with equipment functions and failure modes within the system. Failure information is tabulated in a form such as that given in Fig. 5.1. When completed, the FMEA provides qualitative information which can be useful in making management decisions regarding system safety.

There are some drawbacks to FMEAs, however, because they do not provide accurate probability data nor data on hazard frequency. As normally used, FMEAs concentrate on system components and do not examine system linkages which account for more system failures, and they do not examine the potential for human or operator error within the system. However, there is no reason why safety engineers cannot participate in the analysis and evaluate the human element, nor is there any reason why analysis cannot be extended to include interfacing of system components.

Case History 5.4 A Case for FMEAs

What might have happened if an FMEA had been conducted for the anodizing process in Case History 5.2? The safety team would have consulted an analyst familiar with the electroplating area and the anodizing system and the supervisor to provide some input from the operator behavior perspective (the results may have been tabulated as in Fig. 5.2). ■

HAZARD AND OPERABILITY REVIEW (HAZOP)

The hazard and operability review or HAZOP has been widely adopted by chemical processing plants to identify operation problems. It was originally developed to anticipate hazards and operability problems for new processes where past experience was limited, but has been found to be useful in every stage, from

Figure 5.1 Failure mode and effects tabulation form.

Area: ELECTROPLATING DEPT.

System: ANODIZING

Subsystem: RINSING PROCESS

Component	Failure mode	Effect on Component	Effect on Other components	Effect on Whole system	Hazard degree 0	1	2	3	4	Probability of failure	Detection methods	Compensating provisions or remarks
DIP RINSE TANK	CORROSION	LEAK	NONE	TEMPORARY SHUT DOWN	✓					NIL	VISUAL	—
SPRAY RINSE TANK	CORROSION	NONE	NONE	NONE	✓					NIL	VISUAL	—
HEATER	FAILURE OFF	NONE	WATER NOT HEATED	NO RINSE		✓				SMALL	WATER TEMP. INDICATOR	REPLACE
"	FAILURE ON	NONE	WATER TOO HOT	POTENTIAL SCALD		✓				SMALL	WATER TEMP. INDICATOR, STEAM VAPORS	REPLACE
THERMOCOUPLE	CORROSION	LEAK, CALL FOR HEAT	WATER TOO HOT	POTENTIAL SCALD		✓				HIGH	WATER TEMP. INDICATOR, STEAM VAPORS	BACK-UP INDICATOR, INSPECT
PLUMBING FOR SPRAY RINSE	BURST	LEAK, FLOOD	NONE	POTENTIAL SPRAY, FLOOD		✓				MODERATE TO HIGH	VISUAL	IS PVC OK?
OPERATOR	NO PPE	NO PROTECTION	NONE	NONE			✓			DEPENDS ON TRAINING, SUPERVISION	SUPERVISOR	POTENTIAL CHEMICAL BURNS OR SCALD

DATE _____ ANALYST(S) _____ SHEET ___ OF ___

Figure 5.2 Failure mode and effects tabulation form.

Hazard and Operability Review (HAZOP)

preliminary design to plant shutdown. HAZOP concepts are simple, using brainstorming techniques for a multidisciplinary team with 5–7 members. The purpose of brainstorming is the generation of ideas and typically follows four rules:

- All criticism is withheld until ideas have been completed.
- No idea is rejected; as a matter of fact, radical ideas are encouraged.
- Many ideas are elicited—the more ideas, the better chance of a good one.
- Combination and improvement of ideas lead to better and even new ideas.

We recognize brainstorming as an effective way of encouraging involvement and innovation.

A HAZOP requires detailed plant and equipment descriptions as well as a full understanding of the process and controls, and therefore team members are usually experts in these areas. The team also includes the safety engineer. Each team member is provided with a complete set of blueprints for equipment, flowsheets and equipment manuals. The most common HAZOP uses a *guide-word* approach where guide words are applied to nodes in the process, usually defined by the team leader. A node is a study point in the process or design where potential deviations and their causes and effects can be examined. For example, the application of the guide word "no" to the process parameter "power" results in the deviation no power. Possible causes of this deviation (e.g. operator turns off switch) and the consequences (e.g. product loss) are considered. HAZOP guide words and their meanings are given in Table 5.1. With specific parameters, we must modify these guide words somewhat to determine all deviations; for example, *other than* plus *time* should be *sooner or later*, and *other than* plus *position* should be *where else*.

When we put a HAZOP into practice, we utilize all the management procedures we learned about in Chapter 4. First, we define the purpose, objectives and scope of our study, and then select our team, who organize and carry out

TABLE 5.1 HAZOP Guide Words and Meanings

Guide Words	Meaning
No	Negation of the design intent
Less	Quantitative decrease
More	Quantitative increase
Part of	Qualitative decrease
As well as	Qualitative increase
Reverse	Logical-opposite of the intent
Other than	Complete substitution

Source: AICHE, Guidelines for Hazard Evaluation Procedures, 1985. Reprinted by permission of The Center for Chemical Process Safety of the American Institute of Chemical Engineers.

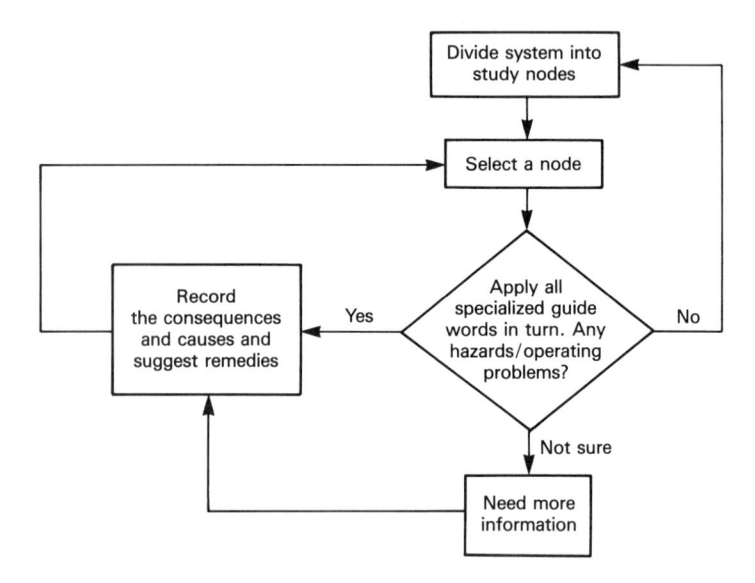

Figure 5.3 HAZOP method flow diagram. (Reprinted by permission of The Center for Chemical Process Safety of the American Institute of Chemical Engineers.)

the team review, and then record and communicate the results to management. This procedure is shown in Fig. 5.3.

There are other modifications of HAZOP than the guide-word method. Knowledge-based HAZOP replaces the guide words with the team's and leader's knowledge and specific checklists of established designs or processes. Creative checklist HAZOP studies were developed to identify hazards early on, based on materials to be used, and to examine the adverse interaction potential with other units in proximity or with the environment. In the line-by-line analysis, each line and vessel of a piping system replace the nodes of a HAZOP. These are examined using the guide-word application. These variations incorporate aspects of the preliminary hazard analyses to HAZOP where experience factors and "what if ?" methodology are incorporated into HAZOP while retaining its structure.

Case History 5.5   A Case for HAZOP

What might have happened if a HAZOP had been conducted for the anodizing process in Case History 5.2? Preliminary consideration of the rinsing process only suggests that the line-by-line experience and creative checklist variations of HAZOP are perhaps superior alternatives for the simple task. Vessel guide-word combinations would not be used because experience indicates consequences would appear in line-by-line analysis. First, let us look at the creative checklist HAZOP:

Hazard and Operability Review (HAZOP)

1. What are the materials?
   - thermocouple
   - tank
   - water
   - stainless steel
   - PVC plumbing
2. What are their hazards?
   - fire? – none
   - explosion? – none
   - toxicity? – none
   - corrosion? – look at thermocouple, because corrosion has been observed in electroplating area
   - radiation? – none

Immediately, we see that HAZOP focuses attention on the thermocouples in the rinse process.

Now, let us look at the line-by-line variation of HAZOP. Our two process parameters where deviations can occur are flow of rinse water and temperature of rinse water. Our HAZOP summary might look like that in Fig. 5.4. ∎

Area: ELECTROPLATING
System: ANODIZING
Subsystem: RINSING

Process parameter	Guide word	Deviation	Consequences	Causes	Suggested action
TEMPERATURE	MORE	HIGHER TEMP.	SCALD POTENTIAL	(1) WRONG SETTING (2) THERMOCOUPLE FAILURE, HEAT ON	(1) CORRECT SETTING (2) REPLACE, INSPECT THERMOCOUPLE REGULARLY
	LESS	LOWER TEMP.	POOR RINSE	NO HEAT	TROUBLE-SHOOT SYSTEM
FLOW	NO	NO FLOW	NO RINSE	VALVE CLOSED	OPEN VALVE
	LESS	LESS FLOW	POOR RINSE	(1) LINE CLOGGED (2) NO PRESSURE	TROUBLE-SHOOT SYSTEM
	MORE	MORE FLOW	SPLASHING	(1) HIGH PRESSURE	MODULATE VALVE OPENING
	OTHER THAN	FLOODING	(1) SCALDING (2) LOSS OF MISSION	(1) BURST PIPE OR VALVE	AVOID PVC

Figure 5.4 HAZOP review summary.

HUMAN ERROR ANALYSIS IN COMBINATION WITH HAZARD ANALYSIS

HAZOPs examine equipment and process operations, but suffer from the same drawbacks as other analytical methods because there is no focus on human error potential. Some textbooks describe human error analysis as a separate topic, but we should try to incorporate human error and human factors considerations into other analytical methods whenever possible on an informal basis. Many approaches have been used in analyzing human errors and in evaluating hazards caused by humans. Nearly all of these approaches are based on task analysis, i.e. what the operator must do during operation of the system. In order to obtain such information, we use the elements of experience and testing learned earlier in this chapter: interviews, job performance, questionnaires, videotapes and even workplace layout and instructions can be helpful. Actual task performance by the analyst has been found to be most useful. The analyst actually experiences problems and can often come up with solutions more readily than if he had listened to someone else describe the problem.

To evaluate errors effectively, we have to record all that we have done in task analysis. Once again, we need to realize the importance of documentation, because it coerces us to organize our thoughts and creates records to benefit others within the company. Documentation can be in many forms: narrative, tabular, graphic (e.g. flow lines or interfacing between units) or even formal reports incorporating all elements. Frequently, error evaluations will be part of a larger analysis, and they must meet the requirements of the larger analysis. Such requirements will dictate the level of effort spent on task analysis.

FAULT TREE ANALYSIS (FTA)

A fault tree analysis is a deductive reasoning method which assumes an undesirable event or consequence and proceeds in a logical manner to examine events or combinations of events which must precede the occurrence of the top event. FTA has been used extensively in government, aerospace and in utility industries. Its value is greatest when the undesirable top event is disastrous to all mankind and thereby easily definable. The fault tree itself is a graphic model that displays the various combinations of equipment faults and/or failures that can cause the top event. The analysts of course must have a complete, comprehensive understanding of the system/plant operation and various equipment failure modes.

One of the objectives of FTA is graphic presentation so that events and their relationships can be visually evaluated; standard symbols have been established for presentation. Figure 5.5 demonstrates how a fault tree is constructed. They can be either qualitative or quantitative; however, quantitative applications are very limited because we rarely know nor can determine the probabilities of an event, and probabilities vary with time in any case.

Fault Tree Analysis (FTA)

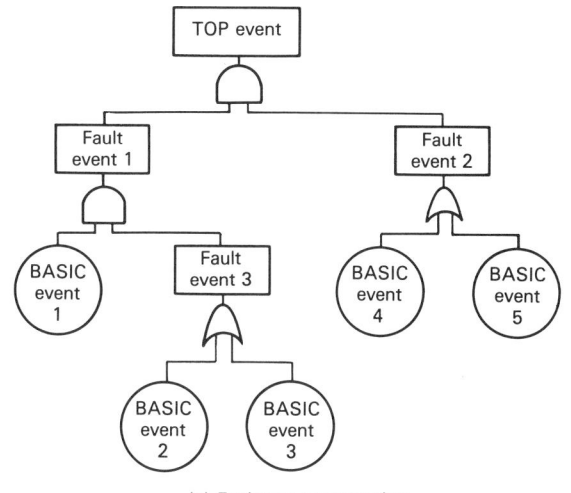

(a) Fault tree construction

OR Gate: The Or gate indicates that the output event occurs if any of the input events occur.

AND Gate: The AND gate indicates that the output event occurs only when all the input events occur.

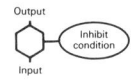

INHIBIT Gate: The INHIBIT gate indicates that the output event occurs when the input event occurs and the inhibit condition is satisfied.

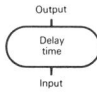

DELAY Gate: The DELAY gate indicates that the output event occurs when the input event has occurred and the specified delay time has expired.

BASIC Event: The Basic event represents a basic equipment fault or failure that requires no further development into more basic faults or failures.

INTERMEDIATE Event: The INTERMEDIATE event represents a fault event that results from the interactions of other fault events that are developed through logic gates such as those defined above.

UNDEVELOPED Event: The UNDEVELOPED event represents a fault event that is not examined further because information is unavailable or because its consequence is insignificant.

EXTERNAL or HOUSE Event: The EXTERNAL or HOUSE event represents a condition or an event that is assumed to exist as a boundary condition for the fault tree.

TRANSFER Symbol: The TRANSFER IN symbol indicates that the fault tree is developed further at the occurrence of the corresponding TRANSFER OUT symbol (e.g., on another page). The symbols are labeled using numbers or a code system to ensure that they can be differentiated.

(b) Fault tree symbols

Figure 5.5 Fault tree analysis.

The first step we take in constructing a fault tree is to define the undesired or top event. Except for potential disasters, this is more difficult than we might think. For example, how many of us would think of the worker being scalded in Case History 5.2 as the top event? In many other cases, the problem is really not suitable for FTA because it is too simple and can be evaluated by other methods. This is actually the situation for our case history because the top event, the scalding of the operator, is immediately attributable to non-use of the personal protective equipment. Therefore, the undesired top event must be carefully selected and properly defined because the entire tree stems from this single event.

When we construct a fault tree, we proceed through levels of fault events which must take place alone or in combination to cause the top event, ending with the most basic contributing causes. At each level, we try to describe precisely what each fault event is, where it is and when it occurs. The fault tree should be completed in levels, and each level should be completed before beginning the next level. It is important that we are systematic and methodical in FTA, but prior experience can be very useful. For example, operating errors which occur within a large plant may have been categorized in past studies and the analyst might simply identify operating errors as the fault event, and then use the triangular transfer symbol (see Fig. 5.5).

When a fault tree is completed, we solve it by determining the number of minimal cut sets, i.e. the combinations of failures that can result in the top event. Most complicated fault trees require computer programs to determine the minimal cut sets. For qualitative solutions, the number of basic events in each minimal cut set is important—the fewer the basic events required, the more important its ranking. A second ranking is used within equal event cut sets—the general rule is to rank human errors above active and passive equipment failures, based upon probability of occurrence. These rankings then can guide our recommendations.

RISK ANALYSIS

Risk analysis is frequently associated with the financial or insurance aspects of conducting business. The methodology of risk analysis, however, can be applied usefully to hazard identification and assessment. Whenever our hazard analysis techniques point to multiple or coincidental hazards, such as equipment failure coincident with a human error, risk analysis can help us to reduce the hazards to an acceptable level. Risk is somewhat intuitive and to what extent risk is worthwhile is personal, difficult to quantify and constantly changing. Nevertheless, risk analysis attempts to determine what a reasonable risk is.

The risk associated with an event is the product of two factors: the *frequency* of occurrence and *severity* of consequences. There are problems which we encounter in the measurement of these values because we can draw quite different

Risk Analysis

conclusions based upon the measurements. For example, the comparison of injuries caused on different machines over one year might be misleading because the machine usage is not considered. A machine which is used half as much as another, but which is associated with the same number of injuries, is certainly more risky to operate. Similarly, severity can be misleading; if we concentrate only on reducing the number of workplace deaths, we overlook the hazards which cause numerous crippling injuries. It is clear, however, that risk can be reduced by reducing either frequency or severity.

Frequency and severity are actually dependent on each other in an inverse way. The more severe an accident, the less frequently it will occur, and vice versa. For example, we find many more industrial lacerations than more serious injuries, such as amputations caused by machinery. This inverse relationship is fundamental to risk analysis because it applies to accidents of all kinds – industrial accidents, motor vehicle accidents (where fender-benders are much more common than fatal accidents), and even natural disasters where minor floods occur more frequently than killing floods, such as that experienced in Bangladesh in 1988.

How do we estimate the risks which can occur in any one job or task? And what risks are we willing to take? There are many ways to estimate risk (frequency data can often be elicited from our company records or from National Safety Council information). One method for estimating severity is to adopt the severity values established by the National Electronic Injury Surveillance System (NEISS), which was established under the Consumer Product Safety Act of 1972. Although oriented toward injuries arising from product use, the nature and severity of injuries differ little from industrial injuries. Table 5.2 lists the NEISS severity values. Figure 5.6 demonstrates how motor vehicle accident data gathered over a 5-year period in Texas clearly show the basic relationships between risk, severity and frequency.

Qualitative methods for severity assessment include the 0–4 rating system of the NFPA (see FMEAs). Such methods are far easier to adapt because rarely

TABLE 5.2 Accident Severity Values Developed by NEISS

Severity Category	Representative Symptoms	Severity Value
0	Incomplete or unnaceptable data	0
1	Mild injuries, dermatitis, sprains	10
2	Punctures, fractures	12
3	Contusions, scalds	17
4	Internal injuries	31
5	Contusions, nerve damage	81
6	Amputations, crushing	340
7	All hospitalized Category 6	2516
8	All deaths	34,721

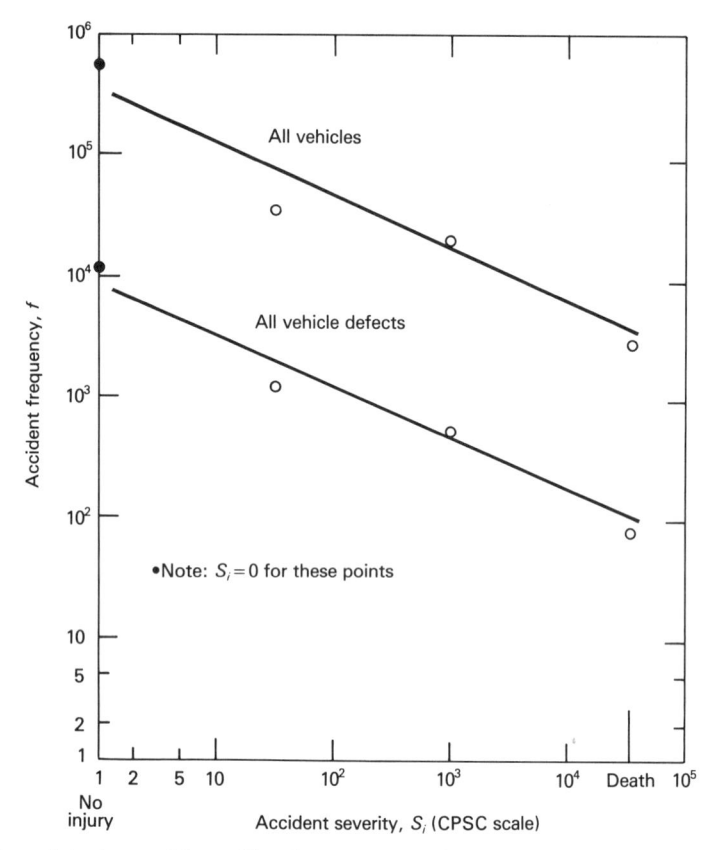

Figure 5.6 Automobile accident frequency–severity distribution (state of Texas, 1971). (From A.S. Tetelman and M.C. Burack, *An Introduction to the Use of Risk Analysis Methods in Accident Litigation*. Reprinted by permission of *Journal of Air Law and Commerce*.)

do we come across such data as that needed for risk assessment in Fig. 5.6. We are still able to evaluate risk qualitatively if we are able to estimate frequency or reliability as well as severity, and then we can set up a matrix to display results for easy interpretation. Chapanis has adapted the qualitative system shown in Table 5.3. Using this system, a risk assessment matrix using frequency and severity is established (Fig. 5.7a). The product of frequency and severity (i.e. the risk) is shown within the matrix (Fig. 5.7b).

The value of this qualitative exercise is that it guides our corrective actions. Chapanis has arbitrarily recommended risks rated 13 or more to be unacceptable, requiring immediate changes or cessation of the activity. Between 9 and 12, careful testing and redesign should be undertaken to eliminate potential problems. Between 5 and 8, technical changes which are easy to implement should only be considered, because serious problems are not expected. Below 5, no changes are needed except those to promote user comfort (see Fig. 5.7c).

Risk Analysis

TABLE 5.3 Qualitative Estimate of Frequency and Severity of Errors

	Probability			Impact
Probability Level	Frequency of Occurrence		Severity Level	Consequence
⑥ Frequent	A typical person is likely to commit this error frequently $>10^{-2}$/day		④ Catastrophic	Almost certain to cause death or severe loss
⑤ Reasonably probable	A typical person is likely to commit this error several times a year 10^{-3}/day		③ Critical	May cause severe injury or major loss
④ Occasional	A typical person will likely commit this error several times during his lifetime 10^{-4}/day		② Marginal	May cause minor injury, minor occupational illness or minor damage
③ Remote	Although a typical person is unlikely ever to commit this error, it is possible for him to do so 10^{-5}/day		① Negligible	Will not result in injury, occupational illness, or material damage
② Extremely unlikely	One may reasonably assume that a typical person will never commit this error 10^{-6}/day			
① Impossible	It is physically impossible for a typical person to commit this error $<10^{-8}$/day			

Source: A. Chapanis (1986). Reprinted by permission.

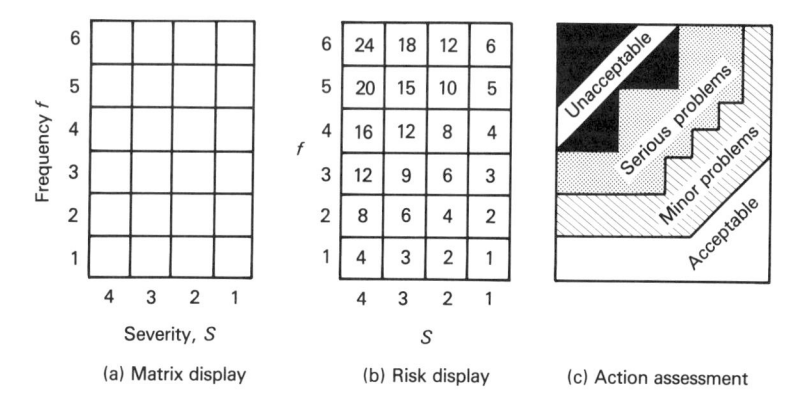

Figure 5.7 Action guide from risk assessment. (After A. Chapanis. Reprinted by permission.)

(a) Matrix display (b) Risk display (c) Action assessment

SUMMARY

The techniques most frequently applied to hazard identification programs have been examined. Each has its own attributes and its own shortcomings and no single method can be applied to all situations. But let us try to put this into perspective. Figure 5.8 demonstrates that the size of a workforce has a direct bearing on injury incident rates, where companies with fewer than 20 employees have good safety records, attributable in large part to intimacy of relationships. As the size of the company grows, however, that intimacy is lost, and hierarchies are necessarily established, with middle management responsibilities expanding. If we recall that full-time safety personnel/total production personnel ratios are typically 1:1000, we readily see there is no room in these companies for a full-time safety professional. This means that safety is most probably relegated to the back burner, with one of the most critical consequences being no hazard identification program.

It would be so much simpler if we could just say establish a hazard identification program and our records will improve. However, we need the impetus to come from management. They should encourage innovative, involved hazard identification procedures based on any one or a combination of the methods discussed in this chapter (or even other less formal methods). When left to the imagination, some very good ideas have been forthcoming from average employees. For example, in a recent questionnaire, a worker recommended interviews, human factors and task analysis by videos in a preliminary hazard analysis, followed up

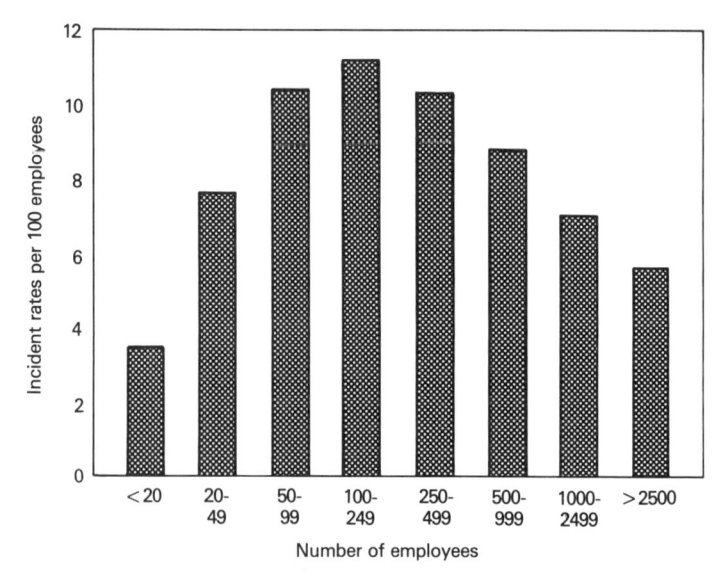

Figure 5.8 Occupational injury rates classified by employment size. (Bureau of Labor Statistics, 1981, 1982.)

by studying the information in a safety circle. What method we use is unimportant, as long as we address this major problem of hazard identification.

FURTHER READINGS

AICHE, *Guidelines for Hazard Evaluation Procedures,* AICHE, 1985.

B. J. BELL, "Evaluating the Contribution of Human Errors to Accidents," *Professional Safety,* **33** (8) 27, 1988.

A. CHAPANIS, "To Err Is Human, To Forgive, Design," *Proceedings of the ASSE Annual Professional Development Conference,* New Orleans, p. 6, 1986.

G. L. PAGE, "Hazard Evaluation Procedures," *Proceedings of the ASSE Annual Professional Development Conference,* Las Vegas, p. 32, 1988.

A. S. TETELMAN and M. C. BURACK, "An Introduction to the Use of Risk Analysis Methodology in Accident Litigation," *Jnl. Air Law and Commerce,* **133**, 1976.

QUESTIONS

5.1. What is the value of recordkeeping in a hazard identification program?

5.2. Why do we have to understand the basics (machines, processes, people and environment) to have effective hazard identification?

5.3. Discuss how safety programs may be more efficient with a hazard identification approach rather than with a conformance to codes approach.

5.4. Define inductive and deductive reasoning, and give an example of each from your own experience.

5.5. Each method of analysis has its own attributes and shortcomings. Focus on one shortcoming of each and recommend how you might compensate for it.

5.6. After reviewing the application of PHA, FMEA and HAZOP for Case History 5.2, what would your recommendation be for future hazard identification procedures in your workplace?

5.7. Explain why risk assessment is valuable in hazard identification programs.

6

Hazard Control

Sound management and effective hazard identification are necessary ingredients, but do not on their own ensure a viable safety program. We have received hints of how we might initiate our program from the topics discussed, from the figures such as the One Minute Manager's Game Plan, the behavioral modification program and even the safety circle operations described in Chapter 4. This chapter explores how sound, participative management can integrate effectively the many parameters into one solid hazard control program.

What are the parameters? Figure 6.1 demonstrates the basic feedback loop we will follow. The first step in our program is meaningful hazard identification, using all of what was discussed in Chapter 5. We must then communicate the information to management, with recommendations so that the proper decision is made. These recommendations should include risk assessment, cost assessments and alternative solutions. In formulating the recommendations, we should keep in mind that the amount of real improvement in hazard control is limited in most instances to engineering or technical controls, which modify the work situation, or management controls, which are aimed at the workers themselves.

There are three basic decisions management can make after reviewing the information and recommendations: they can do nothing, take action to modify the work system, or take action to redesign the work system. The easiest choice, of course, is to do nothing; but stop and think—is nothing being done because it is the proper choice, since little or no risk is involved, or are we procrastinating and enabling the hazard to continue? Doing nothing may also be correct when the cost of the remedy cannot be financed. The cost of remedial action is normally associated with the level of risk. For example, remedial action for low-risk hazards

Hazard Control

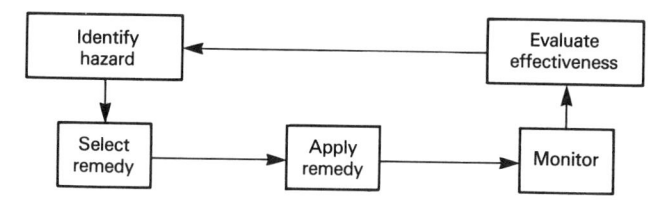

Figure 6.1 Feedback loop for hazard control.

is expensive because they occur frequently; however, the injuries tend to be less severe, and therefore we should address the frequency factors. On the other hand, the costs are also high for high-risk hazards, because the injuries can be severe; therefore our remedies have to address the severity factor. In between these extremes, cost–benefits are maximized (lowest cost) because both frequency and severity are addressed. Figure 6.2 shows a cost–benefit profile, where the high costs of low-risk situations are not warranted and the right management decision is *no remedy*.

Case History 6.1 *No Change: The Right Decision*

Shopping malls have not only changed the shopping habits of most Americans, they have also changed the face of America. Downtown stores have been abandoned because shoppers prefer the convenience of a mall where there is little or no parking problem. Public efforts to reverse this trend have met with mixed success. One frequent solution in the 1980s has been to renovate former department stores and supermarkets to house light manufacturing industries. In many cases, architects have endeavored to retain the ornate decorations of these old buildings while converting the space for new purposes. Such was the case with William Foster Company, whose turn-of-the-century building in downtown Lowell was abandoned in favor of opening a new store in the nearby Dracut Mall. Empty for 5 years, the building was renovated to house both the sales and manufacturing departments of Copy-Rite, Inc., who produce office copying machinery.

Figure 6.2 Cost–benefit profile.

Figure 6.3 Accident stairway at Copy Rite, Inc.

When renovating the building, the architect preserved the ornate stairways, but abandoned the obsolete escalators in favor of an elevator and a modern conveyor system to transfer finished units from the manufacturing and quality control departments in the basement to the shipping department on the main floor. Most workers employed in manufacturing used the stairway for access. During renovation, slip-resistant treads were applied to the steps of the stairway, in accordance with good engineering practice (see Chapter 9). The stairway itself was 54 inches wide and consisted of 14 steps to a landing and 3 stairways, each having 4 steps, leading to the basement floor.

Four months after moving into the renovated building, Mary McCarthy, a 60-year-old employee, tripped when she stubbed her sneaker on the slip-resistant strip at the edge of the landing, falling to the floor before she could grab the handrail. She broke her knee as a result of the fall and has never returned to work, taking a disability retirement.

When the safety committee viewed the scene of the accident (Fig. 6.3), they realized that the handrail on the lower stairways did not project beyond the steps, making it difficult to grasp and in violation of current building codes. However, they also recognized that the injured party was the oldest member of staff employed in manufacturing (average age 28 years), that correction of the hazard would mar the historic beauty of the stairway, and that the publicity of the incident and injury would alert others. They recommended that no changes be made, a decision which has withstood the test of time, for it is 6 years since Mary's fall and no other incidents have occurred. ■

DECISIONS FOR ACTION

When a hazard is identified, action is necessary whether management decides to eliminate it entirely or simply reduces the chance of it causing an incident. Redesign is usually a long-term and expensive decision which often involves the purchase of new and safer equipment. The principles of redesign are presented in depth in Chapter 12 and are not explored further here. Rather, it is important that we emphasize the modifications of the work system. By and large, it is through this alternative that we can make the most impact on hazard control. However, we should exercise caution, because it is as easy to select the wrong remedy as it is to choose one that is innovative and effective.

Fortunately, we can benefit from some general guidelines for control measures. All remedies can be thought of in terms of their placement with respect to the hazard. There are only three possible placements: at the source, at the receiver (the worker) or in the path between the source and receiver. For example, a barrier safeguard which encloses moving machinery or the substitution of a chemical are control measures applied at the source; personal protective equipment and training are control measures applied at the receiver; and remote controls and ventilation are control measures applied in the path. Whenever possible, remedies applied at the source are preferable because they minimize the potential for human error.

Case History 6.2 The Wrong Remedy

PC Assemblers, Inc., is a small company which does job shop assembly work for a number of firms in the Lowell area. Although their name is derived from electronic assembly, a smaller portion of their work is derived from mechanical assembly. They have six pneumatic arbor presses used for the insertion of press-fit parts or for crimping operations. In 1988, April Baldwin had her left fingertip crushed when it was pinched between the ram of the press and the die when the press was inadvertently activated. She has not returned to work as yet and is receiving workmen's compensation.

The workmen's compensation insurer for PC Assemblers came to the plant to examine the machinery which Ms Baldwin was injured on. [The purpose of such an examination is potential third party liability on the part of the manufacturer of the machine (see Chapter 12).] When the representative was shown the machine (see Fig. 6.4), it was explained that the *two-hand* control was not in place at the time of the incident, but was added for safety purposes the next day. However, the representative was also able to view a larger press, activated in the same way as the other machine before the two-hand control was added (see Fig. 6.5).

It was immediately obvious to the insurance investigator that PC Assemblers' remedy was inappropriate. The two-hand control had enabled the machine to conform to OSHA standards, but all the other machines were still in violation of the same standards and subjected workers to the same hazards which led to

Figure 6.4 Accident machine after two-hand control inserted.

Ms Baldwin's incident and injury. For these reasons, the insurance company pressured PC Assemblers to re-examine their safety program. ∎

ENGINEERING VERSUS MANAGEMENT CONTROLS

Although it is convenient to categorize remedies by placement at the source, path or receiver, another differentiation is helpful: the engineering or technical solution or the management solution (not to be confused with management decisions which might encompass all proposed remedies). As was mentioned briefly at the beginning of this chapter, the engineering or technical solution affects the work situation, whereas the management solution involves the worker. Engineering solutions can be found at the source, path or receiver, but management solutions only involve the receiver or worker. For example, personal protective equipment is an engineering solution applied at the receiver, whereas training is a management solution applied at the receiver.

Figure 6.5 Adjacent machine with unguarded foot pedal for activation.

Case History 6.3 Multiple Engineering Remedies

Dracut Technologies, Inc., is a high-tech machine shop for exotic metals used for aerospace applications. Most of the 75 employees are involved in engineering, sales and administration. There are only 15 machinists because most of the work is performed on computer-numerical controlled (CNC) machining centers. Each machinist oversees three machining centers. The work procedure for each task can be summarized as follows:

1. A work sheet stating the name and number of the parts to be machined is prepared and sent to the tool crib.

2. The part name is matched to the machining program and a print-out of the tools and tool heights required for machining is obtained from the computer.
3. The tool crib superintendent assembles each tool into the tool holder and marks the index position for placement in the turret of the CNC machining center.
4. Program and tool holder placement into the machine indexing system is performed by the machinist, who first conducts a dry run (no workpiece), and then repeats the process with a prototype workpiece which is thoroughly inspected by quality control before production is begun.

The machinist assembles the tools only when wear or breakage occurs during production. On these occasions, there is a fixture at each work station to remove the cutting tool and to reset a new one at the correct height. The tool holders are universal, with specific split-ring collets for each tool size which are tightened to the tool holder with a collet nut, as shown in Fig. 6.6. This system, in various forms, has been in use since machining was first practiced.

Avi Nash has been operating CNC machining centers for 14 years and is considered one of the best machinists at Dracut Technologies, both with respect to the quality of his work and his attitudes toward safety. Nevertheless, he lost his eye in 1987 in what at first was considered a freak accident. A drill bit had broken while in use, requiring Avi to replace the drill bit. He placed the universal tool holder in his work station fixture and began to loosen the collet nut. As he did so, the broken drill bit shot upward. Avi jumped back, his safety glasses falling from his face as he did so. As a result, the broken drill bit punctured Avi's eye.

When Avi's accident was being investigated, the tool crib superintendent found drill bits of the same size ejected from the subject collet and tool holder on a number of occasions—the drill bit jumping as high as 18 inches. No other tool holder and collet ejected the drill bit, however. What remedy, if any, should be used?

The safety circle at Dracut Technologies recommended two engineering remedies, both of which would not be expensive and would prevent any such incident

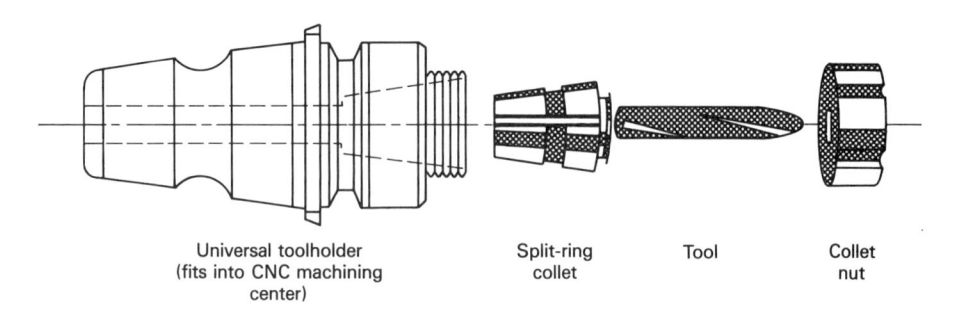

Figure 6.6 Mechanism for inserting cutting tools into CNC machining center.

from occurring again. Their recommendations, based on the severity of the injury (even though it was unlikely to happen again), was to purchase safety glasses which were secured to the head by elastic bands or silicone pads, and to place a hinged Plexiglas barrier safeguard by each workplace fixture. ■

Case History 6.4   Combined Engineering and Management Remedies

In Chapter 5, a single case history demonstrated some of the informal and formal methods for identifying hazards. If we recall the circumstances of the incident and injury, a thermocouple failed by corrosion, giving a false temperature reading, thus causing continuous heating. The worker, unaware of the hazardous situation and not wearing personal protective equipment, turned the PVC valve to start the spray water rinse. The PVC fitting failed, and the worker was scalded seriously.

Because of the incident and injury, remedial action was approved. But what remedies should be selected? The first is simple. Because PVC is an inappropriate plumbing material, the engineering remedy of metal piping and valve was approved immediately. The thermocouple control system was restudied, but an alternative system was not economically feasible. Therefore, a management remedy was settled upon, which specified the monthly inspection of the thermocouples for corrosion, and 6-monthly removal to determine the extent of any damage.

A major concern was the worker's failure to register the overheating. A thermal alert system was looked into, but it too might have fallen victim to corrosion, either failing to alert or giving a false alarm. Again, a combined engineering/management remedy was decided upon. A large-faced thermometer was inserted into the open tank and all staff were retrained to check the water temperature immediately before opening the spray valve.

The final remedy is strictly a management solution, involving the worker's failure to wear the personal protective equipment provided. It was the safety circle which came up with the most acceptable solution, a modified behavioral management effort. Beginning with the supervisors and continuing with all workers, a series of interviews were conducted to establish a baseline. After reviewing the results, a simple retraining program was initiated, starting with the corporate commitment to safety, and the Right-to-Know laws and their purposes. At first, only the supervisors were retrained, but then all other employees were added to the program so as to present a united front. Brief monthly meetings have continued to re-emphasize the issues and cooperation has been exemplary. ■

IMPLEMENTING HAZARD CONTROL MEASURES

The recommendations for hazard control solutions may come from the safety office, the engineering staff, or even the supervisory staff. It should be remembered, though, that management decides what remedy to implement. Our case histories have shown examples of where management has implemented control measures, both engineering and management remedies. Little more can be said, except to urge follow-through and feedback of information. However, what can be done when the management lacks a commitment to safety? Is there some way that control measures still might be adapted? There is, as long as we remember our creativity, commitment and involvement. For example, a crushing injury was caused by a 30-year-old press being considered for replacement. Management was unyielding in its stand not to consider any corrective actions, so no remedies were implemented. The supervisor, undaunted, had a small wooden sign painted with the words:

DANGER

DO NOT TOUCH ROLLS

WHEN MACHINE IS OPERATIONAL

No further incidents occurred before the machine was scrapped.

Innovation and involvement are key factors in the implementation of hazard control programs.

MONITORING REMEDIES

Monitoring is nearly identical to hazard identification itself because the measurement methods used are the same. For example, if the concentration of a chemical solvent is suspected as being above threshold limit values (see Chapter 11), air sampling procedures and chemical analyses are used to identify the extent of the hazard. After applying a remedy such as local ventilation, the same procedure is used to monitor the remedy. We should not be confused, however, into comparing the depth of analysis required in hazard identification with that in the monitoring function. The study required need not be repeated if it is done correctly in the first place. As a matter of fact, monitoring should compare performance to the earlier analysis.

The monitoring function provides us with another type of measurement, that of performance—task and worker performance and staff performance. It also serves to provide us with signals when things are not running smoothly. For example, it is part of the monitoring function to collect the data which become part of corporate safety records. Sudden changes would first be noticed in the monitoring stage.

Where do we place the responsibilities for monitoring? Of course, safety personnel are responsible, but quite often the monitoring function can be integrated with inspections conducted by safety committees or safety circles. This is a good solution because results gathered in these periodic activities can provide the information which makes up our records and which serves as input for the evaluation of our remedies.

EVALUATING EFFECTIVENESS

This chapter began with the feedback loop for hazard control. It was used as a starting point because we should have been familiar with hazard identification from Chapter 5. It is easy for us to misconstrue evaluation then as an ending for hazard control. Evaluation is not the last effort and it is not done after a specific lapse of time. Rather, evaluation is an ongoing and very important decision-making process for the feedback of information, improved hazard identification, etc. There are two aspects to evaluation: assessment and judgment. Assessment is simply the results of measurements made in the monitoring process, such as the incidence rate[1] or severity rate,[2] or it can be more specific, relating to a department or process area.

Once assessments have been made, then the judgment of adequacy of the program can be made. Judgment, of course, encompasses all of our values, including economic, social and political values. Meaningful evaluation must examine our original goals and objectives and how well our program has strived to achieve them. Most important, though, is the feedback of our evaluation so as to facilitate the continued improvement of hazard control programs.

PUTTING IT TOGETHER

It should be remembered that systems safety concepts are based upon the separation of a system into its components, and the examination of all the components and their interrelationships in detail. For the most effective hazard control, we have to be able to integrate what we have learned into a workable program. Such integration must begin by understanding people and their management. To this must be added effective hazard identification and effective controls. Most important, however, we must recognize that no single program can be devised: what

---

[1] Incidence rate (I) is defined as the number of recordable injuries and illnesses (N) per 100 full-time employees per year,

$$I = \frac{N \times 200,000}{\text{Employee hours worked}}$$

[2] Severity rate (S) is defined as the number of lost work days per million hours worked: S = number of days lost $\times$ 1,000,000 = employee hours worked.

is effective in one setting may be a total failure in another. A proper balance has to be found to achieve the best result.

WHOLISTIC SAFETY PROGRAMS

We now review some of the critical points in each of these areas before attempting to formulate a new hazard control program based upon what some people term the wholistic approach. People like to feel successful, appreciated and in some control of their own safety. Management should be innovative, involved, committed to safety and communicative. And there must be a formal approach to hazard identification and evaluation.

Drucker has expressed some ideas in *The Frontiers of Management* (1986) which may help us in forming new, effective safety programs. He recognizes that people decisions are the most important management decisions because they determine the performance of the company. Doesn't the make-up of our safety staff, both full- and part-time, then, determine the effectiveness of our safety program? Staffs are people who analyze and plan, supply knowledge, design policies and give advice. Staffs are fast developing in most organizations and their work should be limited to a few tasks of high priority, a rule most often violated in business. Perhaps we should attempt to limit our safety staff functions, but have larger part-time safety staffs. This would provide support for our workers and not make staff work a career in itself.

Drucker has provided a number of management guidelines for staff selection:

1. Think through the job: job descriptions may last a long time, but assignments change all the time, many times unpredictably.
2. Look at a *number* of people.
3. Think hard about the staffing. The ideal candidate should have a great concern for production and a great concern for people.
4. Discuss it with others.
5. Make sure appointees understand their role.

These guidelines apply to staffing our safety effort as well.

Fragala has described how a comprehensive safety program was formulated with a cooperative management of a manufacturing firm. Three phases of the wholistic approach were implemented:

1. *Safety awareness,* including an educational program on job instructional training, positive behavior reinforcement, ergonomics, cumulative trauma disorders, back injury prevention and job safety analysis techniques.
2. *Safety implementation* by all attendees of the educational program, particularly key managers and supervisory personnel.

Wholistic Safety Programs 135

3. *Safety program preservation* to maintain enthusiasm and energy levels which do not deteriorate with time.

Evaluation of the program as it evolved showed that participation was a key issue (get involved, remember?), with participants sharing their experiences when trying to apply techniques of job instructional training. It was the sharing experiences which helped those peers to overcome problems with application of the concepts. Another key issue identified was the balance necessary between negative behavior reinforcement (rules and regulations) and positive behavior reinforcement, which must be blended into their own management or leadership style. Finally, feedback, through the voice of the plant manager, was considered an important step toward assuring the preservation of the safety program.

In the UK, a movement began in the 1970s to change the emphasis of hazard control programs from over-reliance on "negative regulation by external agencies" toward a system where those who face risk have primary safety control, perhaps better described as self-regulation. The roots of self-regulation were based on those concepts we have espoused—efficient management systems and worker involvement. But these are not ends in themselves, and can only be appreciated in relation to their outcomes—the self-regulation hazard control program. Figure 6.7 illustrates the flow in a self-regulation program.

It would be nice if the flow sequence of the self-regulation program of Fig. 6.7 would work, but in the real world we have work situations which have localized or endemic hazards, no matter what the size of the company is. The framework upon which a workable, self-regulating safety program can be founded depends on the identification of these hazards and use of information on how they might be controlled. Positive intervention is necessary for us to try and control the hazards. Figure 6.8 describes how the hazard sequence should be controlled, with elimination, containment and mitigation the three main options for controlling hazard sequences before they happen.

The second critical feature of the framework for self-regulation is the use of specific technical controls. Technical controls may involve modifying machinery or adding ventilation, changing training procedures, or controlling environmental exposures; no distinction is made between technical and management con-

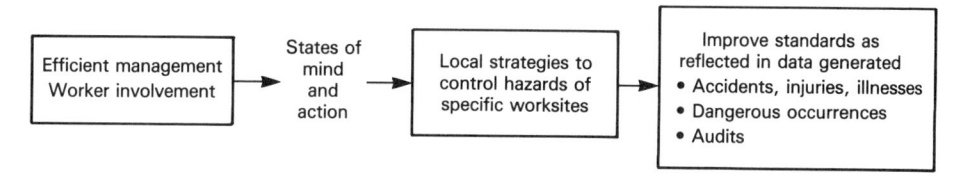

Figure 6.7 Self-regulation hazard control program. (Reprinted with permission from *Omega, The International Journal of Management Science*, II, S. Dawson, P. Poynter and D. Stevens, "How to Secure an Effective Health and Safety Program at Work," Copyright 1983, Pergamon Press plc.)

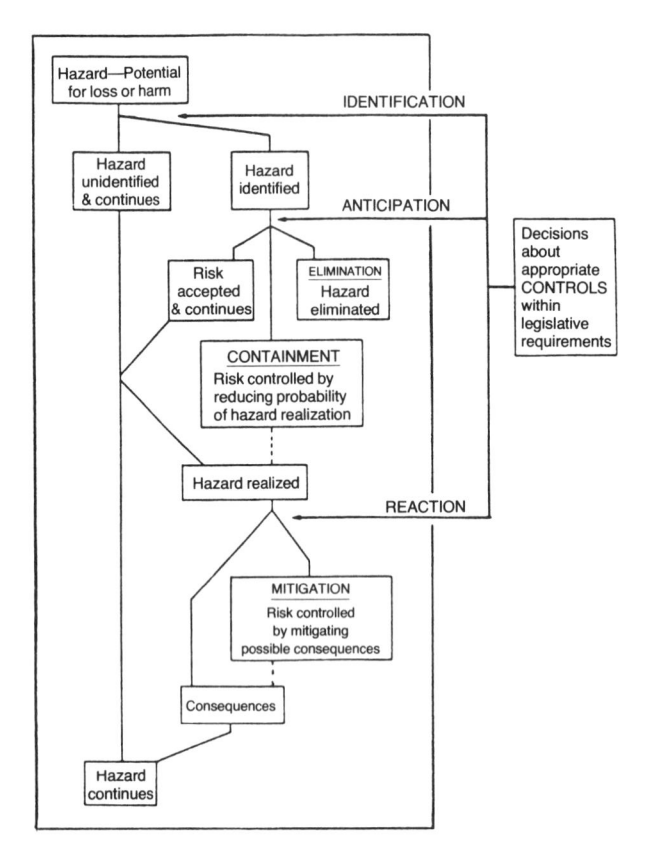

Figure 6.8 Controlling the hazard sequence. (From Dawson, Poynter and Stevens.)

trols for self-regulation programs. However, support and maintenance of the controls are emphasized. In this matter, motivational controls are needed to provide general support—another feature of the framework for self-regulation. The last feature is what, if anything, to do to control risk. Mechanisms for confronting and resolving conflicts over needs, costs and benefits need to be established.

A LAST WORD

There is no one hazard control program which is perfect and we have to find the one which suits us. The following are key points to remember when making a choice:

- perception,
- feedback,
- wholistic, positive approach,
- learn to work with unions,

- make careful people selections,
- try safety circles,
- brainstorm, but limit staff specialties, and
- be the persistent champion for your program.

FURTHER READINGS

S. DAWSON, P. POYNTER and D. STEVENS, "How to Secure an Effective Health and Safety Program at Work, OMEGA," *International Journal of Management Science,* **11** (5) 1983. (Reprinted in *Professional Safety,* **32** (1), 32, 1987.)

PETER DRUCKER, *The Frontiers of Management,* Truman Tally Books, 1986.

G. FRAGALA, "Implementing a Safety Program—A Successful Start," *Professional Safety,* **32** (6), 16, 1987.

QUESTIONS

6.1. Explain the shape of Figure 6.2, the cost benefit profile.

6.2. What can you do as a supervisor in a plant where management lacks commitment to safety?

6.3. Differentiate between engineering/technical solutions and management solutions.

6.4. Explain in your own words the meaning of incidence rate and severity rate. How does their use affect safety program evaluation in a company of 100? In a company of 1000?

6.5. What is meant by a wholistic safety program?

6.6. Do you think that the accident described in Case History 6.2 would have happened if P. C. assemblers had safety circles? Explain your answer.

6.7. Do you agree with the "no remedy" solution in Case History 6.1? Explain.

6.8. Do you think that self-regulation is a way of taking the responsibility for hazard control away from management?

7

Material Handling

The first technical subject to be dealt with in detail is material handling, because approximately 25% of all workmen's compensation claims are the direct result of injuries sustained from the manual handling of materials alone. This is a large proportion even when one considers the number of tasks performed each day. The most common complaint is back pain, particularly lower back pain. It should be made clear from the outset that we have to understand the nature of a potential injury so as to be able to identify the hazards involved and to formulate a remedy.

Material handling encompasses much more than simple manual tasks, however: this chapter covers rigging, (including ropes, chains, slings and hooks), powered industrial equipment (ranging from walkies and fork lifts to overhead cranes), and automated material handling. The injuries sustained with powered equipment are predominantly crushing and/or impact injuries; however, electrocution, burns, falls and property damage are also encountered. Good material handling safety programs apply the general principles of participative management, effective hazard identification and remedies based upon knowledge of the workers and ergonomic applications.

LOWER BACK PAIN

Statistics show that almost 80% of all manual handling claims are for injuries sustained to the lower back, and that 80% of the population experiences lower back pain at some time in their lives. The injuries which lead to lower back pain occur in high-risk industries (construction, mining, transportation and manufac-

Lower Back Pain 139

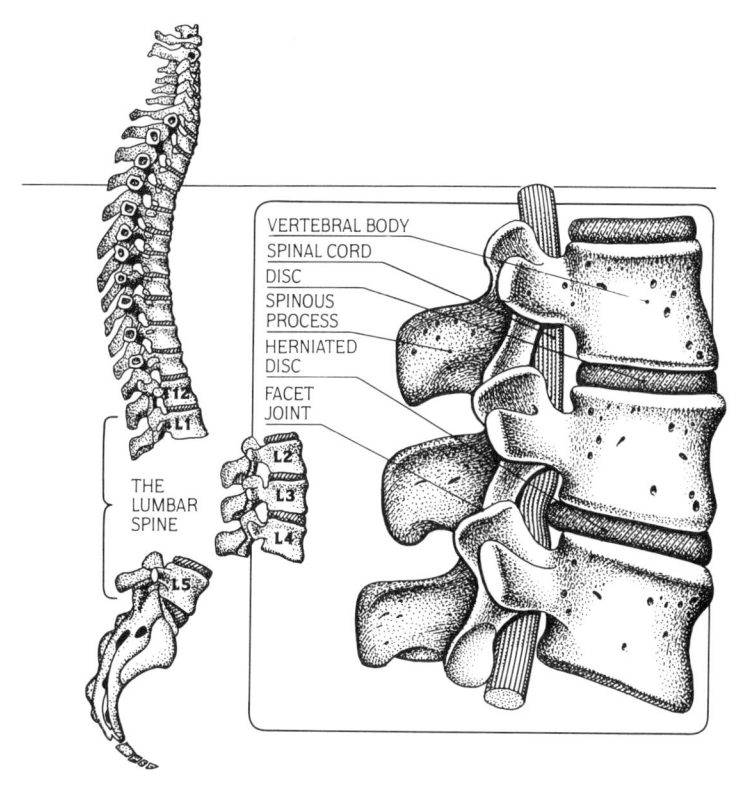

Figure 7.1 Anatomy of the back and spine. (Reprinted by permission of the Bayer Company, Glenbrook Laboratories, Division of Sterling Drug Inc.)

turing), high-risk occupations (laborers, warehousemen, mechanics, nurses), and high-risk activities (lifting, twisting, bending, reaching and falling). The probability of lower back pain recurring is four times greater after the initial episode.

The anatomy of our backs and spines has to be studied in order to begin to understand lower back pain. The spine (Fig. 7.1) is naturally curved, a position which should be maintained at all times by correct posture. It serves to protect the spinal cord, to support weight, and to be movable at the same time. The spine is made up of 24 vertebrae, most of which are separated by discs which act as shock-absorbers. Vertebrae, which consist of the body and spine, surround the spinal cord and are joined at the facet joint. Muscles, ligaments and discs are attached to vertebrae to support the spine. The 24 vertebrae are divided into the cervical spine or neck, the thoracic spine or upper back and the lumbar spine or lower back. Vertebrae in the lumbar region are larger than the others so as to support the weight. The sacrum, the base which supports the back, has several small bones called the coccyx attached to it.

Pain can originate from any part of the back. The ligaments and muscles which hold the bones together and limit motion can be stretched or sprained and can contract in spasms, causing pain with even limited motion. Also, the facet joint can be strained due to overstretching, nerves can be pinched (e.g. by muscle spasms) and the spinal cord itself can be injured due to fractured vertebrae, bone spurs or slipped discs.

The discs, which are firmly attached to the vertebrae, but permit movement by stretching, are made up of annular ligaments surrounding a core of jelly-like substance. However, abnormal forces can cause tears in the ligaments. Tears through the annular ligaments are called ruptured discs, whereas tears within those ligaments are called herniated discs. A slipped disc can result from repeated injuries which permit the jelly to flow out, limiting motion, compressing the facet joint, and causing pain through pressure on the nerves of the spinal cord.

THE RELATIONSHIP BETWEEN LOWER BACK PAIN AND MANUAL HANDLING

Most lower back pain develops gradually and cannot be related to any specific incident. Diagnostic measures are limited and there is little consensus on how lower back pain should either be diagnosed or treated. With respect to industrial compensation, workers who do not perform heavy tasks are better able to tolerate their pain and few man-hours are lost. On the other hand, heavy work which requires excessive exertion can trigger an episode of lower back pain and lost time disability claims are high. There is no exact relationship between workloads and lower back pain and it is now thought that heavy workloads merely trigger the occurrence of lower-back symptoms. If we view lower back pain disability as defined by the compensation claims, however, we find it to be very dependent on the nature of the work.

Workloads have been characterized by different criteria: biomechanical, physiological, psychophysical and epidemiological. Biomechanical studies have demonstrated that the incidence of lower back pain is increased when intra-abdominal pressure, which accompanies any manual handling activity, exceeds 100 mmHg in male workers. Intradiscal pressure studies have demonstrated the importance of lumbar support, backrest inclination and armrests for seated workers. Physiological criteria, such as heart rate and oxygen consumption, have led to recommended weight limitations on the loads workers carry (e.g. 88 lb for adult males). Epidemiological criteria are based upon statistical analysis of the incidence, severity and distribution of lower back pain.

Psychophysical criteria are based on workers who monitor their own feelings of exertion or fatigue in a manual handling task, while controlling one of the task variables (e.g. weight). The other variables, such as frequency, size and distance, are fixed. The psychological aspect to this methodology is significant because lower back injuries have appeared more frequently in workers who perceived their work to be harder. Table 7.1 summarizes a sample of some of the psychophysical studies.

TABLE 7.1 Maximum Acceptable Weights (kg) When Lifting[a]

Workers (%)	One Lift Every:					
	5 s	14 s	1 min	5 min	35 min	8 h
90	10	15	18	24	26	29
75	13	20	23	31	34	37
50	17	25	29	39	43	46
25	20	30	34	47	51	55
10	24	35	40	54	59	64

[a] Males lifting a 36 × 57 cm box to a height of 76 cm above the floor (from Shook, 1983). Reprinted by permission of Dr. Stover H. Snook, Liberty Mutual Insurance Co.

Correlating data such as these with data for other manual handling tasks and field data for lower back injuries, has shown that about 25% of industrial manual handling tasks are acceptable to less than 75% of the workers; half of all lower back injuries, however, are associated with these tasks. In other words, a worker is three times more susceptible to lower back injury when performing a manual handling task that is acceptable to less than 75% of the working population. Therefore, it can be concluded that up to one-third of industrial back injuries can be prevented by designing the task properly.

The National Institute for Occupational Safety and Health (NIOSH) has integrated these biomechanical, physiological, psychophysical and epidemiological criteria for manual lifting (see Fig. 7.2). According to these criteria, we should avoid all manual lifting conditions above the maximum permissible limits (MPL)—the MPL is three times the action limit (AL). Between the MPL and the AL, we should implement administrative action, including selection and training of workers.

MANUAL HANDLING PROGRAMS TO LIMIT LOWER BACK INJURIES

The information from all the studies relating lower back injury to manual handling tasks can be used effectively to reduce their occurrence. Most successful programs incorporate management support, pre-employment screening (despite the inherent diagnostic problems), worker training and fitness, task analysis and ergonomic design, and rehabilitation. Successful training programs have focused on posture, general physical fitness and on worker perception (those programs in which workers have enjoyed participating have been far more effective). Many are modeled on weightlifting programs, where the spine is fixed and strength and flexibility come from the abdominal and leg muscles. Fixating the spine is the purpose of the straight-back, bent-knee lifting method. The correct lifting technique has three elements:

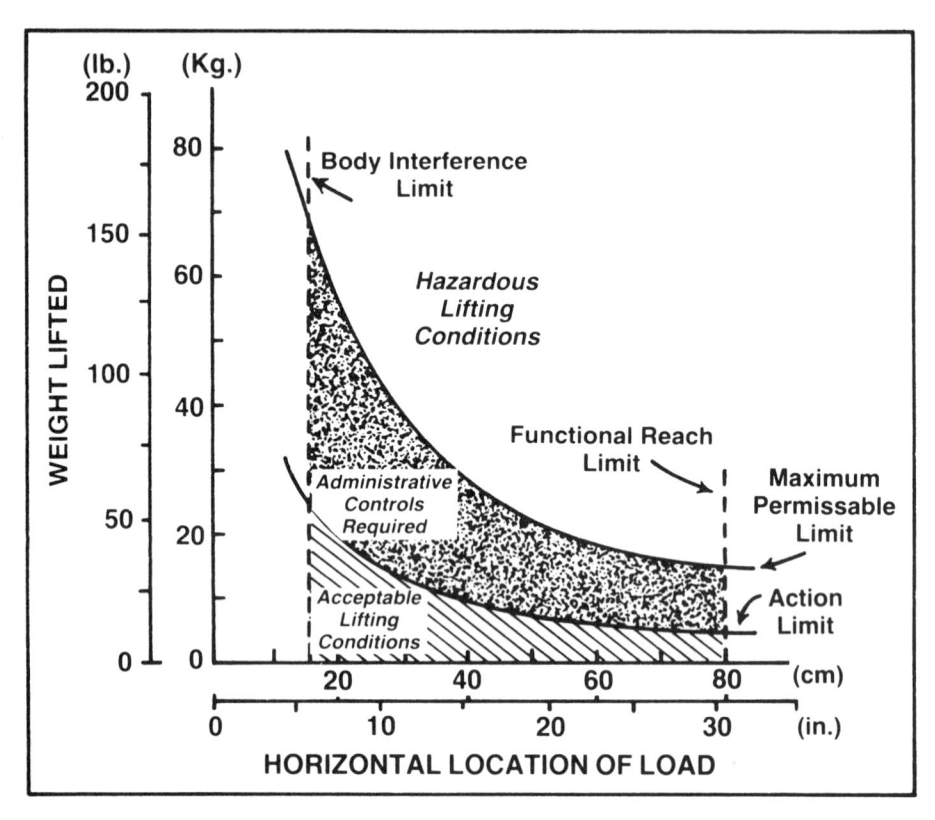

Figure 7.2 Maximum weight versus horizontal location for infrequent lifts from floor to knuckle height. (From NIOSH.)

1. Stand as close as possible to the workpiece, with the knees bent and the back comfortably straight.
2. Lift by straightening the knees to the upright position, without jerking or twisting.
3. Use the shoulders and arms to pull the object close to the body if it is initially some distance away.

For most of us, though, lifting is a simple action which we do not really analyze each time, and we tend to use slightly different methods to suit the specific demands for our daily routine. The thing to remember in *all* lifting tasks, however, is *bend your knees*.

Ergonomic evaluation and task redesign have shown to be effective controls for lower back pain injuries. Task analysis is a key ingredient, involving hazard identification and all the elements involved in time and motion study, as well as the examination of the physical aspects of the task (this is an area where video

TABLE 7.2 Principles of Ergonomics: Manual Handling Task Design

1. *Minimize significant body motions*
   - Reduce bending requirements using mechanical aids
   - Reduce twisting motions, e.g. better layout, use of swivel chairs
   - Reduce reaching motions, e.g. reduce size of workpiece
2. *Reduce object weight/force*
   - Reduce lifting and lowering forces, e.g. using lift tables
   - Reduce weight, e.g. use more containers
   - Increase weight above MPL so that mechanical assistance is necessary
   - Reduce pulling/pushing forces, e.g. with dollies, hand trucks
   - Reduce carrying forces, e.g. with conveyors.

Reprinted by permission of Dr. Stover H. Snook, Liberty Mutual Insurance Co.

analysis should find useful applications). The use of ergonomics and the design of new operations—or the redesign of existing operations—presents an opportunity to improve manual handling safety. The ergonomic principles that must be considered are summarized in Table 7.2.

The rehabilitation of those with lower back disabilities has suffered from the same problems as diagnostic methods, and the great variety of treatments and relative lack of success has been frustrating. Nevertheless, the most successful and cost-effective principles are as follows:

1. Early intervention, without doubting the worker or setting up an adversarial relationship.
2. Conservative treatment for 2–3 months, e.g., bed rest, medication, exercise and education.
3. Follow-up or feedback to let worker know someone cares.
4. Psychosocial evaluation.
5. Early return to work when possible (therapeutic in itself).
6. Vocational rehabilitation when job is incompatible with worker/patient capabilities and task redesign is not possible.

There is general agreement that ergonomic task design is the preferred method for the prevention of manual handling injuries and consequences. Ergonomics has the advantage of being a more permanent engineering solution to the problem, reducing exposure to risk, having less reliance on the worker and reducing medical and legal problems associated with the selection and rehabilitation of workers.

Case History 7.1 A Packaging Problem

Polyethylene terephthalate (PET) is a plastic film widely used in food packaging, such as boil-in-bag food pouches, in photography, in magnetic tape, and in metallized balloons. At Queen Plastics, a competitor of Ballardvale Balloons

(Case History 3.5), metallizing consists of passing the PET sheet through a vacuum furnace where aluminum is evaporated and condenses as a thin, opaque film on the cooler PET substrate sheet. Material handling is automated from the pay-off reel at the entry to the furnace to the take-up reel after metallization is completed. Manual handling has been limited in the past to the removal of 4-foot-long, 10-inch diameter rolls of PET film from the shipping box to a portable horse, a distance of about 30 inches. Each roll of film was wrapped with a loose piece of PET which was taped in three locations with 5-inch long "tape tabs"; loose ends were inserted into the cardboard core.

Fred Ianno and Steve Stadler, both in their early 20s, had several years work experience at Queen Plastics. In 1988, they were lifting a 200-lb roll from the shipping box (on a pallet) to the portable horse. In doing so, Fred's end slipped and began to fall. Reacting instinctively, he stopped the roll from dropping, but injured his lower back, a disability which now appears to be permanent.

Queen Plastics, proud of their safety program, had recognized that the lifting task was above the action limit but below the maximum permissible limit for lifting recommended by NIOSH and had instituted a back training school which both workers had attended. They were perplexed by the incident and spent some time re-examining the task. At first, the safety circle focused on the shipping aspect, but they found that PET was as likely to slip against itself as against Fred's hands. They then examined the rolls of PET supplied by other vendors, discovering that they were taped around their entire circumference, eliminating much of the looseness of the dust cover. Simulated experiments showed that any slipping within the loose dust cover would cease when it tightened, unless the cover failed. The strength of the PET itself, though, prevented this from happening. The final tests conducted showed that the roll which Fred and Greg lifted had to be oriented with the tape tabs at the bottom. When they lifted the roll, slippage occurred between the roll and dust cover, and the tape tabs were too weak to prevent loss of control. Fred saved the roll by catching it, but the force absorbed by his back, however, was an impulse force much greater than the 100-lb static force originally considered in the lifting task.

As a result of the task being reanalyzed, and discussions held with the vendors, Queen Plastics has increased the amount of PET on each roll, thus eliminating *any* manual lifting and using a forklift to transfer all PET rolls to and from the metallization furnace. ∎

MECHANICALLY ASSISTED MATERIAL HANDLING

The ability to move material is vital to all industries, whether manufacturing, storage, delivery or receiving. Although manual handling cannot be eliminated, a variety of mechanical equipment has been developed to make handling safer and more efficient. Accessories such as prybars, rollers, jacks and hand trucks, as well as many specialized tools designed for specific needs, are commonplace.

The hazards associated with these accessories tend to cause crushing injuries to fingers and hands, and therefore they should be identified and controlled by appropriate management procedures.

Mechanically assisted material handling frequently involves raising, lowering and transporting heavy loads over limited distances. Caution is required in these activities to ensure that the load is not dropped and that nobody is struck or crushed while it is suspended. A little knowledge of simple mechanics helps to prevent such incidents. First, the load needs to be attached to the material handling equipment so that the center of gravity does not cause any unnecessary swing when first lifted. Next, the load needs to be securely fastened to the hoist, by tying it or by using slings or hooks. Most importantly, the capacity should not be exceeded by the load. Simple mechanics considerations (Fig. 7.3) show that the angle of slings, ropes or chains affect the size of the load which can be hoisted. Examination of some of the fastening types and materials is helpful when planning material handling tasks.

Ropes are made of both natural and synthetic fibers, and their capacity depends on their size and the material selected. The factors which may affect usage include stretching as well as strength, resistance to moisture, wear resistance and other factors. Synthetic fibers are generally more durable, stronger and more elastic than natural fibers. None of these should be spliced except by approved techniques which ensure that strength is not reduced. Many fibers are woven into web slings which are useful for lifting loads which need the surface protected, e.g. machined metal surfaces. These are easily cut and they are not abrasion-resistant as are wire rope or chains, and therefore they must be inspected properly before use.

Wire rope consists of many strands of wire of small diameter which are braided together in many different fashions to provide a strong, flexible cable. Plow steel, which is a high carbon steel with about 110,000 p.s.i. ultimate tensile strength, is the standard wire used. The breaking load of the cable, of course, depends on the strand diameter, the total number of strands and, to a lesser extent,

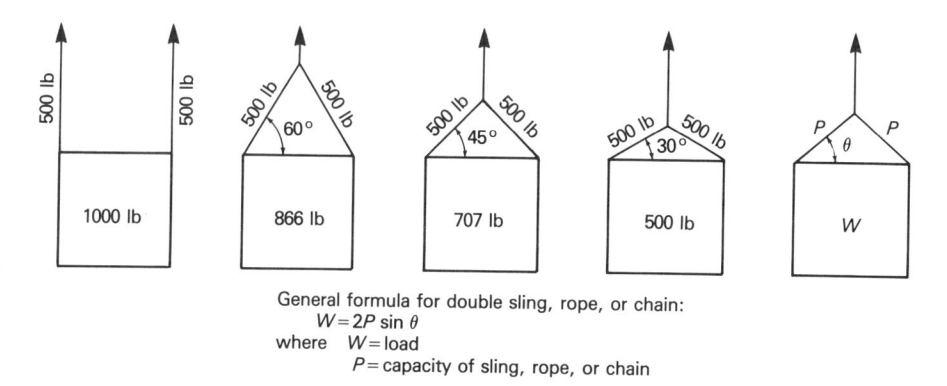

General formula for double sling, rope, or chain:
$W = 2P \sin \theta$
where W = load
P = capacity of sling, rope, or chain

Figure 7.3 Effect of angle on load-carrying ability.

the configuration of the cable. To guard against failure, the actual load or rated capacity is only a fraction of the breaking load, with a minimum acceptable safety factor of 5. As with ropes and slings, abrasion and fracture of the outer strands can occur; corrosion is also a problem with wire rope. Although a number of breaks can be tolerated without loss of wire rope strength, breaks concentrated in one lay necessitate removal from service (lay refers to the distance measured along a cable which a strand makes when wound around the cable axis).

Chains and hooks are typically made of alloy steel, but special-purpose chains are available when resistance to corrosion or other special properties are desired. Because of the contact made with each link during use, wear and deformation are two major concerns for chains and hooks. Many chain failures could be prevented if they were inspected before use. Hooks not only wear, but they can also become deformed, e.g. the hook opening can increase in size.

Any stress that is applied to slings, ropes, chains, etc., is amplified or concentrated at the tip of any crack or cut. This amplification depends on orientation and geometry, but it is greater for long, thin cuts (the phenomenon is shown schematically in Fig. 7.4). Because of their ability to concentrate stress, these cracks or cuts are called stress raisers. They can lead to unexpected failure because the breaking strength is exceeded locally.

Specialized means other than slings and hooks for fastening the load to the hoist (e.g. electromagnets) have been used for many years to lift specific loads. Termed below-the-hook lifters, many of these are undergoing further developments to promote safer material handling operations. Some examples of the more frequently used special lifters are shown in Fig. 7.5. Mechanical lifting devices

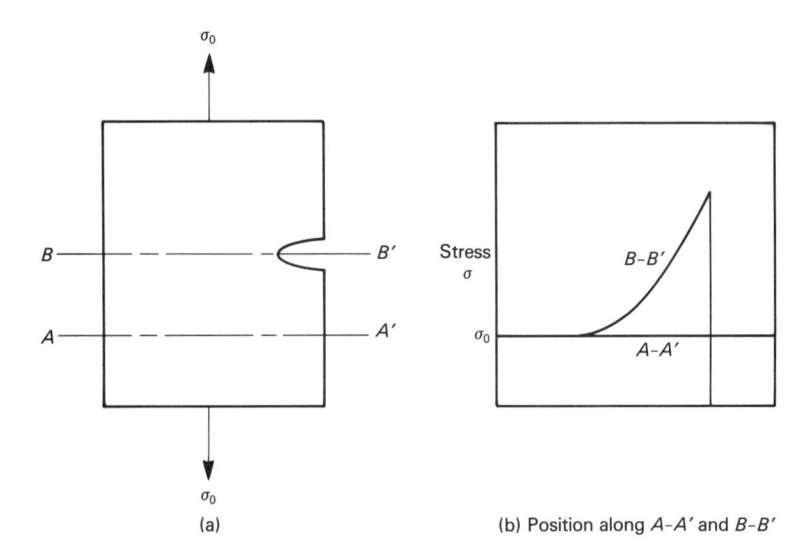

Figure 7.4 Schematic diagram of a cut in a sling (a) and the resulting stress concentration (b).

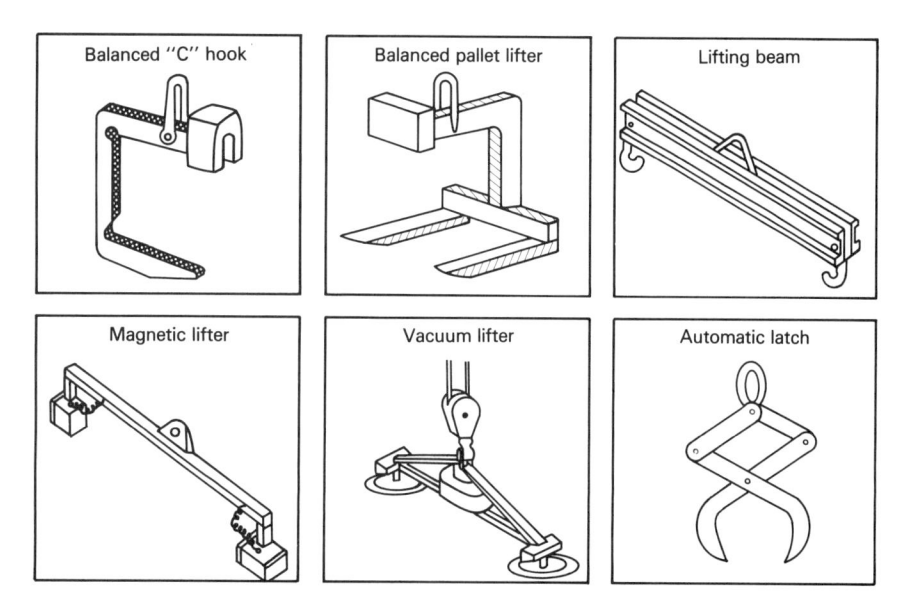

Below-the-hook lifters can increase safety and productivity, particularly when the lifting operation is repetitive. They are a vital extension of the hoist or crane to which they are attached. Too often, users do not fully appreciate that a hoist or crane is only as good as its weakest link.

Manufacturers offer a wide variety of below-the-hook designs to perform different kinds of lifting, stacking, and handling tasks in manufacturing and warehousing. Some examples of the more frequently used lifters are illustrated.

One way to distinguish among these various designs is by the way the lifter attaches to the load. Many designs depend upon some kind of mechanical attachment to the object lifted. Other lifters are surface-contacting devices that connect to the load by the forces created by producing a vacuum or by applying a powerful electro or permanent magnet.

Mechanical lifting such as the automatic latch includes devices that raise the load by pressure gripping it. Structural integrity must be sufficient so that the load is not deformed or, worse yet, collapses.

A load also can be moved by mechanically supporting it. Balanced pallet lifters, lifting beams, and balanced "C" hooks are all load-supporting devices.

Vacuum lifters derive their utility by taking advantage of the pressure differential between atmospheric pressure outside the pads applied to the surface lifted and the vacuum created under the pads. These surface-contacting lifters are effective, for example, with metallic, plastic, and glass sheets, with rolled products, such as paper, and with other loads. Magnetic lifters work well when the surface to be lifted is ferromagnetic.

Figure 7.5 Common lifters used with cranes. (Adapted from *Modern Materials Handling*, November, 1987. Reprinted by permission.)

include drum turners and automatic latches, both of which grip the load by applying pressure to it as it is lifted. Of course, the load must be strong enough so as not to deform or collapse in the lift. Balanced pallet lifters and balanced "C" hooks move materials by supporting them from beneath. Vacuum lifters are effective in lifting sheet materials; they take advantage of the pressure differential between atmospheric pressure and the vacuum created beneath the lifting pads in order to grip the workload. Electromagnets work effectively, but are limited to lifting ferromagnetic workloads only.

Case History 7.2   Better to Miss the Brass Ring

The advent of cable TV has meant that cable lines have had to be installed as for the telephone and electric utilities. John Mackyn joined New-Wal Cablevision as a cable installer 8 months before his accident and injury. The cable stringing method used by New-Wal employed a pole-to-pole support cable wrapped with two signal cables. The wrapping was carried out using a lasher, which was pulled along near ground level. In order to pull the lasher, a wire rope was clipped to it in two places; this rope passed through a ring, and a polyester fiber tow-rope pulled by the New-Wal Cablevision truck was clipped to the ring.

When the crew began work on the day of the incident and injury, the ring could not be found. John went to the local hardware store and bought a ring of the same size (1.75-inch i.d., made of 0.25-inch brass). The ring was stamped Taiwan. When the new ring was put in place, the cable installation began. After 1 hour, the ring broke, and the clip on the end of the flailing tow rope struck John in the face, breaking his cheek and nose, narrowly missing his eye.

New-Wal's safety circle was asked by management to address this incident and injury. In turn, the circle asked Dalzell Associates to conduct some tests on similar rings purchased after the incident at the same hardware store. Dalzell Associates found that these brass rings became less deformed than other brass rings or even a plated steel ring before failing, and that they broke at load values of less than 50% of other brass or steel rings. Their poor performance was attributed to a lower zinc content and the use of a cast-metal rod, identifiable by their microstructure (Fig. 7.6), which shows extensive coring or non-uniform composition due to the casting process.

The circle used this information to recommend that cable installation crews be made aware of the problems encountered when purchasing any parts used in loading, and that New-Wal's purchasing specifications for replacement parts be re-examined. ■

PALLETIZING THE WORKLOAD

Lifting, lowering and transporting loads for limited distances is encountered in processing, shipping or receiving, and storage. Performing these tasks efficiently and safely is made easier by palletizing the workload. Manual operations have largely been replaced by safer and more productive automatic palletizing. Box or container palletizing is the most popular method; bags and sacks being used to a lesser extent. This is true because bagged products are not always stable and handling is more of an art than a science. Stacking boxes is much easier and we now possess technology that allows the interlocking of layers, whether on pallets, sheets, or even without either. Loads can be banded with steel or plastic strapping, tied with rope or even stretch-wrapped.

Palletizing the Workload

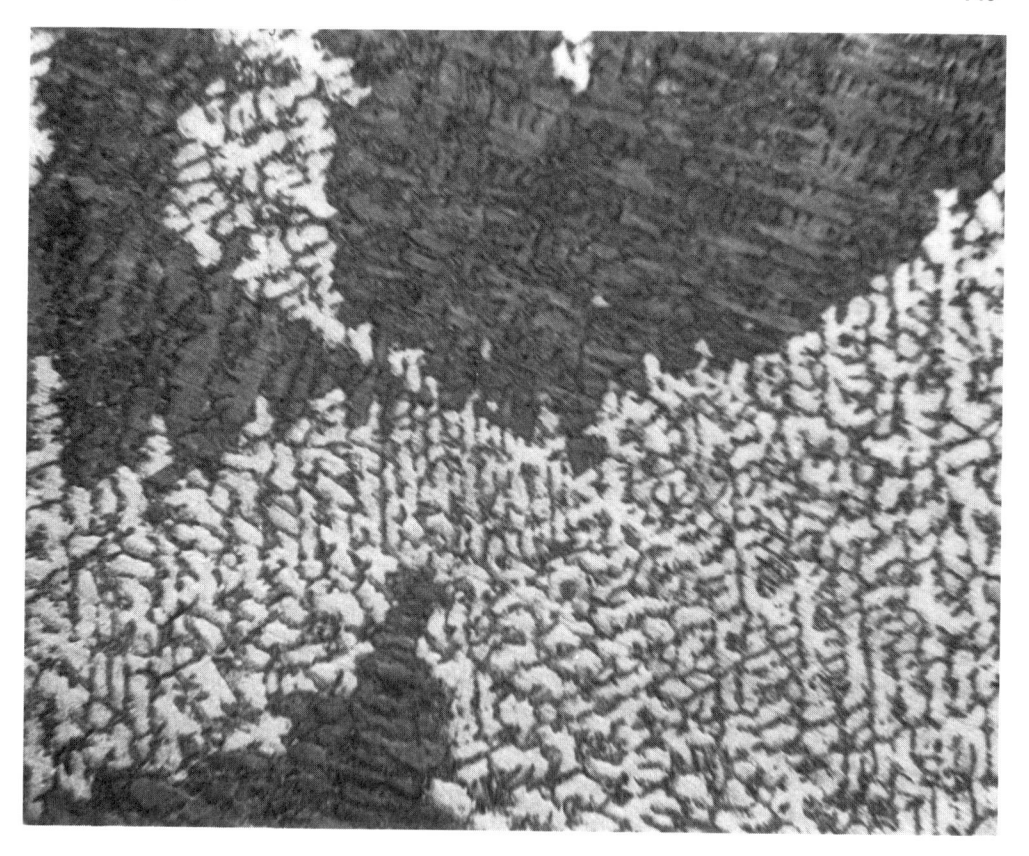

Figure 7.6 Microstructure of Taiwan brass ring (100×).

Automatic palletizing motivated by increased productivity has frequently improved handling safety. For example, unlike rectangular boxes, square boxes do not interlock, and thus stacking is ineffective. This has resulted in lost production, messy clean-up and some minor lacerations in bottling operations. In the Fort Smith, Arkansas, plant of Hiram Walker and Sons, safety and productivity were both improved when they installed a system based on the application of a temporary adhesive to lock square boxes together when moving them and for storage. The boxes are easily separated, without destroying the box surface or even the graphics.

Automation in palletizing includes fixed units such as the Alvey 2000, used for beverage canning and bottling operations (Fig. 7.7.) This high-performance unit palletizes 120 cases (24 each) per min from a single infeed conveyor. Automation also includes flexible systems, including powered manipulators, automated guided vehicles and robots which can be programmed for many different palletizing tasks. Automation, therefore, introduces many safety improvements,

Figure 7.7 High-performance palletizer, Alvey 2000, is for beverage canning and bottling operations. Palletizes 120 24/12 cases per minute from a single infeed conveyor. This is one of the biggest and fastest palletizers available. (Reprinted by permission of Alvey Conveyor, Inc.)

but also new, more subtle and complex safety problems. Good systems safety concepts are needed to solve such problems.

POWERED INDUSTRIAL VEHICLES

Whenever we think of productivity, flexibility and maintenance in materials handling, powered industrial vehicles dominate our thoughts. For distribution and storage systems alone, we find numerous lift-trucks such as stockpickers, pallet trucks and the most popular fork-lift truck. Other appliances used in the construction industry include backhoes, front-end loaders and even earth-moving equipment. Safety goes hand-in-hand with higher productivity and flexibility, and therefore modern vehicle production adopts vastly improved ergonomics as well as utilizing new technologies (e.g. hydraulic, electric and electronic components) to make them safe. Statistics reveal, however, that injuries suffered from fork-lift accidents alone have been increasing about 65% faster than total industrial accidents. Before we look at how to reverse this trend, let us look first at some of these powered vehicles.

Palletizing is the first step in a storage or warehousing operation. The storage systems need to be designed with safety of personnel in mind, taking into consideration such factors as floor loading, anchoring and strength of shelving, and the potential uses of fork-lifts. Stockpickers or order selector trucks are very popular for warehouse storage and distribution. Most of these have elevating platforms which raise the operator with the pallet. (Fig. 7.8) As the lift rises, the capacity falls, e.g. at 30 feet the capacity is usually about 1500 lb. These stock-

Figure 7.8 Latest stockpicker series from Raymond is completely redesigned to make the operator's life as easy as possible and productive over a fast-paced picking shift. Controls are convenient, easy to use and electronically or electrically connected for fast, smooth response. Anti-fatigue floor mat is replaceable as individual tiles, side compartment arms flip up for easier rack access, and tether line is mounted outboard to give the operator more freedom. (Reprinted by permission of Raymond Corp.)

pickers operate in aisles which are only a few inches wider than the vehicle itself. The picker must use fall protection equipment and traffic patterns must be planned beforehand to promote both productivity and safety. Guidance systems along the aisles and computers help to accomplish this. Some units even have on-board realtime interactive data terminals to relay new instructions or even messages to the picker.

Pallet trucks account for about 25% of all industrial trucks sold each year and they are used primarily for transporting palletized materials. Over short distances, walkie pallet trucks are mainly used, whereas walkie/riders are used over longer distances and for continuous operations. Pallet trucks offer the advantage of maneuverability because of their short lengths and tight turning radii. Many of them have end controls, where the operator rides in a standing position on the end of the truck or walks beside it (speeds of 4–7 mph are typical). Fig. 7.9 shows the control handle for one of these units.

The main problems experienced when operating pallet trucks are crushing injuries caused to people caught between the truck and fixed objects or by the truck wheels themselves. To prevent these injuries, all of the systems safety principles need to be utilized, particularly management involvement through appropriate training and effective hazard identification.

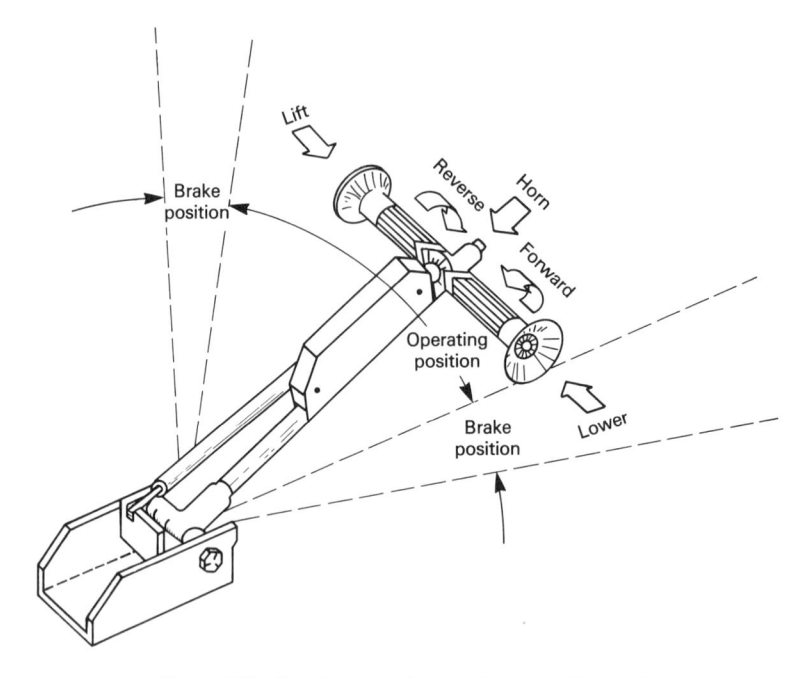

Figure 7.9 Handle control operation for pallet truck.

Case History 7.3   A Case for Training

Best Sales distributed powered industrial trucks for several manufacturers. Dexter Best, son of the founder, arranged to demonstrate a low-lift, walk/ride pallet truck to the foundry department at General Industries in Midport. General Industries already used two pallet trucks, but they were interested in replacing the older one which was no longer dependable. Dexter had only been with the company for 2 months, and he was not able to answer many of the detailed and technical questions asked of him by the General Industries personnel. His lack of training and unfamiliarity with the product led to a hands-on demonstration by several employees.

John Royce was fairly new at General Industries, and he had already developed a reputation in the finishing department for being careless and inattentive. During the demonstration, Dexter permitted John to operate the pallet truck. While operating the vehicle, John managed to place the truck into reverse so that the vehicle jumped forward, running over and breaking his right foot and ankle.

The immediate response of General Industries was to reject the pallet truck and to dismiss or, at a bare minimum, reassign John after he had recuperated. The incident was then presented to the safety circle at General Industries. After analyzing the problems the team came up with the following recommendations for management:

1. Cease business relations with Best Sales (Best Sales is no longer in business).
2. Funnel all purchases of powered industrial equipment through the safety office.
3. Enforce training requirements for *all* operators of powered industrial equipment.
4. Include differences in response of different vehicles, since examination of the subject pallet truck had revealed much faster acceleration than existing pallet trucks. ■

Fork-Lift Safety

The fork-lift (e.g., Fig. 7.10) represents the workhorse of all powered industrial vehicles, fulfilling the requirements of many broad assignments in all aspects of lifting and transporting materials. The very capabilities which provide this versatility, however, contribute to the hazards inherent in fork-lift operations. Perhaps the best-known fork-lift hazard is instability; the center of gravity is affected by the truck design, size and shape of the load, load placement on the forks and the height to which the load is lifted. But only a small proportion of fork-lift accidents actually involve the vehicle overturning; most accidents occur when a fellow worker or the load itself is hit.

Other characteristics of fork-lifts pose difficulties for operators. Fork-lifts are generally steered by the rear wheels rather than by the front wheels, and they steer easier when loaded than when empty. In addition, the load is carried in front, so that the mast/load assembly interferes with visibility. These features contribute to operator error, and training is a must to reduce errors and incidents, as well as conform to OSHA regulations. Some companies have adopted improved training and retraining programs for fork-lift operators, plus the rigorous enforcement of safety rules. Although these companies also point to no fork-lift accidents during short-term studies, the approaches are traditional and lack the innovation and commitment required in today's safety management.

Figure 7.10 Fork lift. (Reprinted by permission of *Modern Materials Handling*.)

Kaiser Aluminum and Chemical Corporation[1] has taken a more modern systems safety approach to fork-lift safety. Beginning with a NIOSH study and analysis of their own 5-year fork-lift accident experience that substantiated the NIOSH study, Kaiser categorized the types of fork-lift operations where accidents occurred as:

- *The task,* e.g. most accidents occurred in backing up.
- *Activities preceding the task*—decisions about the upcoming task, how and when it would be performed.
- *Activities remote from the task*—mainly vehicle factors and maintenance.

Kaiser also adopted training and enforced fork-lift safety rules, but used management commitment to safety and systems concepts to come up with their present program which includes the people, vehicles and environment of the fork-lift materials handling procedures. Planning for these operations was expanded to *involve* engineering, purchasing, production, maintenance and fork-lift operators in group planning of tasks and traffic patterns. Focusing on non-operator as well as operator behavior led to better training programs and better communication between operators, co-workers and pedestrians. By comparing the fork-lift trucks with alternative, ergonomically safer vehicles presently available, and examining maintenance practices, traffic patterns and illumination, Kaiser now has more effective control procedures. For example, the purchasing department was able to arrange a favorable leasing arrangement for new, more ergonomically designed fork-lifts which were also more versatile than existing units scheduled for replacement.

A separate approach to improved fork-lift safety demonstrates that we must be flexible, that what works for one company may not work for another, and that we can all learn by communication. A NIOSH study[2] of warehouse fork-lift safety and implementation of a behavioral modification program (see Chapter 4) focused on the operator training aspects and human error reduction only. After a detailed specific task hazard analysis was conducted to model a training program, instructional sessions were devised to introduce the topic, to demonstrate incorrect ways of handling the situation and, finally, to specify the *correct* procedure to adopt. When reinforced with performance feedback, group performance goal setting, peer group modeling and management support, this study not only showed dramatic reductions in individual error rates, but that improved performance is also maintained long after the program ceases.

We have benefited from these studies in a subtle way. Both the NIOSH and Kaiser studies indicated that most fork-lift error-induced accidents occur when backing-up. The behavior modification study showed that errors while driving in

---

[1] See Akamatsu and Grand (1984).
[2] See Cohen and Jensen (1984).

reverse were reduced dramatically in one warehouse, but they were unchanged in a second. It was revealed that the operators in the second case resisted change because the adoption of correct procedures caused them to breathe noxious fumes from the propane-powered fork-lifts (battery-operated fork-lifts were used in the first warehouse). This information has led to an ergonomic redesign of exhausts for current propane-operated vehicles.

Case History 7.4 An Unplanned Task

Randy Older has been an employee at Mill Stream Lumber Co. since it first opened for business in 1962, working as a material handler and recognized by everyone as the safest and most competent fork-lift operator. He was often called upon to transfer orders of lumber which arrived about twice monthly by railroad freight cars. He was therefore familiar with the palletizing procedures that used disposable, inflatable dunnage bags to stabilize loads within freight cars during transportation (see Fig. 7.11).

When unloading lumber from the freight cars, Randy would open the doors, cut the metal strapping around the entire workload, and then use the fork-lift to transfer the individual pallets.

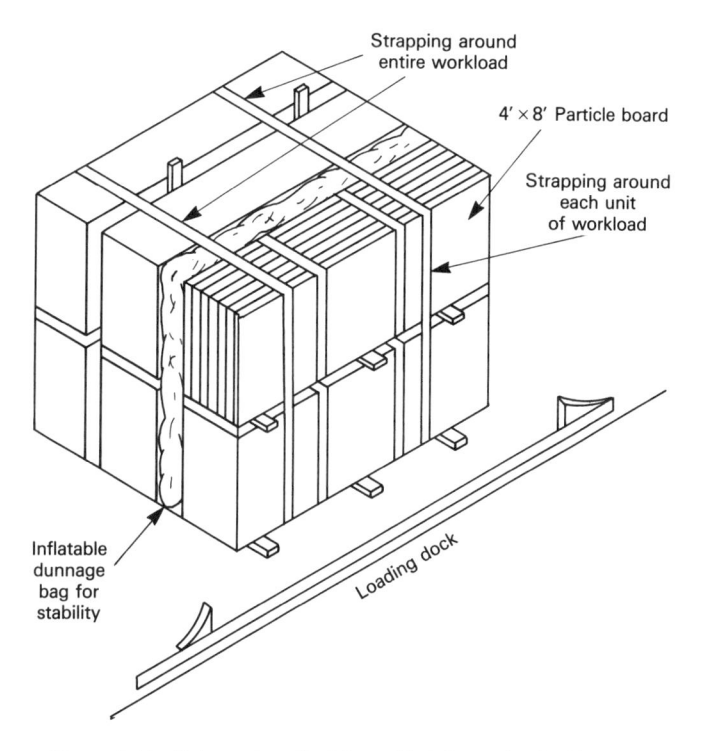

Figure 7.11 Schematic palletizing of freight car load of lumber.

In 1987, Randy was asked to unload particle board from the freight car at the loading dock. He drove his fork-lift to the freight car, parked it and broke the seal on the freight car doors. He slid the right door open but saw that the left door had a particle board unit leaning against it; the individual unit strapping on it was broken. He therefore used his fork-lift to move the workload back about 6 inches, just far enough to let him open the door, and everything appeared normal. Randy then cut the strapping around the entire workload so that he could proceed with the unloading. He moved only about 4 feet toward the fork-lift when the entire load collapsed, striking him and throwing him forward, pinning his left leg, which was crushed so severely that it needed to be amputated.

A reconstruction of the incident showed that the dunnage bag had been cut when the strapping was placed over the entire load, deflating it. During transportation, the workload, particularly the inner units, had shifted, and would have collapsed except for the all-around strapping. Randy was unaware of the internal instability because the external units were normal. As soon as the strapping was cut, however, the load was ready to collapse.

As a result of Randy's incident and injury, the use of dunnage bags for stabilizing workloads between freight car doors has been modified. Today, dunnage bags parallel to the doors can only be used external to the workload. ■

Cranes

The use of cranes differs from other material handling procedures used in industry because cranes are usually confined solely to in-processing material handling. Although there are many types of cranes, wall or boom cranes, trolley hoists and overhead cranes are most often used. Very serious or fatal injuries have been identified when:

- being struck by the hook or load,
- being struck by a dropped crane block,
- being caught, for example, between the load and building structure, and
- falling from cranes or boarding ladders.

Victims are not only operators or hookers, but include maintenance personnel.

Investigations have shown that workers do not understand and follow established safety procedures. In other words, this is another example of where innovative, committed safety management can intervene, and this is exactly what Interlake, Inc., did. Interlake is a steel producer which uses about 150 overhead cranes with capacities up to 150 tons (see Fig. 7.12). Using systems safety concepts to identify the real risks, safety routines were developed and slide presentations were selected because they permitted better interaction with the workers during the presentation. Initially, three presentations were developed (one each for operators, hookers and maintenance workers)—and then their performance was ob-

Figure 7.12 Overhead crane. (Reprinted by permission of Modern Materials Handling).

served. It became apparent that certain critical safety aspects had to be repeated to all groups, e.g. the proper use of hand signals. Today, retraining presentations are held annually and feedback on worker behavior is essentially continuous because of the increased understanding and communication established among all workers by the program.

AUTOMATED MATERIAL HANDLING

Today's manufacturing technology emphasizes productivity and optimization of inventories. Without proper material flow through manufacturing, inefficiency and possible chaos result. One of the most common errors in materials handling systems in manufacturing is back-tracking, which confuses operations and increases handling times, costs and the risks of handling accidents. A simple process layout can avoid this problem: the total distance travelled can be reduced by simply improving the flow pattern (see Fig. 7.13).

Besides improved plant layouts, we should also look at the benefits of automated handling between work stations for productivity and safety. There are three basic types of automated processing line: fixed, programmable and flexible. Fixed automation, which is probably the oldest automated technology, is specialized, has a high initial investment cost, is highly productive and is inflexible when accommodating product changes. Certain specific operations or series of operations are performed on the product, with both the operation and transfer from station-to-station automatically controlled. In a synchronous system, a transfer machine or rotary index moves parts to succeeding stations, and the next cycle

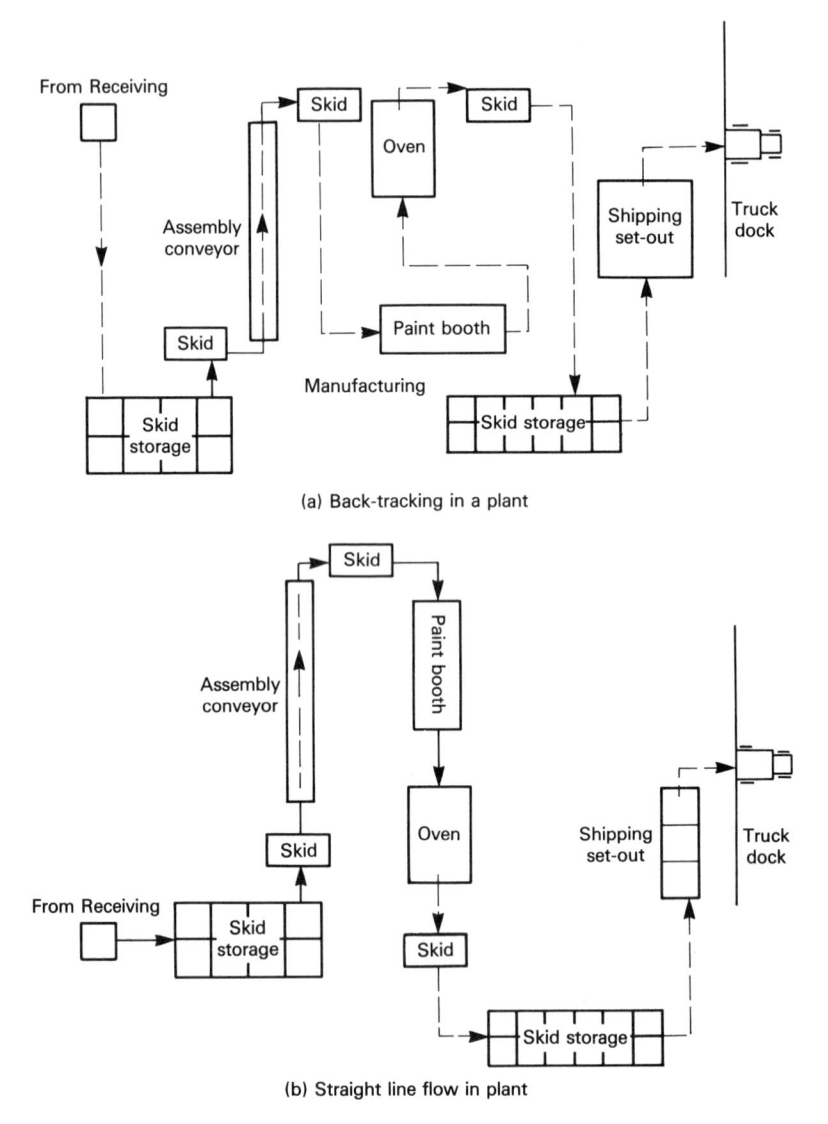

Figure 7.13 Inefficient back-tracking and efficient straight-line flow through processing in a plant.

begins. The time for each cycle is controlled by the slowest operation and any breakdown halts all operations. Non-synchronous systems do not suffer from this problem because a process inventory is maintained at each operation. Both straight-line flow (Fig. 7.13) and parallel flow systems are popular.

With programmable automation, programs are developed to instruct the machine what to do. Any change in the product requires a new program. There are many numerically controlled and computer numerically controlled machines which are popular, e.g. lathes, machining centers, drills and stamping machines. Programmable machines are best suited to batch production and are most flexible for product configurational changes. They represent high capital investment, but they are not as productive as fixed systems.

Flexible automation is a relatively recent development which has evolved for medium production volume and relative flexibility. Flexible automation can be distinguished from programmable automation because past programs can be changed and physical set-ups can be changed with no loss in production time. Computer-integrated manufacturing (CIM) and the factory of the future symbolize flexible automation.

There are many arguments for and against automation. Two key arguments in favor are increased productivity and reduced costs. The most common arguments against automation are worker retraining or replacement and a reduction in the labor force. Automation affects safety in two ways, both of which involve perception. Removing a worker from a hazardous operation only reduces the frequency of contact, not the hazard, giving the perception that the automated operation is safer when it is not necessarily so (Case History 7.5 demonstrates this). The arguments against automation also reflect other safety issues: as long as job losses are perceived, the job becomes stressful. The effects of stress and how to cope with it are covered in Chapter 11.

Case History 7.5   Automation Needs Ergonomics

When ice cream is first made, it is soft and must be hardened in a freezer. After pouring the soft mixture into a half-gallon container and shrink-wrapping six containers together, the automated system at Kost Creamery uses a series of roller and belt conveyors to transfer the product from the processing room to the hardening rooms (maintained at $-20°F$). Here, the packages are collected until the system is filled. Unloading, palletizing, and storage is then automatic.

Joe McGentry was a "chest man" in the refrigerated unit of Kost Creamery. In 1989, an alarm went off which alerted Joe to a jam in one of the conveyors. In order to reach the jam, he had to climb a fixed ladder and enter a narrow walkway to the lifter mechanism which stands the plastic wrapped unit upright while transferring it from the infeed conveyor to the collection system. One unit was jammed in the upper (no. 3) conveyor system (Fig. 7.14). He knew that he had to turn the switch off before releasing the jam, but he reached over and shut off the switch under the large #3 in Fig. 7.14, which is the wrong switch. Due to this human error, the lifter continued upward rapidly when the load was released, amputating Joe's finger.

Kost Creamery had been proud of their safety record prior to Joe's injury and charged their safety circle with evaluating the problems. Their comprehensive study found fault with the design and installation of the machines (e.g. the switch controls for two different machines were adjacent to each other), and with man-

Figure 7.14 Conveyor system.

agement and themselves over poor hazard identification. Although automation reduces the frequency of worker involvement with machinery, safety is not really improved unless safeguarding principles are adhered to. ∎

THE FUTURE OF MATERIAL HANDLING

The current emphasis is on flexible manufacturing, rather than factory automation, because many products are not suited to automation and they neither generate the long-term demand nor sufficient profit to warrant the investment necessary for automation. Participative management, pride in workmanship and better working relationships are being credited for gains in productivity and quality, the same qualities that have been espoused for improved safety.

Another in-process material handling concept is Just-In-Time (JIT) manufacturing. JIT is a total systems approach which conveys the idea that capital, equipment and labor are made available only in the amount required and at the time required to do the job. A JIT manufacturing system is simple to operate,

TABLE 7.3 Basic Principles in Just-In-Time Philosophy

1. Each worker or work unit is both a customer and a supplier; each has a right to safe operation.
2. Customers and suppliers are an extension of the manufacturing process.
3. Continually seek the safe path to simplicity.
4. It is more important to prevent problems and hazards than to fix them.
5. Obtain or produce something only when it is needed.

Adapted From R. T. Lubben, *Just-in-Time Manufacturing*, 1988. Reproduced by permission of McGraw-Hill Book Company.

flexible and competitive, and it has been applied successfully to such diverse activities as electronics and automobile manufacturing. Because the development of high-quality processes and products is the responsibility of an entire company, JIT manufacturing includes all responsible functions within a company, not just production. Safety personnel must ensure effective hazard identification and control as well as integrating safety into JIT management.

Let us now look at how safety can be integrated with the philosophy and goals of JIT. Table 7.3 lists Lubben's five basic principles, modified to integrate safety wherever needed. In JIT, the manufacturing system is continually optimized and integrated by removing the excess inventory, equipment and labor to force production-related problems to the surface so that they can be dealt with. The purpose of this is to achieve trouble-free production and optimum productivity. Over the years, certain elements have been identified which depress productivity; chief among these is poor design. Because of this, two of the basic goals of JIT manufacturing involve design:

1. Design for optimum quality/cost and ease of safe manufacturing.
2. Minimize the amount of resources expended in designing and manufacturing a product.

In a JIT system, then, there is constant pressure to use the fewest resources. Safety personnel have to be involved when the questions "What is required?" and "Why is it required?" are answered.

A LAST WORD ON MATERIAL HANDLING

Materials handling, both in-process and warehouse handling, can be performed safely. To develop this, the cooperation, involvement and dedication of management, and technical and safety professionals are required. A recent series of articles in *Modern Materials Handling* characterized five basic steps in developing successful materials handling systems:

1. Assess current operations and establish improvement goals.

TABLE 7.4 Material Handling System Selection Factors

Here are ten "comparative features" for getting the "best fit" between handling options and company needs. When choosing between systems, you will want to keep all ten in mind—each to a degree determined by your company's needs.

Flexibility	The capability of accommodating predictable changes, such as different rates or routings.
Controllability	The capability of responding to or predictably carrying out specific directives.
Adaptability	The capability of accommodating unanticipated changes, such as new flow paths.
Reliability	The capability of performing the desired missions, often called "up time."
Manageability	The capability of providing usable information to decision makers (human or electronic) and to produce timely and accurate transaction data.
Ease of implementation	The degree of difficulty and the amount of effort required for installation.
Maintenance factors	The cost of keeping the system running after installation.
Safety	The degree to which workers' well-being is jeopardized while operating or working near the equipment.
Integrability	The ease with which the components can be made to function together compatibly.
Ease of planning	The amount of time and effort, based on available skills, required for design work.

Source: Modern Materials Handling (April 1987). Reprinted by permission.

2. Identify and specify requirements.
3. Select suppliers and acquire equipment.
4. Install the system.
5. Test for design operation.

Safety personnel have to be involved in all phases to ensure successful and safe materials handling systems. When we realize this, the system selection features (Table 7.4) take on added meaning for *all* features.

FURTHER READING

M. AKAMATSU and E. V. GRAND, "Forklift Safety Involves More Than Just Forklifts," *Professional Safety,* **30**(1), 15, 1984.

T. H. ALLEGRI, *Materials Handling: Principles and Practice,* van Nostrand Reinhold, 1984.

J. B. BENSON, "Control of Low Back Pain Using Ergonomic Task Redesign Techniques," *Professional Safety,* **33** (9) 21, 1987.

H. H. COHEN and R. C. JENSEN, "Measuring the Effectiveness of an Industrial Lift Truck Safety Training Program," *International Safety Research,* **15**, 125–135, 1984.

D. E. DICKIE, *Rigging Manual,* Construction Safety Association of Ontario, 1975.

M. P. Groover, *Automation, Production Systems, and Computer Integrated Manufacturing,* Prentice-Hall, Englewood Cliffs, N.J., 1987.

R. T. Lubben, *Just-In-Time Manufacturing,* McGraw-Hill, New York, 1988.

Material Handling Engineering, published monthly by Penton Publishing, Cleveland, Ohio.

F. E. McElroy, ed., *Accident Prevention Manual for Industrial Applications: Engineering and Technology,* 8th Edition, National Safety Council, Chicago, Ill., 1980.

Modern Materials Handling, published monthly by Cahners Publishing, Newton, Mass.

National Institute for Occupational Safety and Health, *Work Practices for Manual Lifting,* NIOSH Publication No. 8I-122, Cincinnati, Ohio, 1981.

M. H. Pope, J. W. Frymoyer and G. Anderson, *Occupational Low Back Pain,* Praeger, 1984.

S. H. Snook, "The Design of Manual Handling Tasks," *Ergonomics,* **21,** 963, 1978.

S. H. Snook, "Workloads," *Proceedings of an International Symposium on Low Back Pain and Industrial and Social Disablement,* Back Pain Association, 1983.

QUESTIONS

7.1. Describe the correct method for lifting an object. Is this effective in reducing injuries? Explain.

7.2. Wire rope, chains, slings and hooks are commonly used to mechanically assist in material handling. Assuming you are responsible for safe operation using these items, describe the key features of your program.

7.3. What are the main causes of operator error in forklift operation? How would you address their correction in a forklift training program?

7.4. Assuming you are responsible for safety in a warehouse with different stockpickers available, describe key features of your program.

7.5. Describe the safety problems imposed by overhead cranes in the workplace.

7.6. How does automation affect safety?

7.7. Describe the three types of automation.

7.8. How does J.I.T. affect safety?

8

Machinery Safeguarding

Do you know how many different kinds of machines are used in the workplace? If anyone attempted to answer this question, we would treat their responses lightly. And do you know how many different kinds of machines are defective and still used in the workplace? We should again treat lightly any answer to this question, even though it should be zero in order to conform to OSHA regulations. The fact is that there are many machines in use today which are unsafe to operate. There are also many reasonably safe machines that are used for improperly designed tasks which make *the task unsafe,* or improperly trained employees add an element to a task and/or machine which turns the manufacturing system into one which is ergonomically unsafe.

The principles of safeguarding have not changed dramatically in recent years, and accidents resulting in injuries have not been reduced simply by citing them. As a matter of fact, machine injuries account for some 10% of all industrial injuries; moreover, these injuries are among the most severe. This chapter explores what we know about machine safety, what we *should* know about the machinery, the task, changeover processes and maintenance, and then explores the integration of participative management to point the direction in which we must go so as to reduce significantly the number of accidents and injuries caused on machinery.

HAZARDS AND MACHINE MOTION

The Industrial Revolution brought about a relationship between mankind and machinery which has rivaled interpersonal relationships and has endured to the present day. The relationship is a stormy one, where machines have become

Hazards and Machine Motion

extensions of ourselves, bringing out some of our best qualities, but also fostering frustrations, anger and even hatred. The hazards associated with machinery are not new, but neither are they necessary, even in today's advanced machines. There are a number of basic motions and actions which can create hazards in machinery, and these must be either eliminated by design wherever possible or protected against in other ways.

The hazardous actions we should all be familiar with are as follows:

1. *Rotating motions* – includes rotating shafts or flywheels which can snag anything (e.g. loose clothing) with which it comes in contact, and screw mechanisms, including drill bits and augers which can not only snag, but also lead to pinch hazards.
2. *Reciprocating motions* – where the direction of linear motion between fixed limits is reversible. The increasing use of robots has demonstrated the hazards of these motions, e.g. crushing against fixed objects.
3. *Crushing or compression actions* – these actions can cause crush injuries, but also include shearing, pinching and bending actions which can lead to amputation. A special category of crushing or compression action is the in-running nip point of any machine where parts rotate or move relative to each other.

Statistics tell us that the most hazardous types of equipment are woodworking machinery and power presses and that the most common form of injury involves fingers and hands which become caught in the moving parts of a machine.

The hazards of machine moving parts can also be classified by the location or purpose within the machine. Point of operation hazards are usually distinguished from other hazards within the machine. For example, Fig. 8.1 illustrates three common point of operation hazards caused by shearing, cutting and bending motions. Other hazards, termed pinch points or nip points—to distinguish them from point of operation hazards—occur with the transfer of energy, motion or material guidance (Fig. 8.2).

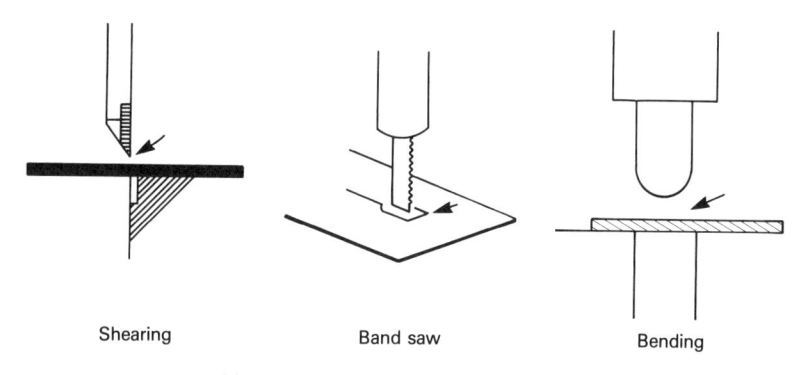

Figure 8.1 Point of operation hazards.

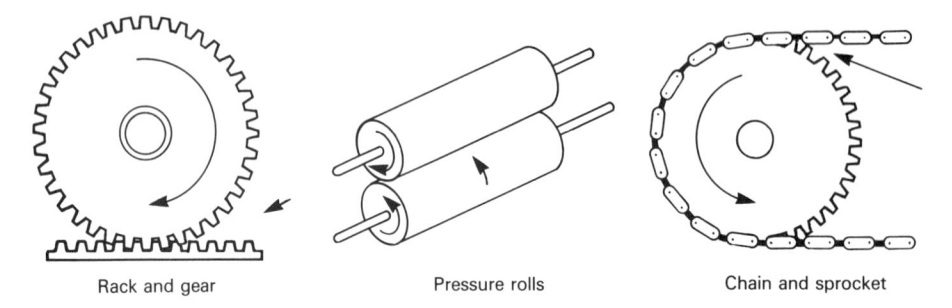

Figure 8.2 Pinch point hazards.

Machine designers familiar with their functional requirements must also be aware of human factors as well as the aforementioned hazards in order to design a reasonably safe machine. Although we know that certain motions are hazardous, the mere absence of motion does not make it safe. A common problem encountered is the inadvertent activation of equipment. Machine designers *must* consider the needs involved in the interaction between the operator and the machine. Anthropometry, the maximum functional zone of the operator and controls must be considered when designing a machine, so as to prevent coincidental hazards.

Like designers, safety personnel too must be aware of the hazards involved with machines and recommend changes to equipment and processes that are felt to be unsafe. Some machinery hazards identified in human factors studies include:

1. Push-type starter controls that do not have extended sleeve guards and unguarded foot-pedal controls permit inadvertent activation.
2. Line-of-vision controls are needed at multiple start-up stations on large machines to prevent activation when workers are at risk of injury.
3. Complete control over activation is necessary to prevent other workers being injured by the actions of the operator.
4. The design of push sticks and feeding tools should take into account anthropometry and the task for which they are to be used.
5. Hand-holds should be provided where machines are climbed on or over.
6. Two-hand controls must be designed so as to prevent override.
7. Safety stop buttons must be within the reach of the operator's free hand when the other is caught in the machine. These must be larger than all other buttons and left unguarded. Trip wires or bodyguards can also be used to reduce the severity of injuries.
8. Control buttons and visual displays should be color-coded. Also, the layout of the controls should be logical and should not violate stereotypical coloring codes.

Principles of Safeguarding

PRINCIPLES OF SAFEGUARDING

Understanding machine hazards is only the first step in ensuring the safe operation of a machine. The second is safeguarding *by design,* which includes human factors aspects as well as providing mechanical safeguards. Safeguards are simply means to prevent access to the hazard. They are usually of four types:

1. Barrier guards.
2. Interlocking guards.
3. Automatic guards.
4. Remote guards.

Any of these must be integral to the machine (and not afterthoughts), well-constructed and tamper-proof, able to permit task performance, easy to maintain and do not introduce new, but different hazards. Methods are not new and have been described in OSHA regulations, ANSI specifications, professional society publications, and safety manuals (summarized in Fig. 8.3).

For reasons of effectiveness and cost, the most commonly used form of safeguarding is barrier guards which enclose the hazard. In most cases, manufacturers provide permanent barrier safeguards for point-of-operation and for power transmission components. However, in many instances, where older equipment has been modified for use in new applications, well-meaning employers have made and installed various types of barrier safeguards. In these cases, it is most important that a safeguard is adequate. Figure 8.4 illustrates the relationships between guard location and opening which will prevent entry of hands or fingers into the hazard area.

Barrier guards limit access, whereas other safeguarding methods control either the machine or the worker actions. Interlocking guards, for example,

LIMIT ACCESS	CONTROL MACHINE FUNCTION	CONTROL EMPLOYEE ACTION	CONTROL MACHINE/ EMPLOYEE	WARNINGS AND INSTRUCTIONS

• Barrier guards
  1. Fixed
  2. Adjustable
	• Interlock devices
	• Photoelectric devices
	• Automatic material handling
	• Barrier guards on controls
	• Presence-sensing devices
	• Proper fixturing
		• Restraints
		• Pull-back devices
		• Feeding devices
			• Two-hand controls	• Awareness barriers
				• Color coding
				• Warning labels

Figure 8.3 Safeguarding point of operation or pinch point hazards.

168 Machinery Safeguarding Chapter 8

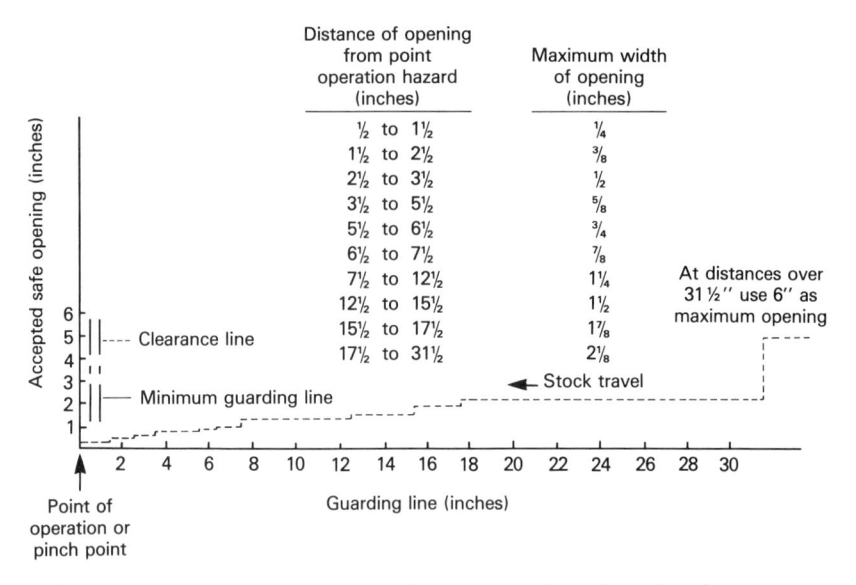

Figure 8.4 Maximum openings permissible between edges of guards and top surface of feed table at various distances from point of operation (ANSI B11.7).

Distance of opening from point operation hazard (inches)	Maximum width of opening (inches)
½ to 1½	¼
1½ to 2½	⅜
2½ to 3½	½
3½ to 5½	⅝
5½ to 6½	¾
6½ to 7½	⅞
7½ to 12½	1¼
12½ to 15½	1½
15½ to 17½	1⅞
17½ to 31½	2⅛

At distances over 31½″ use 6″ as maximum opening

control machine action by preventing operation when the guard is not in place, and must be incapable of removal while the machine is in operation. Interlocking guards are commonly used in molding operations where access to the point of operation is necessary from time to time. It is common to use a spring-loaded switch for interlocks, such switches being depressed for circuit continuity. A second form of interlock is an electric eye or photoelectric device, consisting of a beam of light which deactivates the machine if interfered with.

Other forms of safeguards which control machine action include the sleeve guards for push-button controls, automatic material handling methods (e.g., vibratory feeding devices), presence-sensing devices and—something which is too often overlooked—proper fixtures for workpieces.

A popular safeguard used for many punch presses is the two-hand control. This safeguard controls both the machine action and the worker, forcing the worker to place his hands on the control buttons to activate the machine. As mentioned earlier, the position of the control buttons must be such that both hands are required for activation and the control buttons must be fail-safe, usually ensured by proper electrical circuitry.

Those safeguards which control the employee alone include restraints which limit access to the hazard, automatic pull-back or push-back devices which sweep away from the hazard when the machine is activated, and special tools. These latter safeguards include push sticks, long-handled tongs, and other types of hand-feeding devices. Care must, of course, be taken that such devices are available for use.

TASK FACTORS

The hazards of working with machinery are well known and have not changed significantly as the machines themselves have become more sophisticated. Safeguarding principles have not changed appreciably in recent years either. The practical application of safeguarding has increasingly been enforced, either through enforcement of OSHA regulations or because of increased manufacturer concerns for user safety. Nevertheless, statistics do not suggest any reduction in either the frequency or severity of injuries occurring on machinery. In attempting to understand these truths and convinced that we can reduce the number of these incidents, we have to look elsewhere to achieve this goal.

Many hazardous conditions have not been safeguarded against simply because they have not been recognized as hazards. It was shown in Chapter 5 that hazard identification is not easy and requires our collective ingenuity, dedication and proper study. In other words, we have to look at task factors and operator factors. In many cases, the machinery might remain the same, but the tasks they are asked to perform and the workers who operate them constantly change. Therefore, safety personnel not only have to keep abreast of the times, they have to provide feedback.

A task can be defined as a recognizable unit of a job, i.e. a unit that involves performance from beginning to end with a visible outcome. If we consider machinery safeguarding problems only, a task represents the smallest unit of a system, and therefore we should be able to apply systems safety concepts to hazard identification for each task. In other words, the operator and the environment must be considered as well as the machine. We must learn to apply the methods for identifying hazards described in Chapter 5 to each task and build the hazard control procedures into the standard process descriptions for that task. Let us look at the first case history with this in mind.

Case History 8.1 Press Brake Accident

A press brake is a basic precision power press which is versatile and capable of high production rates because of interchangeable dies. Safety requirements for press brakes have been established by the American National Standards Institute in ANSI B11.3. The safety requirements include the training of operators, die setters and maintenance personnel. The hazard in a press brake is the point-of-operation crushing hazard which occurs when the dies are closed. It is difficult to design a standard barrier guard for a press brake because of the variable die sizes which promote its versatility. Therefore, task analysis is essential for its safe operation.

Christopher Maloney, a 23-year-old press brake operator, has been employed at Bradford Machine Co. for 3 years. He worked in the machine shop and sheet metal shop before being trained to operate the 15-ton press brake shown in Fig. 8.5. The training consisted of both formal and on-the-job sessions with his

Figure 8.5 Press brake.

supervisor and group leaders. Most of the job requirements required crimping of pre-threaded inserts into punched holes of electronic hardware. However, he was asked to use a "V"-shaped die for bending 90° V-brackets from 1.5-inch square steel sheet, 0.062 of an inch thick. This was a one-time task that would take no more than 100 man-hours to complete, and therefore no special safety precautions were taken.

In order to bend the pieces, he had to take each piece (lightly coated in oil), place it on the die against the stop, hold it between his right thumb and index finger, release it and remove his hand, and then depress the foot-lever to activate the downward movement of the press slide. One of the pieces slipped from his hand as he was placing it into position and he did not remove his hand quickly enough when his foot depressed the activating pedal. The die closed crushing his thumb and index finger.

The supervisor, group leaders and safety officer at Bradford Machine met to determine what caused the incident and what might have prevented it. After reviewing the injured party's statement, inspecting the press brake to determine that performance was standard, and analyzing the specific task, the general consensus was that the cause of the incident and injury was task-related. The size of the workpieces required the operator's hands to be much closer to the point of operation hazard than before. This task variable had been overlooked because of the short-term nature of the task and safety concerns had been compromised, causing the hazard to be unidentified until the incident occurred. ∎

OPERATOR FACTORS

It is common to find operator error listed as the cause of an injury sustained on machinery. However, as shown in Chapter 3, although human errors can never be eliminated completely, through understanding and commitment we can reduce the causes of human error. This means that listing human error as the cause of an accident is really an escape clause which does little or nothing to identify the hazards when using machinery.

Human errors should be considered as potential hazards in machinery operation, but ways should be devised to reduce the potential for them to occur. For example, breaks during tedious operations, the observance of operations with appropriate supervisory guidance, and the discussion of potential error situations in training sessions might be beneficial.

Let us now look at another press brake injury, unrelated to the task or machine, but caused by human error.

Case Study 8.2   Press Brake Accident

Roger Felbaum is a 67-year-old retired school janitor who receives both Social Security benefits and a small pension. He worked part-time for Hanover Fabricators, specialists in fabricating hardware for electronic equipment. He was only able to work 15 hours a week so as to remain eligible for his Social Security benefits and was assigned to Hanover's 25-ton press brake.

Jack Dame, the foreman and scheduler for the press brake area, did not instruct Roger on die changing procedures, telling Roger to contact him whenever a run was finished and the die was to be changed. Six months went by with no problem. One day, however, when the foreman had gone to the tool room to get the die needed for the next task, Roger began removing the set screws from the upper platen. When he got to the last one, he loosely held the die up with his left hand. When the screw came out, the upper die dropped just enough to pinch his hand, making him jump and inadvertently depress the foot pedal control, thus activating the machine and crushing his hand.

The safety circle at Hanover Fabricators, which included the Vice President, concluded that the incident and injury were indeed caused by operator error, but that more should have been done to recognize the potential hazard when employing Roger. Although the circle recommended that part-time employees should not be assigned to the press brake, they also initiated a review of the training programs in the press brake department with the intent of formalizing training and retraining methods. ■

LEARNING TO RECOGNIZE HAZARDS

What can be done to recognize machinery hazards before accidents and injuries occur? It seems clear that systems safety engineering concepts must be adopted. First, commitment, innovation and direction from management are necessary if our efforts are to be successful (these were examined in detail in Chapter 4). We need now to focus on those managers who have day-to-day contact with the machines, tasks, and workers. This means responsibility extends to the line supervisor and the safety professional. Good communication between these two key personnel is essential to recognize hazards, but assistance can be solicited from safety circles as well.

The tools at our disposal for hazard recognition have been described in Chapter 5, i.e. preliminary hazard analyses, FMEAs, HAZOPs and systems analysis. One important difference, but a very real difference nevertheless, is that we are dealing with real machines, real tasks and real people. Too often we think of safety generalities and fail to relate these to safety specifics. Case History 8.3 demonstrates what dealing with real machines should entail.

Case History 8.3   Leather Band Buffer

Suede leather is the major product of Ducharme Tanning Company. One of the final manufacturing processes is a sanding operation on a band buffer to produce hides of uniform thickness. The band buffer is basically a specialized belt sander using a 4-foot wide abrasive belt which abrades the leather hides as they are passed by the high-speed belt on a conveyor. Carl Ducharme, the President of Ducharme Tanning Company, recently purchased the second-hand band buffer shown in Fig. 8.6 and requested the safety circle to address any problems in the safe operation of the machine. Greg Maider, the group facilitator, prepared the circle's report, summarized as follows:

Machine Factors

The potential hazards of this machine include an in-running nip point at the friction rollers which move the hides past the abrasive belt; guarding of this hazard area is

Figure 8.6   Leather band buffer.

essential to prevent fingers or hands coming into contact with the moving abrasive belt. Abrasive belts create dust and are known to fail occasionally; injuries can occur if belts disintegrate at high speeds and guarding of the belt is essential to prevent anyone from being struck by flying parts. It is the opinion of the safety circle that barrier guards on the machine provide adequate protection, but that a sign should be added to direct operators not to run the machine unless the guards are in place because the guards have to be removed to change belts and are not interlocked. A second sign is also recommended to require eye, face and respiratory protection to prevent injury from particulates produced.

There is also a barrier guard at the entrance to the in-running nip point of the machine. However, for the distance from the guard to the hazard, which is 4½ inches, an opening of ⅜ inch is recommended. The present guard extends the width of the machine, but the opening is ⅞ to 1⅛ inches; this should be adjusted.

Ergonomic considerations have focussed on the control panel, [shown in Fig. 8.7], and the method for engaging or disengaging the pressure rolls. The control panel violates ergonomic design principles because the start and reverse buttons are oversized while the stop button is protected by a side shield and inconveniently located between the two operating buttons. The safety circle recommends the start and reverse buttons be replaced with side-shielded buttons and an oversized red stop button be installed on the control panel. Another factor of concern is the 70-inch distance from the stop button to the farthest in-running nip point. The maximum horizontal distance with both arms extended is only 59 inches, so the machine could

Figure 8.7 Control panel for leather band buffer.

not be turned off by the operator for all potential hazard positions where fingers or hands could be trapped. The safety circle recommends that this situation be corrected by two methods—first, install an emergency stop button or bar directly above the operator's normal position and to incorporate into the training program the use of the foot lever which disengages the pressure rolls. Such maneuvering could eliminate all serious injuries, but might be readily overlooked in an emergency if not included as part of the operator training.

Task Factors

The operator of the band buffer stands on a raised platform. Leather hides for sanding are placed at the rear on a wood cart, so the operator must turn, pick up a hide, turn back facing the machine, place the hide beneath the barrier guard, feeding it into the in-running nip point until friction drives the hide forward against the high speed abrasive belt. Task variables include the grit size of the belt which in part controls the amount of leather removed, variable thickness of the hides (particularly for initial abrasion) and the separation of the friction rolls. These variables must be controlled to assure proper feeding of the hides.

Operator Factors

It is the consensus opinion of the safety circle that all operators must be trained in both task factors and band buffer safety. In order to best facilitate the training, the circle recommends video-taping the operation, discussion of the video with safety and operational personnel and utilization of the video after editing for training of all new operators.

Respectfully submitted,

Greg Maider
Safety Circle Facilitator ■

SOME REALITIES IN HAZARD RECOGNITION

Many hazardous conditions are not controlled because they are not recognized as hazards. For example, in an operation where pressure rolls were used to apply liquid glue (similar to the way in which paint rollers apply paint), a barrier safeguard was provided on the in-running nip point during application. Normal maintenance cleaning procedures using solvents and a wiping action, however, involved reversing the rolls, thus reversing the position of the in-running nip point which was unguarded. Accidents and injuries could occur during this normal maintenance procedure.

Recognizing hazards is also difficult with new machinery. For example, workers tend to be so absorbed in the setting up of a new machine that they neglect operator safety. Other problems include the transfer of material on to or off of a machine, controls located where the operator's view of the activity area is impaired, flexible controls, machine operation for routine and non-routine maintenance, inadequate illumination, emergency stops, and worker stressors. In each of these cases remedial action is not difficult. In some cases, the hazards, once known, can be easily eliminated, such as poor illumination and elimination or relocation of controls and emergency switches. By and large, however, we have to employ proper training and retraining techniques.

Case History 8.4   A Case of Similar, but Different Machines

Derek Nelson, an 18-year-old high school graduate, was employed for 1 month at S & R Brush Company, manufacturers of toothbrushes, scrubbing brushes and specialty brushes. He worked in the plastics department where various plastic handles were fabricated on 10 injection molding presses: 5 of them were identical, having a 75-ton capacity. Three 30-ton presses were also purchased from the same manufacturer. All of these presses had interlock guards, which would deactivate the machines if removed. Horizontal guards had to be removed when plastic parts became stuck in the dies and had to be freed; in such cases an alarm would sound, the guard would be slid open, and the operator would reach in from above to free the piece. Once the guard was back in place, the machine had to be turned on.

The last two injection molding machines in the department, 12-ton presses, were not fully utilized. These two machines appeared the same (except for size), but actually differed in performance, particularly the interlocked barrier safeguard. On these machines, once the barrier guard was replaced, the machine would immediately restart unless the emergency button had been pressed before the interlock was removed. Also, the interlock switch was located on top of the slide bar, which required the operator to reach from below to free the parts that became stuck.

When the alarm sounded on the small press, it was the first time Derek had to free a part on this size press. He removed the guard, reached in from the top, inadvertently activating the interlock switch. The machine moved, trapping his fingers in the hot mold as it closed. Derek's accident and injury are easy to understand with 20/20 hindsight, but we must be able to recognize the hazard in order to provide adequate remedies. It seems clear that such cognizance is only achievable through management innovations in areas such as training and increased warnings on machinery, plus active evaluations of hazards of machines and processes by safety circles. ∎

SAFEGUARDING, AUTOMATION, ROBOTICS AND FACTORIES OF THE FUTURE

The need to safeguard machinery for operator protection has been combined with the need for increased production to provide the unique remedy of automation. Automation eliminates the presence of workers in hazardous areas and provides faster manufacture of products with more precision and more uniformity. Although automation is not new, the machinery for automation is unique for each product. Therefore, the problems in safeguarding are the same as those already discussed in this chapter. A new concept has made its way into the workplace in the last decade, however, that challenges our safety considerations, one that has combined automation with programmable versatility. This concept, based on robotics, is being termed the workplace of the future, a workplace which is based on computer-integrated manufacturing (CIM).

The workplace of the future is controlled by computers, including the design concept (computer-aided design or CAD), production scheduling, manufacturing processes and controls (computer-aided manufacturing or CAM), inventory controls, maintenance scheduling, and even accounting and purchasing. The integration of these functions into a flexible manufacturing system (FMS) is what CIM is all about.

The advanced factory, with programmable robots at the heart of the system, is smaller, cleaner and quieter than was expected. As workers have been separated from the hazardous tasks, one would expect safety records to be significantly improved. Historically, however, the introduction of automation has increased the number of worker injuries (due in large part to the poor hazard identification with new machinery) and the same has been true with robotics. The majority of injuries which have occurred have been caused by contact with the moving parts of a robot.

Although there are similarities between robots and conventional machinery, three major differences can be identified which are the concern of safety personnel: speed of movement, predictability of movement and hazard zones. In conventional machinery, hazard zones may be difficult to recognize, but they are fixed with time. The main difference between conventional machinery and robots is that a robot can be programmed to do different jobs and to react to changes in the process, even making decisions from a limited number of choices.

Lessons learned in safety engineering have been applied to robot safety, particularly the human factors aspects and systems safety approaches. Robot safety must include the usual considerations of man, machine, environment and the interface behavior, but it must also consider software. Robot hardware design should be based on sound engineering practices, especially those safeguarding principles which have traditionally been used where man interacts with the machines during operation. Emergency stop switches, for example, must appear on the control panel, but they must also be added to the pendant used in the teach mode where the operator may be moving in the robot's work envelope. Of course, the robot's movements are slowed considerably in this teach mode, again by design. From the operator viewpoint, workers replaced by robots have proved to be retrainable, and little, if any, socioeconomic impact has been experienced where CIM has been adopted. Of course, comprehensive instruction and operation procedures must be incorporated into thorough training programs.

Although this section was intended to be a view of the future of safeguarding

Safeguarding, Automation, Robotics and Factories of the Future

LEVEL I is the workstation perimeter
LEVEL II is within the workstation
LEVEL III is adjacent to the robot arm

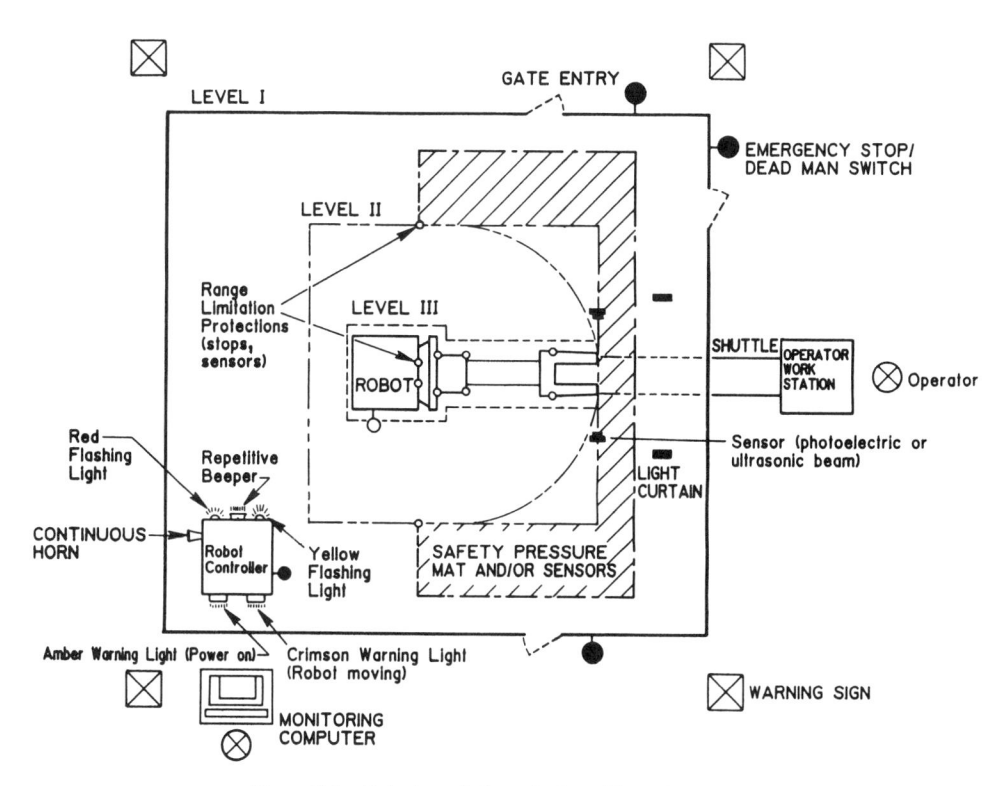

Figure 8.8 Robot workplace design. (From B. J. Jiang.)

and not a comprehensive treatise of robotics safety, Jiang has summarized a systems procedure for robot safety, using the workplace design shown in Fig. 8.8. This summary, assuming appropriate technical needs are met, is reproduced here.

Robot Selection

- Manufacturer's Safety Specifications
  1. Mechanical
  2. Electrical
  3. Software
- Test and Evaluate
  1. Start-up
  2. Emergency handling

3. Software
  4. Sensory devices
  5. Dynamic performance
- Task Design
  1. Keep operator outside work envelope, except in teaching mode
  2. Set up check procedures for malfunction
  3. Provide comprehensive training
- Workplace Design
  1. Install interlocked perimeter fences
  2. Use warning signs
- Maintenance
- Programming

FURTHER READING

H. A. AKEEL, "Intrinsic Robot Safety," *Professional Safety,* 29 (12), 27, 1983.

AMERICAN INSURANCE ASSOCIATION, ENGINEERING AND SAFETY SERVICE, *Machinery and Equipment Safeguarding Manual.*

J. K. BLUNDELL, *Machinery Guarding Accidents,* Hanrow Press, 1983.

D. A. COLLING, "Machine Safeguarding and Safety Management," *Professional Safety,* 28 (8), 26, 1982.

B. JIANG, "A Systematic Procedure for Robot Safety," *Proceedings of the ASSE Annual Professional Development Conference,* New Orleans, p. 245, 1986.

F. E. McELROY, *Accident Prevention Manual for Industrial Operations,* 8th Edition, National Safety Council, Chicago, Ill., 1980.

H. McILVAINE PARSONS AND G. P. KEARSLEY, "Robotics and Human Factors: Current Status and Future Prospects," *Human Factors,* 24 (5), 535, 1982.

S. W. MEAGHER, "Designing Accident-Proof Operator Controls," *Machine Design,* June 23, p. 80, 1977.

S. W. MEAGHER, "Machine Design Mechanisms of Injury," *Professional Safety,* 25 (2), 19, 1979.

NATIONAL SAFETY COUNCIL, *Guards Illustrated,* 4th Edition, National Safety Council, Chicago, Ill., 1981.

J. M. ORLOWSKI, "Engineering Aspects of Guarding of Machinery and Equipment," in *Products Liability* (eds L. R. Frumer and M. I. Friedman), Matthew Bender, 1975.

OSHA REGULATION, 29 CFR Part 1910, Sub-Part 0, *Machinery and Machine Guarding.*

OSHA, Pamphlet 2227, *Essentials of Machine Guarding.*

R. D. POTLER, "Safety for Robotics," *Professional Safety,* 29 (12), 18, 1983.

L. B. TRUCKS, "Standards for Robotics Injuries—Legal Aspects," *Proceedings of the ASSE Annual Professional Development Conference,* New Orleans, p. 261, 1986.

QUESTIONS

8.1. Name and define three hazardous actions or motions created by machinery.

8.2. What are the four types of safeguarding? Which are the most common and why?

8.3. How can you reduce the potential for operator error in machinery operation (a) if you are the designer and (b) if you are the employer?

8.4. Using your own experience, give an example of a machinery safeguarding hazard which is (a) design-related, (b) task-related, and (c) operator-related.

8.5. Examine the role of management in identifying the hazards associated with machinery.

8.6. What hazard identification procedures are suited for identifying specific production machinery hazards?

8.7. Explain why safeguarding considerations are different for automated machinery and for programmable machinery.

9

Falls

The National Safety Council estimates that there are 200,000 to 300,000 disabling injuries in work-related falls annually. These falls occur on the same level, from higher elevations or on stairways. This chapter examines why falls occur and how we can prevent them. It will be shown that the hazards which lead to falls are not simple and are not always easily identified, and that prevention requires all the creative efforts of safety management, both in understanding human behavior as well as human factors.

What is meant by falls on the same level? We can fall by slipping or tripping: in a slip, traction is lost between the shoe and the walkway surface; a trip, on the other hand, involves loss of balance when your foot encounters an obstruction or high friction and your body motion continues while your foot is momentarily stationary. This chapter looks at friction, ambulation, design, construction and the maintenance of walkway surfaces.

Falls on stairways are different from falls from higher elevations, even though 80% of stairway falls occur when descending. In looking at stairway falls, we will examine not only stairway design and ambulation, but distractions, lighting and even shoe factors.

These types of falls are common to retail businesses, business offices, and other public areas as well as in industrial settings. Almost 2 out of 3 work-related falls from higher elevations, though, occur in construction or in manufacturing. These include falls from ladders, scaffolds, roofs and even elevators. All of these will be considered in this chapter.

SLIPS AND FALLS

A slip is something we have all experienced, but studies of what actually happens when we slip have been fragmented, with only a few researchers addressing all aspects of the phenomenon. Early efforts concentrated on defining and measuring the slipperiness of walkway surfaces. How difficult this was is perhaps best demonstrated by the "performance" definition proposed by Miller (1984):

> A "slippery" work surface is that combination of (1) a host transient surface (e.g. a shoe), (2) an agent structural surface (e.g. a floor) and (3) contaminant conditions (e.g. water or oil), all of which together have the propensity to cause the initiation and/or promotion of sliding between the host and agent during the performance of actual or anticipated tasks.

Although the majority of us believe the tendency to slip is closely related to the coefficient of friction, there are differing opinions on how measurements should be made, what causes inconsistencies in the data and which coefficient of friction is involved in walking, turning, climbing, pushing and pulling. For many years, it was thought that the frictional force was proportional to the load or pressure of one against the other, was independent of the contact area and was independent of sliding velocity. Although the first two relationships are generally true, we now know that friction is dependent on sliding speed and that μ, the coefficient of friction (the ratio of friction force to load), may vary by as much as 50% according to the speed of motion. We therefore consider two coefficients of friction: static μ_s for surfaces at rest, and dynamic μ_k for surfaces in motion. We also know that each varies as well. For example, μ_k decreases as the sliding speed increases and μ_s depends to some extent on the length of time the surfaces have been in contact.

A great deal of effort has been spent determining how to make μ reproducible and how to measure it accurately. Today, we use only the drag-type or articulated strut-type tests (Fig. 9.1) for measurements. Measurements of μ have been made between various shoe materials and many different floor surfaces, sometimes using foreign substances such as water between them.

Measuring the coefficient of friction helps us to identify only two of the factors which can cause slipping, i.e. the shoe factor and the floor factor. Using these two factors, we have been able to define working surfaces as either slip-resistant ($\mu_s > 0.5$) or slippery ($0.5 > \mu_s$). This acceptance is based on level surfaces and employees walking with no load.

Two other factors significant in slipping are ambulation of the individual and task-related factors. Each of us has a characteristic walking gait, determined by age, sex, anthropometry and even personality, but all of us have a general bipedal gait pattern which can be characterized by a stance or support phase and a swing phase. In the stance phase, our body's weight is supported on one foot in contact

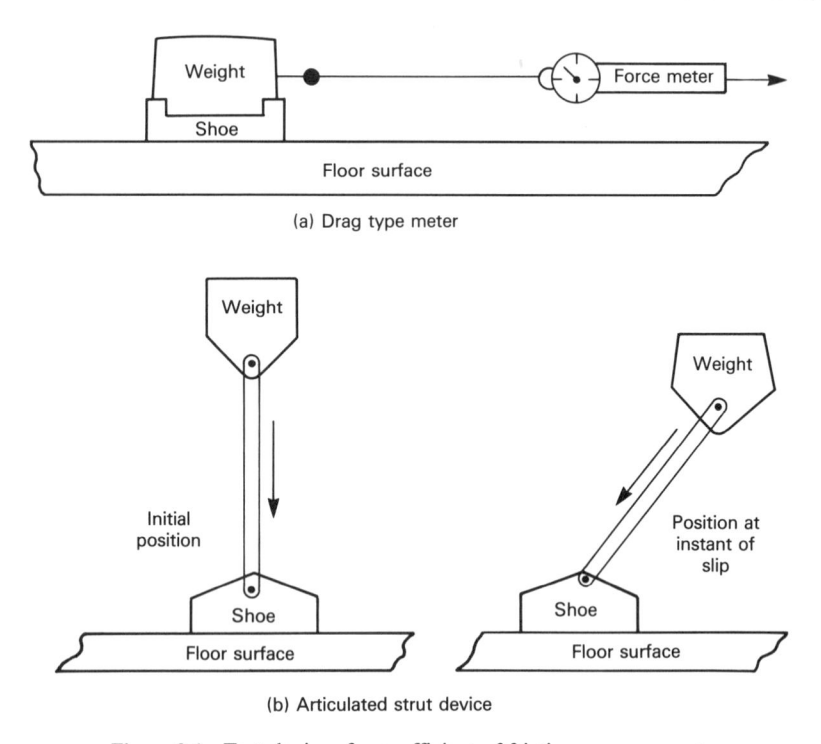

Figure 9.1 Test devices for coefficient of friction measurements.

with the walking surface. In the swing phase, our free foot swings ahead of the supporting foot to propel our center of gravity forward, and to contact the walking surface so as to become the supporting foot. The walking gait cycle begins when the heel of the swing foot contacts the walking surface; the heel then rocks forward to bring the foot into full contact with the walking surface and to support our body weight as the other foot pushes off at the toe to begin its swing phase. As the heel touches the floor, our body is unstable, and is dependent on sufficient frictional force to permit us to regain our body balance. It is at this time when a slip-resistant surface is essential to prevent a slip and fall.

Task characteristics also affect the potential for slip and fall incidents because of the effect on the characteristics of the force on the individual. Forces exerted while lifting, walking and turning during task performance vary, thus causing the traction between shoes and walkway surface to vary as well. Specific task factors that influence the forces include the speed at which the task is performed, loading, types of motion (i.e. pushing, pulling, lifting), frequency and changes in direction of movement. All of the factors which affect slipperiness of walkways are summarized in Fig. 9.2.

Prevention of Falls Caused by Slips or Trips

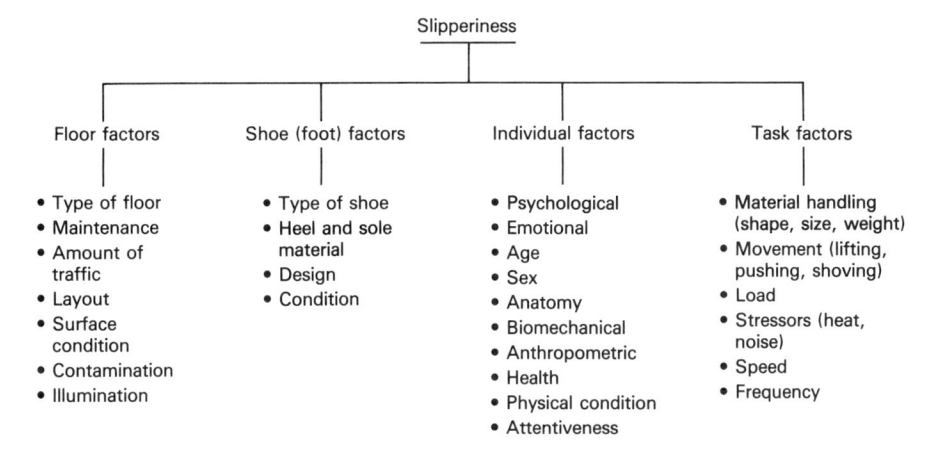

```
                            Slipperiness
        ┌───────────────┬───────────────┬───────────────┐
   Floor factors    Shoe (foot) factors  Individual factors   Task factors
```

- Type of floor
- Maintenance
- Amount of traffic
- Layout
- Surface condition
- Contamination
- Illumination

- Type of shoe
- Heel and sole material
- Design
- Condition

- Psychological
- Emotional
- Age
- Sex
- Anatomy
- Biomechanical
- Anthropometric
- Health
- Physical condition
- Attentiveness

- Material handling (shape, size, weight)
- Movement (lifting, pushing, shoving)
- Load
- Stressors (heat, noise)
- Speed
- Frequency

Figure 9.2 Factors that influence slipperiness of walkways.

TRIPS AND FALLS

Falls on level surfaces are not only the result of slipping, but can be caused by tripping as well. There are two types of tripping: the first is caused by some obstruction which blocks the normal gait in the swing phase, causing a fall when we are unable to regain our balance; the second is caused when there is a sudden increase in friction, interfering with the gait when the swing phase is changing to the support phase. The differences between trips and slips are important when we consider the nature of the injuries caused by falls on the same level: in slips, our legs slide forward and we fall, injuring our back or buttocks; in trips, our center of gravity moves ahead of the leg which is momentarily stopped or slowed down, and we are more likely to injure our arms, elbows, knees or even our heads when we strike the floor.

The factors which determine tripping hazards are nearly identical to those which affect slipperiness. For example, floor factors, such as change in material, maintenance (e.g. missing tiles or carpet fold) and amount of wear from traffic, influence both the hazards of tripping and slipping. The same is true for the individual factors and task factors given in Fig. 9.2. Even shoe factors affect tripping hazards, e.g. ladies' high-heels can catch in small depressions on the walkway surface.

PREVENTION OF FALLS CAUSED BY SLIPS OR TRIPS

An evaluation of the factors which cause falls by slipping or tripping is important only if we are able to apply that knowledge to the prevention of future incidents. We have to remember also the lessons of systems management, human behavior and hazard control. For example, a program designed to eliminate high-heeled ladies' shoes would be ill-advised. Therefore, the emphasis needs to be placed on the floor and task factors, with only motivational recommendations for shoe and individual factors.

When addressing floor factors, we can divide our efforts into two stages, i.e. construction and maintenance. In the construction stage, we should install slip-resistant floor surfaces and use proven materials which are easy to clean and will wear well. A meaningful maintenance program should not only include cleaning and repairing, but also a training program for employees to alert them to the hazards. If the hazards are not eliminated immediately, warnings should be posted as to the existing temporary danger.

Training should also play an important role in our task analysis. For example, local vacuum systems are common for table-saw cutting operations, but are not 100% efficient. Sawdust has caused many falls around a table-saw, some of which have led to blade lacerations and amputations when the worker was trying to regain his balance without falling. Only training can overcome such a hazardous condition. We should design handling tasks with fall hazards in mind, just as we consider potential back injuries in the design.

Case History 9.1 Slipping Injury in Print Room

When John Tuholski retired from the military, he found a job at New-Wal Computer operating an offset printing press. The press, manufactured by a German company, uses 19 × 25½ inch paper. It is a large press, occupying a floor space of about 7 × 12 feet and it is about 7 feet high. In order to operate the press, there are a number of slip-resistant steel surfaces for standing purposes at various locations. Beneath the press is an oil drip pan which is 6 feet 7 inches long and 3 feet 9 inches wide.

John is 5 feet 4 inches tall and weighs 135 lb. The task he performed required putting a plate into the press (Fig. 9.3) 4 feet 6 inches from the floor. John sat on a drop-seat for this procedure. He stood on the same drop-seat when adjusting the ink flow. In order to descend, John would use a grab-bar on the press and step down. On one occasion, his left shoe slid on the concrete floor because of accumulated oil on his shoe as he descended (see the position of his feet in Fig. 9.3). The amount of oil was minor, though, and it rapidly dissipated. Although he did not fall, he wrenched his knee so severely when the sliding abruptly stopped that he underwent knee surgery the next day. He was incapacitated for 2 months.

Immediately following the incident, the New-Wal safety circle met to determine why it had occurred. After careful analysis, it was agreed that members had been too lax when reviewing task analyses for hazard identification. Proper attention would have revealed the exposure of workers' shoes to the oil and the hazard, if identified, could have easily been eliminated in the construction stage. A permanent cover was therefore added to eliminate any possible contact with oil in the drip pan and regular scheduling of task analyses was resumed. ■

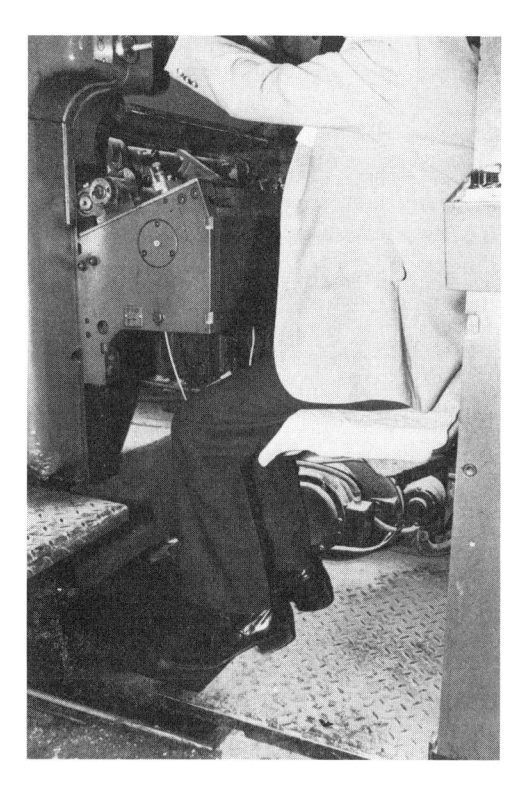

Prevention of Falls Caused by Slips or Trips 185

Figure 9.3 Operator position for putting plate into printing press.

Case History 9.2 Tripping Incident on Freight Elevator

Jim Coluccini, an overweight 19-year-old laborer, had been employed by Commercial Furniture, Inc., an office furniture manufacturer. The distribution center where Jim worked was on the fourth floor, and much of his work involved manual labor.

On the date Jim was hurt, he and three others were unloading a delivery. They placed the five boxes of tables and six boxes of chairs in the old freight elevator and ascended to the fourth floor, with Jim operating the electrical controls. When unloading the elevator, Jim passed the boxes out to the others, but carried one out alone when no one had returned. He lifted the large box and backed from the car, tripping on the floor edge because the car was not leveled. In tripping, he fell backward over one box in the aisle and the one he was carrying fell on top of him. As a result of his back and rib injuries, Jim has been unable to return to work.

Although the Workmen's Compensation carrier has examined the elevator, David Dreskin, the President of Commercial Furniture, asked Dalzell Associates to reconstruct the incident in order to prevent any recurrence. Dalzell Associates used the systems approach, examining the worker, the elevator and the environment. Although they closely scrutinized Jim because of his weight condition and the fact he was the only worker who tripped when exiting the elevator, Dalzell Associates concluded that Jim's only contribution to the incident was temporary inattentiveness. Similarly, Dalzell Associates found the working environment at Commercial Furniture to be safety oriented, with a posted safety policy, safety posters changed monthly, visible training efforts for manual lifting and mechanically assisted equipment for much of the material handling.

When attention was directed to the tripping hazard, itself, the focus became the levelling mechanism of the elevator, which was one commonly used for freight and commercial elevators. When the elevator car was commanded to travel to a specific floor, it traveled at a single speed until an electrical brake was applied prior to reaching that floor. The car coasted to the approximate floor level, with levelling actually dependent on a permanent magnet switch located on the top of the car. This switch is activated at the same time as the electric brake, forming a high magnetic field which attracts an angle iron vane attached to the proper position in the elevator hatchway. The nature of this magnetic levelling device indicated that the exact car position would be dependent on the load. Dalzell Associates therefore measured the levelling at the fourth floor entry to Commercial Furniture. They found it to be independent of the origin because of the single speed, with the average difference in car and building floor level influenced by load as depicted in Fig. 9.4.

Whereas the load on the elevator at the time of Jim's incident was estimated to be 1350–1600 lb, the elevator car level was probably $\frac{1}{4}$ to $\frac{3}{8}$ of an inch below the building floor level, creating a variable tripping hazard. In a continued effort to prevent repetitive incidents, Mr Dreskin reviewed the Dalzell Associates report with his staff, and several alternative remedies were discussed. It was decided that elevator modifications were not economical and were not within corporate jurisdiction as a tenant anyway. Therefore, a simple warning was posted at the elevator entry and training programs were altered to alert all employees to the potential tripping hazard. In addition, all tasks involving oversize packages were assigned to two workers or mechanical handling. ■

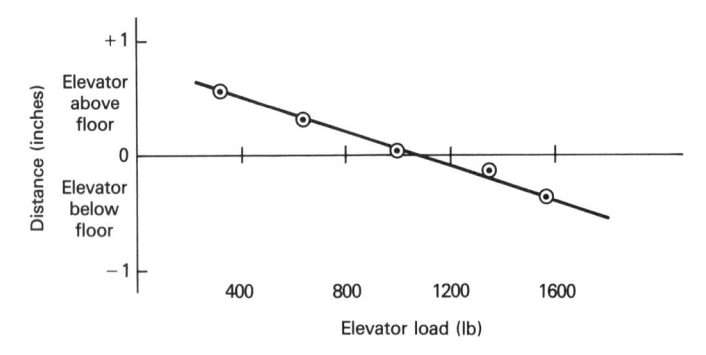

Figure 9.4 Influence of load on elevator leveling.

STAIRWAY FALLS

Falls to lower levels can also be caused by slipping from a scaffold, ladder, stairway or ramp. A slip is not the only cause of these falls, however. Although stairway falls are recognized as a leading cause of occupational injuries, there was little research in this area until the 1970s.

To understand the hazards of stairways, we need to differentiate between the angles of inclination for the methods of climbing and learn a little about stairway geometry. Ramps or inclines have a continuous change of height with horizontal distance traveled, whereas stairways and ladders have a discontinuous change. Ramps or inclines are useful up to 20° where the slope is too steep to remain stable. Stairways are useful from 20° to 50° and angles from 50° to 90° require ladders.

A stairway system is comprised of the riser which is the vertical distance between two adjacent stairs, the tread which is the surface we walk on, the nosing which is the projection of the tread beyond the riser, landings where stairway systems change direction or where long stairways are separated, and handrails usually located at the side and above the stairway. Approximately 80% of all stairway accidents occur when descending. This is not surprising because a fall while ascending often causes no injuries, because forward momentum is stopped by the higher stairs. Falls while descending, however, can cause serious injuries through the repeated impact on the sometimes sharp nosings of industrial stairways. When we add to this the fact that overall employee exposure to stairways is much greater than exposure to other elevated work surfaces, it becomes apparent why we need to reduce stairway falls. Although we will not look at the problems of falls on ramps because of many similarities (particularly in remedies) to stairway falls, we have to remain cognizant of the potential, as the number of handicap access ramps in industrial settings is increasing.

An analysis of workmen's compensation claims has shown that stairway falls occur most frequently in three distinct areas of high exposure to stairway use or to new and unfamiliar stairs. For example, accidents frequently occurred at entrances and exits where unique problems are faced, i.e. there is increased traffic flow because of funneling at entrances and exits, haste due to lateness on entering or eagerness on leaving, abrupt changes of visual environment and an abrupt change in level, surface materials and conditions (e.g. snow, ice or rain). Stairway falls also occurred more frequently in offices and manufacturing areas and at field locations where stairways are used only once or infrequently.

These studies identified four broad categories of events in stairway falls: (a) design conditions; (b) environmental factors; (c) inherent user characteristics; and (d) task performance factors. The most prominent design-related problem was stairway materials; almost 90% of falls occurred on metal, concrete, carpet and brick surfaces and doorways opening abruptly on to staircase tops or bottoms. In ascending, nosing projections and open risers were design factors involved with falls. Environmental factors are normally associated with housekeeping, but can also include slip-resistant materials on walking surfaces, proper drainage conditions, illumination and visibility of steps. We have long been aware that distinct orientation edges for stairway systems are inherently safer.

User characteristics include the design, condition and/or maintenance of shoes as well as obvious handicaps, e.g. age, sex and group ecology (i.e. being alone or with others). Chief among shoe factors in stairway falls is women's high-heel shoes, which can catch on the smallest of tripping hazards. User performance errors are involved with almost half of all stairway falls. While the activities of climbing or descending stairs are commonly taken for granted and are considered as specific tasks, completion of other tasks such as manual handling requires continual information processing as well as complex balancing. Inattention or preoccupation with such matters as not spilling hot coffee have caused many falls on stairways.

As in all accident phenomena, multiple causation for stairway falls is common. One area that has received some attention is the effect of certain changes in the immediate visual surroundings which can trigger abrupt changes in our behavior. Visual distraction is more complex than originally thought, but the basic concept was based on our behavioral dependence on the spatial arrangement in our surroundings, a dependence which in itself is very complex. Studies of visual distraction on stairways showed, however, that incident rates were lower in the presence of certain visual distractions, but that they were higher for those who were *actually* distracted; in other words, those who resisted distraction may have exercised caution because of the abrupt change, thus reducing their risk, whereas those who were easily distracted had no resistance and exercised less caution.

Steps have been and continue to be of interest to psychophysical research. Although most codes have maximum riser heights of up to about 8 inches and variations of 3/16 inch, and minimum tread widths of 9–10 inches with permissible variations of 3/16 inch, many codes have adopted the seventeenth-century universal formula:

$$2r + t = 24 \text{ to } 25$$

where r is riser height in inches and t is tread width in inches. Recent trends have been to the so-called 7–11 stair which fits into this universal code and avoids high incident rates associated with higher risers. When comparing psychophysical comfort and records for low incident rates, however, it appears that stairway systems should be designed with a maximum riser height of 6 inches and a minimum tread depth of 11 inches. This suggests the optimum stairway angle is 28° or less.

Handrails have also been the subject of specialized studies, both from stairway fall incidents when handrails were present and from human factors analysis. Handrails serve four fundamental purposes: (a) to guide us, (b) to provide stability, (c) to serve as a pivot when changing direction as on a landing, and (d) to serve as a grab-bar to regain stability after a stumble or slip. Most building codes specify that handrails should be 30–34 inches above the nosing, be able to support 200 lb, project at least 1½ inches and preferably 3½ inches from the wall, and permit handrail spacings of 88 inches. Anthropometric studies indicate that handrail heights of 36–38 inches are far better for 90% of the adult population and that, at these heights, handrails should be spaced about 48 inches apart to be within reach of every user. Graspability was also considered and round handrails of 1¾-inch diameter were selected for achieving the best power grip. It is interesting that statistics show that the incident rate is higher for stairway systems equipped with handrails, but that severity of injuries is significantly reduced by the presence of handrails.

We can simplify stairway fall prevention technology by remembering three necessary conditions for safe stairway use:

1. Steps that can readily be seen.
2. Treads large enough to provide adequate footing.
3. Reachable and graspable handrails.

In order to prevent falls in the future, we should design new stairway systems which reduce the chance of human error and make the stairway systems more forgiving should such errors occur. On existing stairways, we may be able to do little about tread and riser height, but can add slip-resistant surfaces, paint to provide orientation factors, insist upon a sound housekeeping policy, and provide proper handrails: in other words, apply all of our persistent dedication and innovative management principles.

FALLS FROM LADDERS

Ladders are familiar to all of us, a major reason why falls occur from them. Because of our familiarity, we misuse ladders and do not treat them as tools designed to make our tasks easier. For this reason alone, our safety programs should include training on ladder safety.

There are three main types of ladders which we encounter in industry: fixed ladders, extension ladders and stepladders. There are also three types of ladders, based upon working load capacity:

Type	Duty Rating	Working Load (lb)
I	Industrial-heavy	250
II	Commercial-medium	225
III	Household-light	200

The most common causes of falls from non-fixed ladders have been identified with the lack of stability and sliding, including lateral sliding for extension ladders. Most of the standards for ladders therefore address means to prevent these situations, calling for proper support and proper inclination angle or secure spreaders in the case of stepladders, plus cautions to avoid over-reaching or placement near doors opening toward the ladder. The origin of many standards, however, is obscure, and few research studies have examined ladder safety.

For example, the directions for properly setting the angle of inclination for an extension ladder, according to ANSI Standards for portable wood or metal ladders, are to place the ladder base a distance equal to ¼ the working distance of the ladder away from the base for the vertical support; this angle is 75.5°. Psychophysical studies showed that the mean angle of inclination selected by volunteers unaware of this standard was 71.9°. A total of 90% of us can set this angle approximately by standing erect with toes touching the ladder base and arms placed straight out with hands grasping the ladder. Other standards probably have evolved from human factors considerations, such as the requirement for fixed ladders to be at least 4½ inches from a vertical wall. This distance prevents slipping because our weight would be supported by only the toes of our feet.

Although most ladder falls are attributable to misuse and not to failure of the product, we can find an exception with portable aluminum stepladders, which conform to the ANSI A14.2 safety regulations. This regulation specifies step bending tests, top loading compression tests, front, side and rear stability tests, and a torsional test of 45 ft-lb. In many cases, examination of aluminum stepladders after someone has fallen from them has shown the spreader bar to be deformed due to bending in the horizontal plane, even when the spreader bar was locked and the ladder properly stabilized. In order to determine why spreader bars bend, causing the ladder to fall as well as the user, human factors studies were used in combination with strain gages applied to a 5-foot Type II aluminum ladder. Previous studies of extension ladders had shown that: (1) the body trunk remains parallel with the ladder on ascending, (2) most people use the hands to grasp the side rail while ascending, and (3) when working from the top part of the ladder, the body trunk is generally parallel to the working surface and the legs are leaning against the ladder. With a 5-foot aluminum stepladder, most people grasp the top cap with their hands while climbing; if hurried or careless, the rear side rails can lift, altering the base support. While working from the ladder, the legs are leaning against the ladder and the body trunk is generally parallel to the ladder as for extension ladders, but one hand grasps the side of the top cap. In extending the other hand, a twisting motion is imparted.

Strain gages were mounted on the front and rear side rails and on the spreader bar of a 5-foot aluminum stepladder. Strain was measured with a 160-lb person on the ladder, first as a static load and, second, twisting the top cap without disturbing the base support. The stress calculated from the strain was:

Location	Stress (p.s.i.) Static	Stress (p.s.i.) Twist
Front side rail	400	5800
Rear side rail	370	2400
Spreader bar	580	6400

The increase in stress when twisting combined with the lack of design rigidity makes aluminum stepladders susceptible to collapse when improperly supported or when a worker over-reaches. Wood stepladders are much less susceptible to collapse caused by spreader bar bending because of the increased strength of galvanized steel spreader bars.

An effective industrial safety program for ladders should include training and retraining which explains, for example, why we should not overextend while working on a stepladder, should coordinate a good inspection and replacement program for ladders and provide careful task analysis to ensure a ladder is the right tool. Because we are all so familiar with ladders, it is a good idea that *all* employees be provided with training in the safe use of ladders.

Case History 9.3   An Unnecessary Ladder Fall

Dick Peterson is a 37-year-old journeyman's electrician, a trade he has practiced for 10 years since graduating from vocational high school and serving with the military for 8 years. For the last 3 years, Dick has worked for Newton Electric, installing and renovating commercial, industrial and residential electrical services. Last year, extensive renovations were made to the office of the Movers Unlimited, Inc. warehouse. On the day of Dick's accident and injury, he was laying cable for an auxiliary switch for the overhead door used by the moving vans.

The overhead door was 13 feet wide and 13½ feet high and was operated by a motor to the right of the roller guides at the top of the door and immediately above the wall switch. When activated, the door would open fully or close fully, i.e. it would not stop at intermediate positions. The control panel for the motor was adjacent to the motor about 1 foot from the wall opening.

Dick knew that there were no other controls to activate the door other than the switch visible to him, and so he did not de-energize the circuit. In order to lay the Romex cable, he extended the 24-foot wood ladder above the door and stapled the cable to the wood doorframe. When he reached the control box, he cut the cable and stripped the insulation, then decided to make the connections before descending. In opening the control box and attempting to remove the fuse with his screwdriver, however, Dick inadvertently short circuited the switch, thus activating the motor to open the door. When the top of the door pushed the ladder away from the wall, Dick grabbed on to the roller guides, but had to drop when the door rollers approached his hands. Dick was out of work for 10 months with a broken ankle and broken femur.

Dick's boss, Jim Newton, was initially irate, thinking that Dick had been stupid not to de-energize the door *before* working on the auxiliary switch. The workmen's compensation representative, however, sat down with Jim and explained the concepts behind the behavioral safety program her firm highly recommended. As a result of that meeting and Jim's desire to prevent future accidents, Newton Electric now adheres to a comprehensive training and retraining safety program which emphasizes not only safety specific to electrical hazards, but human factors, human behavior and human error. ■

FREE FALLS

An unimpeded fall from a known elevation to a known impact is known as a free fall. The most common industrial free falls are those by workers from scaffolding or roofs and falls from balconies or landings not having adequate railings. Most industrial falls occur in construction and manufacturing areas and occur to workers under the age of 35. Slipping or loss of footing prior to the fall is the most frequent cause of the fall.

Free falls can be prevented by the use of personal fall protection equipment, but systems concepts including dedicated management, continual training, effective hazard identification and selection of proper equipment for the task are even more important. Perhaps the last case history in this chapter best exemplifies this importance.

Case History 9.4   Elevator Shaft Fall

Cornish Industries is a condominium association with small member companies which have renovated an old textile mill in Lowell. Both stairways and a freight elevator service all floors and the basement. Each member company has renovated their own area, but Cornish Industries is responsible for the stairways and elevator. When first formed, Cornish Industries was aware that the hatchway elevator did not meet elevator safety codes because there was no interlock safety gate, and therefore they contracted with an elevator service company to replace the cable release system with push-button activation, to repair the electromechanical door interlock system which was inoperative, and to fireproof the hatchway as well as add internal safety gates.

Parts were on order, but had not been delivered 6 months after the contract was signed. Most member companies in the meantime were working to maintain their production as well as complete renovations. Simms Computer Service, who assembled printed circuit boards for a major computer manufacturer, occupied one-quarter of the third floor of Cornish Industries. Their manufacturing area could be entered opposite the elevator door. John Vail, a 19-year-old trainee at Simms, had no experience or training on the elevator—in fact, he had never used it since he began work.

On the day of his incident and injury, John was working near the open doorway of the manufacturing area when a delivery man entered and asked his assistance in getting the elevator car to the first floor. The elevator door was closed, but John knew that the door mechanism was released through a hole someone had put through the wallboard. When he opened the door, however, the elevator was not there, but at the 4th floor. He did not know that the elevator would not operate because his door was open and the electromechanical device prevented activation. In leaning into the shaft to pull on the cables, John fell about 20 feet to the pit below, breaking his leg and suffering internal injuries. John has yet to return to work after 10 months.

At a meeting of the Trustees of Cornish Industries following the incident, it was decided that the condominium association could not afford to delay formulation of a safety program until all renovations were completed. First, the hole in the wall was covered and instructions were posted on each floor that doors were to be left open until the elevator renovation was completed and, when inside, how to compensate for the malfunctioning electromechanical device. In addition, a letter was sent to each member company requesting them to circulate elevator training information to all their employees. In the meantime, each trustee was asked to draft a realistic safety program for the association for the next meeting. Today, that program has been refined and is before each company for approval. ■

FURTHER READING

P. C. BUCK AND V. P. COLEMAN, "Slipping, Tripping and Falling Accidents at Work: A National Picture," *Ergonomics,* **28** (7), 949–958, 1985.

BUREAU OF LABOR STATISTICS, Bulletin 2195, "Injuries from Falls from Elevations" (June 1984), and Bulletin 2214, "Injuries Resulting from Falls on Stairs" (August 1984).

H. H. COHEN, J. TEMPLAR AND J. ARCHEA, "An Analysis of Occupational Stair Accident Patterns," *Journal of Safety Research,* **16,** 171, 1985.

J. A. EDOSOMWAN AND T. M. KHALIL, "Accident Prevention in Slips and Falls—A Comprehensive Approach," *Professional Safety,* **27** (6), 30, 1981.

C. H. IRVINE, "A Human Factors Approach to Slippery Floors, Slippery Shoes and Ladder Dangers," *22nd Annual Meeting of the Human Factors Society,* 1978.

C. H. IRVINE AND M. VEJVODA, "An Investigation of the Angle of Inclination for Setting Non-Self-Supporting Ladders," *Professional Safety,* **23** (34), 7, 1977.

J. M. MILLER, " 'Slippery' Work Surfaces: Towards a Performance Definition and Quantitative Coefficient of Friction Criteria," *Journal Safety Research,* **14,** 145, 1984.

J. C. NELMS, "The State of the Art of Fall Protection," *National Safety News,* p. 75, March 1978.

J. L. PAULS, "Review of Stair-Safety Research with an Emphasis on Canadian Studies," *Ergonomics,* **28** (7), 999, 1985.

S. I. ROSEN, *The Slip and Fall Handbook,* Hanrow Press, 1983.

S. S. SZYMUSIAK AND J. P. RYAN, "Prevention of Slips and Fall Injuries," *Professional Safety,* **28** (6 and 7), 1982.

J. TEMPLAR, J. ARCHEA AND H. H. COHEN, "Study of Factors Associated with Risk of Work-Related Stairway Falls," *Journal of Safety Research,* **16,** 183, 1985.

QUESTIONS

9.1. List the major categories of falls discussed in this chapter. What experience have you or others you know had in these categories?

9.2. What is the main difference between a trip and a slip?

9.3. Figure 9.2 lists factors that affect the "slipperiness" of a surface. Which of these can be applied to the falls discussed in this chapter?

9.4. What are the three necessary conditions for safe stairway use? In order to prevent falls in the future, what might be done to existing high-risk stairway systems?

9.5. What factors should be emphasized in a comprehensive program designed to reduce slips and trips. Briefly explain how your company can proceed with the program.

9.6. This chapter describes some factors in ladder design and suggests that all employees should be *trained* in the safe use of ladders. Choose one of the three types of ladders and list steps on how to set up and use this ladder safely.

10

Fire Prevention and Protection

Fire is more than just of casual interest to safety engineers because of life safety, property damage and business interruption concerns. Safety engineers are already doing a good job, because industrial fires comprise a high percentage of total national fires, but have not led to extensive loss of life. There is large room for improvement, though, and the principles needed to accomplish this are covered in this chapter. Fire, however, is no different than our other safety concerns, and therefore we will look at some case histories which will demonstrate once again the difficulties in fire hazard identification and why application of the basics is so important.

Fire is, first of all, a chemical reaction where a fuel material is oxidized very rapidly, giving off heat. The most common type of fire is characterized with a flame, which is the result of burning gases. This simple fact demonstrates how complex fires really are because liquid or solid fuels must be vaporized to form the gases which form the flames. Thus the vaporization of liquids or the pyrolytic decomposition of solids to form flammable vapors is important if we are to understand fire. In order for a fire to occur, several conditions must be present. We must have (a) a combustible fuel, (b) a source of oxygen and (c) heat. These conditions have become known as the fire triangle, and all of the elements must be present or no fire will occur.

There are complications, of course, such as the rapid expansion of burning gases, smoke and toxic fumes generated, and other factors which we will explore in this chapter in order to learn how to prevent or control fires. We will also have to examine other factors, such as how fires spread, and how to detect and extinguish fires to minimize or eliminate damage. One of our major concerns is the ignition of fires, which can frequently be caused by electrical sources. Electrical hazards and disaster and emergency preparedness planning will be included in this chapter.

CHEMISTRY OF FIRE

Our understanding of fire begins by looking at the components of the fire triangle. Fuels are materials which will oxidize and produce heat. We must recognize what properties of fuels are important in this manner. For example, all fuels do not begin to burn at the same temperature and all fuels do not provide the same heat. Fuels must be heated to a temperature, called the ignition temperature, where combustion can first occur; because heat is evolved in the process, the reaction can continue without any additional external heat input. The ignition temperature is not accurate because of chemical differences in the fuel and varying fire conditions which limit or expand the availability of oxygen. An example of a chemical difference in the fuel would be the difference between green wood and seasoned wood for camp fires.

There are other temperatures which can be accurately measured, such as the flash point of combustible and flammable liquids. The *flash point* is determined by raising the temperature of the liquid slowly until the vapor pressure above the liquid reaches the proper mixture with oxygen to be ignited when exposed to a flame. Only a flash occurs if the flame is removed because not enough heat is generated. However, when the liquid continues to be heated above the flash point, the vapors are produced at a rate which maintains the proper mixture with oxygen and fire can be sustained even if the ignition flame is removed. The minimum temperature at which fire can be sustained defines the *fire point*.

Fires can also be started with no external ignition source. The temperature at which burning occurs automatically is termed the autoignition temperature. The autoignition temperature is usually much higher than the flash point for flammable or combustible liquids. However, the difference is not large for solid materials which are combustible, but are not ordinarily considered to be fuels, such as plastics. An example of autoignition which is well known, but not common, is spontaneous combustion. In spontaneous combustion, a chemical system is present where an exothermic reaction occurs at ordinary temperatures. The most common example of such a system is the drying oil in oil-base paints. Hardening in these paints occurs by oxidation of the oils which gives off heat. When heat is not allowed to dissipate, the temperature rises, in turn speeding up the reaction. When the autoignition temperature is reached, flames erupt.

There is another fundamental ignition characteristic which is so well known that we sometimes fail to recognize it in another form. In order to start a camp fire, we all use kindling wood before we add large logs because kindling ignites readily. The reason for this is that chemical reactions occur at the surface and kindling has more surface area exposed for its volume than a log. When this simple principle is extended to heating fuels, we atomize heating oils so that those droplets which have the right vapor/air mixtures can be ignited by a spark. Further extension can demonstrate how materials such as metals, which we do not ordinarily consider to be fuels, can actually lead to very hot fires when divided into fine powders which have high surface area/volume ratios.

Fuels are ordinarily classified as solids, liquids or vapors, judged by the normal condition in which we find the fuel. Vapors burn rapidly because they mix readily with oxygen and burning expands the gases which are produced. There are few gases which do not burn, either because they are chemically inert, like helium or argon, or because they are already oxidized to the highest state, such as carbon dioxide or steam. But even those vapors which burn do so only when mixed with air in certain quantities, determined by the stoichiometry of oxidation. For example, natural gas can be ignited only if there is at least 4.7% but less than 15% of the gas in the air. These limits of flammability are called the lower and upper limits and are identical to the explosive limits. We must remember that fires and explosions differ only in the containment of the rapidly expanding gases produced in combustion.

Flammable limits are also applicable to liquids and solids, but they must first vaporize or decompose to form the gases. However, when liquids are atomized to form mists, or solids are suspended in air as fine dusts, specific limits of flammability of the mixtures are exhibited. Definite hazards of explosion or fire exist in sawmill or granary operations because of flammability.

Other properties of solids and liquids are important to determine whether ignition will readily occur or not. These properties are derived from thermodynamics, but are not accurately determinable because of variable conditions of different fire conditions and fuels. Three properties which point out the amount of energy required to heat fuel and form flammable vapors are specific heat, heat of fusion or decomposition, and heat of vaporization. The higher these values, the more heat is absorbed and the material is considered less flammable. The most common example of the importance of these properties is water. Water is our most effective cooling agent for extinguishing fires because it absorbs the most energy in heating and vaporization than any other extinguishing material.

The most important thermodynamic property of fuels, however, is the amount of heat which is generated during combustion. We will see in Case History 10.5 how the amount of heat generated was critical even though the fire point was very high for the material. Heat is described in different ways, but the most common units are Btus, calories (cal) and Joules (J). The British thermal unit or Btu is defined as the amount of heat required to raise the temperature of 1 lb of water 1°F, measured at 60°F. Calories are defined similarly for metric units and Joules represent the SI unit for heat. Useful conversion factors are:

- 1 Btu = 252 cal
- 1 Btu = 1055 J

TABLE 10.1 Combustion Properties of Common Flammable Gases

Gas	Specific Gravity	Heat of Combustion (Btu/ft^3)	Autoignition Temp. (°F)	Limits of Flammability Lower (%)	Limits of Flammability Upper (%)
Natural gas (methane)	0.6	1050	900	4.7	15.0
Butane	0.79	670	—	6.0	17.0
Acetylene	0.91	2499	581	2.5	81.0
Propane	1.52	2516	920–1120	2.15	9.6
Hydrogen	0.07	325	752	4.0	75.0
MAPP gas	1.48	2450	850	3.4	10.8
Anhydrous ammonia	0.60	386	1204	16.0	25.0

- °F = 9/5°C + 32
- 1 lb = 454 g

The combustion properties of common gaseous, liquid and solid fuels are listed in Tables 10.1, 10.2 and 10.3. These properties have been considered when defining the classifications of fires.

- Class A fires consist of ordinary materials such as wood, paper, plastics and textiles.
- Class B fires involve flammable liquids: a flammable liquid has a flash point below 100°F, whereas liquids with higher flash points are referred to as combustible liquids.
- Class C fires pertain only to live electrical equipment; electrically caused fires are not Class C fires.
- Class D fires involve combustible metals.

TABLE 10.2 Combustion Properties of Common Flammable Liquids

Liquid	Heat of Combustion (Btu/lb)	Flash Point (°F)	Autoignition Temp. (°F)	Limits of Flammability Lower (%)	Limits of Flammability Upper (%)
Gasoline	20,700	−45	495	1.4	7.6
Alcohol, methyl	9,600	54	725	6.0	36.5
Kerosene	19,800	110–165	490	0.7	5.0
Acetone	13,200	0	1000	2.6	12.8
Methyl ethyl ketone	14,600	30	960	1.8	10.0

TABLE 10.3 Combustion Properties of Common Solids

Solid Material	Heat of Combustion (Btu/lb)
Wood (pine)	9,000
Wood (oak)	7,200
Charcoal	14,700
Coal (anthracite)	13,800
ABS plastic	15,200
Nylon 6	13,300
Polyethylene	20,000
Polystyrene	17,800
Polyurethane	10,300
Polyvinylchloride	7,200
Aluminum	13,300

PRODUCTS OF COMBUSTION

When fuels are burned, the products of combustion cause many hazards which can alter the manner in which the fire spreads and how it is combatted or hazards which affect any personnel in the vicinity. All combustion products represent a rapid expansion of gases, further expanded by the heat. Venting of fires is an important consideration in firefighting and in planning means of egress from a building in a fire situation.

Besides the expanding gases and heat, though, other hazards exist from combustion products. These include smoke and gases, many of which are toxic. Common toxicants in fire gases and their effects are listed in Table 10.4. Other fire gases, which are non-toxic, but dilute the oxygen levels in air, are CO_2, H_2S, NH_3 and H_2O. Oxygen normally comprises about 21% of air. When oxygen is consumed by combustion and replaced by fire gases, the oxygen level can be critically reduced, dependent on the ventilation of the fire. When oxygen levels fall below 16%, we are slowed considerably, but become euphoric at the same time, not able to recognize the danger. At levels below about 12%, life cannot be sustained.

Smoke is produced in fires where oxidation of carbonaceous materials is incomplete. The generation of smoke will depend on the fuel, the availability of oxygen and on ventilation. Alcohol, for example, produces no smoke, wood a light gray smoke, while only hydrocarbons produce dense, black smoke. A fire burning in a confined space without adequate oxygen will produce more smoke than a well-ventilated fire and the color will be darker. The effect of the fuel or color, however, is much stronger and cannot be reproduced by inadequate oxygen supply to the fire.

TABLE 10.4 Toxicological Effects of Fire Gases

Toxicant	Sources	Toxocological Effects	Estimate of Short-term (10-min) Lethal Concentration (ppm)
Hydrogen cyanide (HCN)	From combustion of wool, silk, polyacrylonitrile, nylon, polyurethane, and paper	A rapidly fatal asphyxiant poison	350
Nitrogen dioxide (NO_2) and other oxides of nitrogen	Produced in small quantities from fabrics and in larger quantities from cellulose nitrate and celluloid	Strong pulmonary irritant capable of causing immediate death as well as delayed injury	200
Ammonia (NH_3)	Produced in combustion of wool, silk, nylon, and melamine, concentrations generally low in ordinary building fires	Pungent, unbearable odor; irritant to eyes and nose	1000
Hydrogen chloride (HCl)	From combustion of polyvinylchloride (PVC), and some fire-retardant treated materials	Respiratory irritant; potential toxicity of HCl coated on particulate may be greater than that for an equivalent amount of gaseous HCl	500 (if particulate is absent)
Other halogen acid gases (HF and HBr)	From combustion of fluorinated resins or films and some fire-retardant materials containing bromine	Respiratory irritants	HF: 400 HBr: 500
Sulfur dioxide (SO_2)	From materials containing sulfur	A strong irritant, intolerable well below lethal concentrations	500
Isocyanates	From urethane polymers pyrolylis products, such as toluene-2,4-diisocyanate (TDI), have been reported in small-scale laboratory studies; their significance in actual fires is undefined	Potent respiratory irritants; believed the major irritants in smoke or isocyanate-based urethanes	100
Acrolein	From pyrolysis of polyolefins and cellulosics at lower temperatures (400°C)	Potent respiratory irritant	
Carbon monoxide	From incomplete combustion of wood, oil coal, etc.	Interferes with oxygen-carrying capacity of bloodstream	

Case History 10.1 Propane Fire in Warehouse

In recent years, it has become economic for many industries to store records, unused equipment and other items at central warehouses which provide pick-up and delivery services. Storage warehouses have been established in many unused mill buildings in economically depressed areas. One such multi-story building was leased by STOR Warehousing in Lowell. In order to accommodate their customers' needs and utilize all areas of the building, STOR purchased four propane-operated fork-lift trucks. Extra propane tanks for the trucks were stored on the ground level in a separate room in conformance with regulations. The storage area, however, was poorly ventilated, and when a leak developed in one storage tank, a fire and explosion occurred which caused extensive damage.

When Dalzell Associates were asked to reconstruct the incident, they focused attention first on the fuel loading. No fuels other than propane were found to be involved; fire from the leaking tank provided sufficient heat to surrounding tanks that the liquid propane expanded, until pressure was sufficient to deform and burst the tank. This phenomenon is commonly referred to as a BLEVE, the acronym for Boiling Liquid Expanding Vapor Explosion. It occurs for any liquids contained and heated above boiling point, but each explosion at STOR added further fuel to the fire. Figure 10.1 shows exploded tanks from STOR after the fire was extinguished.

Dalzell investigators were perplexed initially because there were no sources of ignition in the storage room. However, this led them to examine the exits from the storage room. One unused door led to the loading platform and an overhead door. The investigators immediately focused on the usual activities performed in the loading area. Interviews with employees quickly led to the reasons for the incident. Propane, the vapor density of which is 50% greater than air, leaked beneath the door to the loading dock and beneath the closed overhead door. STOR

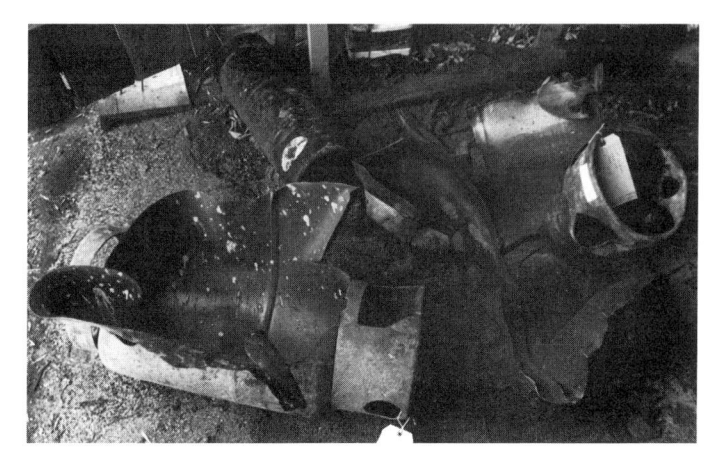

Figure 10.1 Bleve failures of propane tanks.

Warehousing enforces a strict NO SMOKING policy in the warehouse and it was learned in the interviews that employees who smoked commonly went outside by the loading docks at break time. The first explosion reportedly came during the afternoon break, ignited by a burning match, which flashed back to the storage room, much like the action of a lit fire. Unfortunately, the tank which initially leaked (and therefore the cause of the propane leak) was never determined.

When confronted with the investigation results, STOR Warehousing extended the NO SMOKING ban to all company property and contracted to have an exhaust system installed for the storage room. ■

FIRE DEVELOPMENT: SEVERITY AND DURATION

The protection of life is of primary importance in fire situations, and therefore we must be aware of the hazards and how they change as a fire develops. Fires develop over time and the environment deteriorates as smoke and heat build up to endanger life. Since there are a great number of variables involved, many of which are not controllable, we can only generalize about the process of deterioration of the environment and increasing hazards to life safety. Initially, when fuel, oxygen and heat combine, there is little or no hazard. When ignition occurs, most fires develop slowly so the rate of hazard increase is slight. Eventually, however, the fire intensity will increase sharply, building up smoke, heat and the dangerous products of combustion.

Life safety concerns imply that we want to avoid exposure to harmful levels of the products of combustion. If we examine the hazard development as fire develops (approximated in Fig. 10.2), there is a time interval between the first detection and the critical level of fire development which represents hazards to life safety. All actions to preserve life safety must be undertaken in this time interval.

Another way of looking at Fig. 10.2 is in terms of the severity and duration of a fire rather than hazard and fire development time. We will find it useful to examine some of the many factors which contribute to their behavior.

Ignition

The ignition source, along with the fuel first ignited, will determine the initial fire development. For example, if we drop a lighted match against charcoal lighter fluid, fire develops rapidly; a lighted cigarette, however, dropped into a sofa corner may smolder for hours before the sofa bursts into flames. We can conclude correctly that an ignition source really is dependent on the fuel and presence of flammable vapors in the right mixture.

Fire Development: Severity and Duration

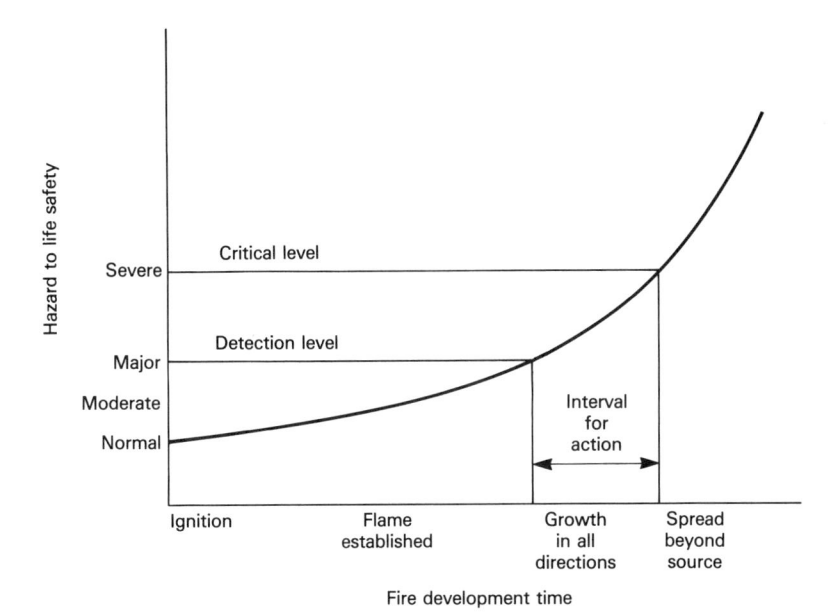

Figure 10.2 General effect of fire development on hazards to life safety.

Flame Spread

Flames spread along the surface of the original fuel which was ignited, dependent on the properties of the fuel and the supply of oxygen. The moving flame heats adjacent unburned fuel, adding more flammable vapors and increasing the flame size; thus, we see that flame spread is also a function of the fuel. For example, a flame will spread across the surface of gasoline in seconds, whereas the process of burning might be halted across a solid wood floor unless heat is supplied from some other fuel. Locating the first fuel to burn is very important, because flame spread to nearby combustibles will be quite different from flame spread by exhaustion of the original fuel ignited. In general, flames spread faster upward than they do horizontally or downward due to heating by the hot combustion products flowing upward and outward from the fire. This causes the characteristic V-pattern which we often use to determine how a fire has spread.

Case History 10.2 Fire Spread in Cold Storage Warehouse

The Henniker Cold Storage Food Warehouse was designed and built in 1958. It is a four-story windowless brick building, with the upper three floors used for frozen food storage; the temperature in these areas is maintained at 0°F. The ground level area houses the offices, a windowless cold storage area maintained at 40°F, and shipping and receiving for this cold storage area.

Insulation of the windowless building is exemplary. The only modification which had to be made was the addition of 3-inch layers of corkboard to the ceiling of the cold storage and office areas to prevent heat transfer to the frozen food sections above. Above the cold storage area, however, sufficient moisture became entrapped in the open spaces of the cork and froze. As the water froze, the resulting expansion caused some of the cork to loosen and fall; this problem was solved early by sealing the cork layer with a thin layer of aluminum foil, adhesively bonded to the cork. The system worked well for 25 years.

Henniker is in the center of New England's apple orchards and the cold storage area is actively filled with apples throughout late September until December. It was during this time that the elevator service for the frozen food floors was being modified. The elevator shaft was located over the cold storage area and the shaft formed a small room with a low (4-foot) ceiling within the cold storage area.

While the elevator was being modified, a hole was cut through the ceiling of the elevator shaft, using a jackhammer to break up the concrete. The welders then used an oxyacetylene torch to flame-cut the reinforcing bars still in the hole. Molten slag falling into the small cold storage room started a small fire in some papers lying on the floor. The fire extinguisher in the shaft was not charged, creating momentary panic, but one worker saw another extinguisher below in the small room. As he dropped through the hole and went toward the extinguisher, the whole room was suddenly engulfed in flames. The worker went out through the door and hurried from the cold storage area, alerting other workers in the shipping and receiving area.

The fire in the cold storage developed very rapidly, spreading through the entire area at the ceiling level in a matter of seconds, then subsided and was easily extinguished with just a little water. When it was all over, the aluminum foil had fallen, covering the crates of apples, and there was soot and water covering the room. However, the apple crop was almost unharmed. Only the top apples in the top crates were charred. Even the cork insulation material was charred only to a depth of about $\frac{1}{2}$ inch.

The investigation of this fire focused on the flame spread which was unusual. First, the welder dropped into the small room believing there was no impending danger, but the room was engulfed in flames. However, he managed to escape the small cold storage room and wider area. Shortly afterward, the whole cold storage area was engulfed. The only path of the fire to be traced was the typical V-pattern of the pyrolysis on crates directly outside the door of the small room where the fire began.

Knowing that flame spread is primarily dependent on the fuel, the investigation focused on the fuel available near the ceiling height in both the small room and the larger storage area. The slowdown in flame spread allowing the worker's easy escape occurred when the small room fire was burning slowly, forming the V-pattern in the crates. At the ceiling, flammable vapors had to have been present in the right mixture, but there was no clue as to what they might be. No known

Fire Development: Severity and Duration

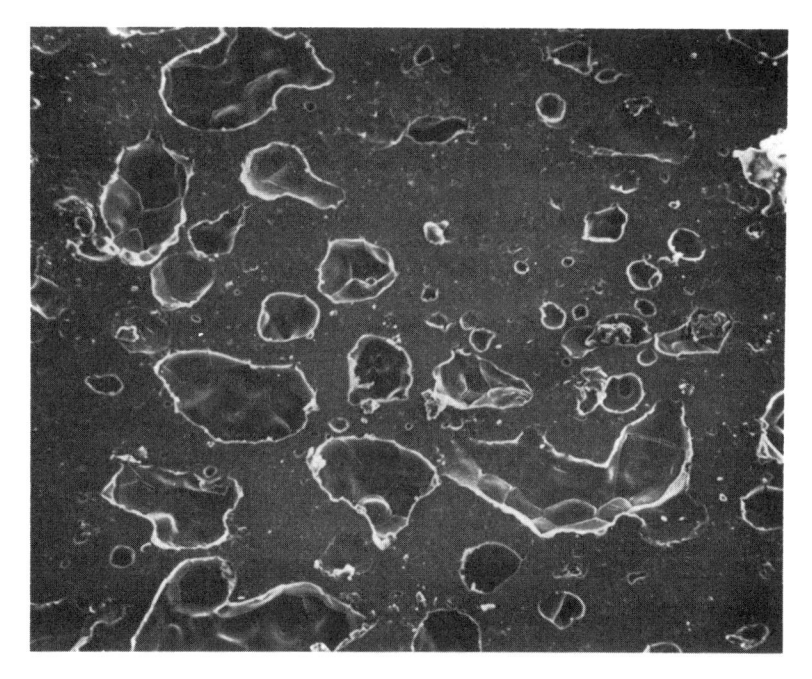

Figure 10.3 Melted pattern of aluminum foil surface adjacent to cork insulation (150X SEM micrograph).

flammable vapors were attributable to the apples and no other source could be identified.

One investigator focused on the aluminum foil, questioning why *all* of the foil should have fallen when the duration of the fire was short and other damage was minor. Her examination of the foil in a scanning election microscope provided the beginning point for conclusive studies. Figure 10.3 illustrates the surface of the aluminum foil which had been adhesively bonded to the cork insulation. This micrograph shows melting of the aluminum, with the pattern of the melted and resolidified areas the same as the spacing in the cork insulation.

The investigator retrieved a piece of the cork insulation from an area unaffected by the fire, submitted it to a chemistry laboratory where the sample was heated and the vapors which evolved were analyzed. These vapors included toluene, benzene and xylene, all extremely flammable when in the right proportions with air.

The investigation concluded that the rapid flame spread occurred when the flammable vapors were first exposed to ignition—initially in the small room—then slowed as it moved to the larger room and burned upward to the ceiling and through the aluminum foil, then rapidly through the cork, melting and dropping the entire aluminum foil facing. The foil had prevented normal dissipation of flammable vapors evolved from the cork insulation for many years, permitting accumulation to the flammable limits. ■

EFFECT OF ENCLOSURE AND HEAT TRANSFER IN FIRE DEVELOPMENT

Once a fire has been started, its spread beyond the source is determined by the enclosure within the building. The fire spread rapidly in Case History 10.2 but it did not develop to cause major damage because the storage room and its contents never became heated and involved in the fire. In a larger fire, however, the effect of the enclosure will become a factor because it controls the oxygen source through ventilation and it traps hot combustion gases in the upper portion of the room. This is a result of heat transfer under fire conditions. Heat must be transferred to other materials for the fire to spread and can take place by convection, conduction or radiation.

Convection is simply the movement of heat by hot gases, either the combustion products or gases heated by conduction in contact with the fuel. Convection usually spreads fire upward and away from the fire. Conduction is the transfer of heat through a material from hot to cold areas; how fast heating or cooling occurs is widely variable, dependent on the type of material. Metals are excellent heat conductors, whereas concrete and insulation materials are poor heat conductors. Radiation is the transfer of heat by electromagnetic waves from a source to a solid within line of sight with the source. For example, heat is radiated from the sun and warms the solid earth, with no heat loss in the space traveled between source and solid. Flashover is a means of fire spread where items in a room not in direct contact with the original flames suddenly can burst into flame due to high radiation.

When a fire is contained within a room where the oxygen supply is adequate, the fire will continue to grow until all the fuel is consumed. Such a fire is limited by the amount of fuel, and is therefore called a fuel-controlled fire. When all fuel is burning, but is slowed because it becomes starved of oxygen, we have a ventilation-controlled fire.

While convection, conduction and radiation are the conventional methods of fire extension, other methods can extend a fire as well. For example, light fuel material (e.g. newspaper) can be caught up in convection currents and deposited, still burning, elsewhere where it can ignite other fuels. Aluminum cookware has been known to spread a fire when it has melted and flowed or fallen to a lower area. The potential to extend a fire due to moving material which is either flaming or heated to high temperatures should not be overlooked.

Case History 10.3   A Challenge to the Professional Fire Investigator

The Paul K. Rogers Building in Lowell was the site of the senior high school in 1930, but was abandoned in favor of the new Lowell Central High School in 1968. When the resurgence of industry began in the 1970s, developers remodeled the Rogers Building. The remodeling was carried out to suit the new tenants. It catered mainly for office space (e.g. medical, legal, dental and insurance offices), but there were a few retail jewelers, a computer software systems company and a manufacturing company which had its assembly operations sited there.

Recently, there was a fire which destroyed much of the second and third floors of the center and right wings of the symmetrical building. The left wing and lower floors were unaffected. A dentist's office, the computer software systems company and the assembly area of the manufacturing company were involved. The remodeling was significant in the investigation of the fire, and therefore sketches of the third floor layout and the elevated views are reproduced in Figs. 10.4 and 10.5 (the letters in these figures indicate the key locations in the fire and investigation).

When fire was first reported, it originated from an office in an adjacent building at 7:58 A.M. by a worker who sighted flames at location A. At 8:01 A.M., an employee of the computer software systems company reported two small closet fires at locations B and C. Hand fire extinguishers were used to try to quash the fire, but this had to be abandoned and the areas were evacuated. Fire erupted from the building only at position K.

The initial reports of three separate fires plus the apparent separation of the

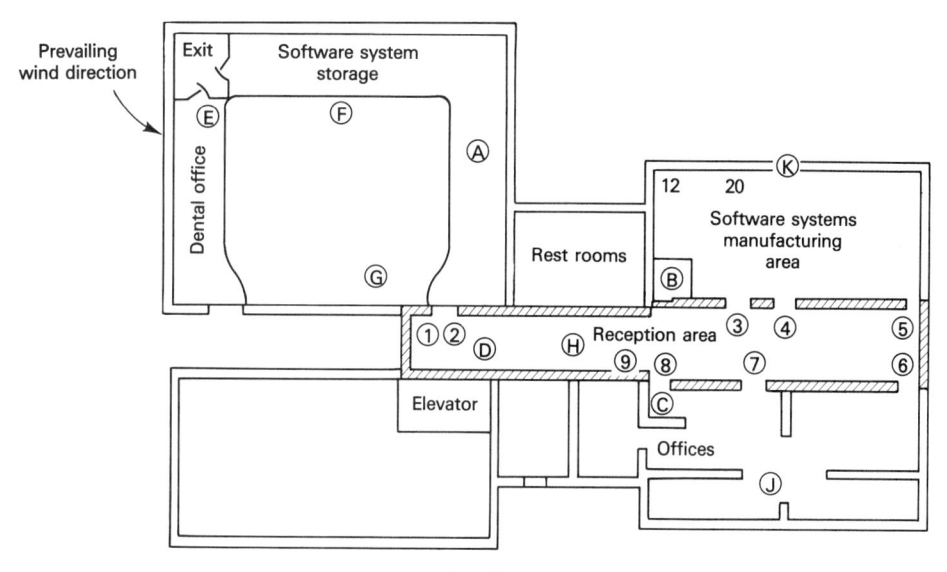

Location
A: Flames sighted from adjacent building
B, C: Closets where fires discovered
D: Location of small room on 4th floor (see Fig. 10.5)
E, J: Location of highest heat in area
F, G: Location of distorted steel
H: Roof vent
K: Flames erupted from building

Figure 10.4 Third-floor plan of Paul K. Rogers building.

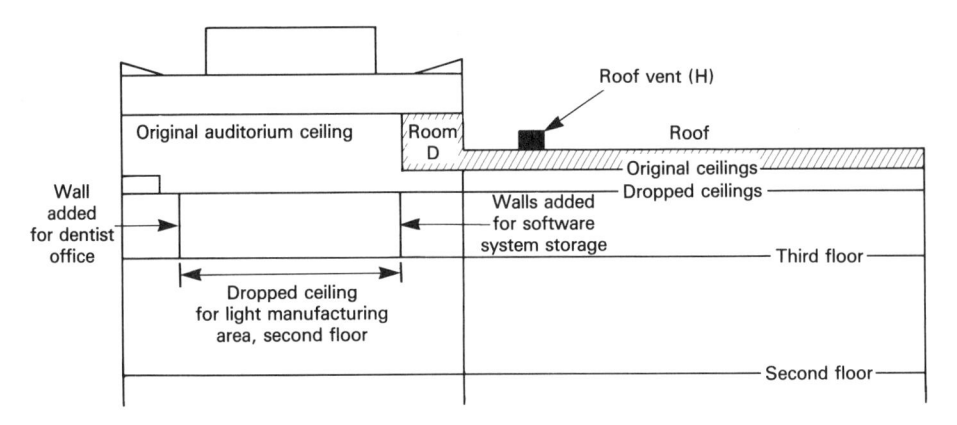

Figure 10.5 Elevation plan of areas in Paul K. Rogers building which involved remodeling of original auditorium.

areas by a firewall (later proved erroneous) led to an investigation of arson which confused and limited the investigation. There was no evidence of any incendiary origin, and clean-up proceeded. No accusations were ever made and the investigation stalled for a long time, but interest was renewed because of insurance claims and counterclaims among the occupants' and the proprietors' insurance representatives, all involving liability of the losses. All parties agreed to abide by the investigation results of Dalzell Associates, whose efforts started long after clean-up had been completed.

The first important discovery made by Dalzell employees was a room adjacent to the original auditorium just below the roof level, which provided potential communication among the three fire areas. This room is marked D in Figs. 10.4 and 10.5. In this small room there were two openings to the ceiling area of the original auditorium, 2 × 2 feet and 3 × 3 feet square. There was also a 2 × 12 foot crawl space leading to the east wing. The discovery of this led to the theory that a fire could have spread through the openings from the auditorium into room D and through the plenum to the east wing, starting the smaller fires (when discovered) at locations B and C. For this theory to be proven, however, it would be necessary to explain why the original fire was not discovered earlier than the smaller fires, what temperatures were reached in the different areas, why dust and an old cardboard box in room D were not pyrolized or consumed and why the fire did not vent through the roof vent (marked H in Figs. 10.4 and 10.5) before spreading to the east wing.

These problems required extensive measurements, sampling of materials, laboratory testing and interviews. Measurements showed the total volume of the auditorium to be 200,000 ft^3; this was dead space hidden by dropped ceilings above the manufacturing area on the second floor and the walls and dropped ceilings of the original balcony on the third floor. By comparison, the plenum and room D was about 5000 ft^3 and the open areas of the east wing accounted for 10,000 ft^3. With such volume differences, a fire could develop extensively in the auditorium with no awareness before spreading into the east wing through the plenum. The temperature reached during the fire was estimated from microstructural studies of steel wires used to suspend the dropped ceilings. The highest temperature was found at location E in the auditorium and at locations C and J in the east wing. Another indication of the temperature was taken from the distortion of the steel beams which supported the original plastered auditorium ceiling. Such distortion is caused by the contraction of steel when heated above the temperature where austenite is formed (about 1300°F). These temperatures suggested the fire started in the north-west corner of the auditorium, with hot gases accumulating near the ceiling, but moving in the direction of points F and G where distortion occurred. Point F is explainable because firefighters broke the skylight above to vent the fire. Point G, however, indicated heat flow toward room D.

The highest temperatures occur in ventilation-controlled fires. This type of fire is easier to extinguish, and the actual fire was controlled in a short time by the firefighters. This fact is significant because during development and spread from the auditorium, the hot gases would have been oxygen-depleted, and would have been unable to ignite the cardboard boxes in room D or dust in the plenum. Oxygen depletion was confirmed by measuring the heats of combustion of the dust from the plenum and similar dust from the west side (unaffected by the fire), and the cardboard box from room D and an identical box found in the west side. The results showed a lower heat content in the samples from room D and the east plenum, suggesting distillation of vapors by exposure to high heat, not experienced by the samples in the part of the building unaffected by the fire.

The only remaining question to be answered was why venting did not occur through the roof vent H. News media photographs showed heavy dark smoke from the east wing, but only light colored smoke near the roof vent and heavy soot deposits were found in the plenum east of the roof vent, but not west of it nor inside the vent itself. In addition, bolt holes were found in the gutter of the roof vent. All of these discoveries suggested that the roof vent was blocked at the time of the fire. This was confirmed by the firm which had installed the air conditioning for the computer software manufacturing company. During installation, the make-up air was obtained from the space above the dropped ceiling, but the air conditioning system could not be balanced until the roof vent was closed.

Thus it was concluded that there was only one fire, originating in the closed area of the original auditorium. Other aspects of the investigation divulged numerous wiring code violations, suggesting an electrical origin to the fire. The proprietor of the building was found liable for 90% of the damages, and the computer software manufacturing company for 10% because the roof vent had been blocked during their remodeling. ■

FIRE PREVENTION AND PROTECTION

Fire prevention is paramount, but fire protection has to be planned as well because no program is perfect. Fire prevention programs which are successful incorporate all the creative aspects of management, commitment and good hazard identification, i.e. the essence of systems fire safety. The best prevention programs are built upon innovation, tailored to the needs of each individual company, and avoidance of the elements which combine to form the fire triangle. This means being aware of the fire characteristics of all materials in a plant—the fuel loading—and potential ignition sources, rather than controlling them by avoidance or training. Although we will be creative in our approaches, perhaps some pitfalls can be avoided if we look at one high-tech company's experience.

Case Study 10.4 A Hydrogen Explosion

Microchip Processing, Inc. (MPI) is a small high-tech company located in Dracut. They employ 45 people in research and development of thin-film materials for government and electronic applications. When a hydrogen explosion occurred in the research laboratory, where a chemical vapor deposition experiment was being conducted, extensive damage was caused. The preliminary report issued 4 days afterward demonstrates that MPI was knowledgeable of the hazards of the chemicals, and was concerned first and foremost with life safety. We can best understand what happened by reading that report:

Preliminary Report on Accident at MPI on March 22nd

Location: Research Laboratory
Time: 2:10 P.M.
Activity: The accident occurred where a chemical vapor deposition experiment was underway to deposit silicon. This experiment uses hydrogen (H_2) and trichlorosilane (Cl_3HSi) gases. These chemicals are contained in cylinders located under a hood in the lab. Four hydrogen cylinders containing 200 ft^3 of H_2 at about 2000 psi were attached to a manifold under the hood. One tank containing about 25 lb of liquid Cl_3HSi was also under the hood and connected to the same experiment through a separate line. Tanks of hydrogen selenide (H_2Se) and hydrogen sulfide (H_2S) were also under the hood, but were closed at the tanks and not in use. All tanks were chained in place.

Chronological Events of March 22nd

In the morning, Brian Bullard and Brian Camasso, research technicians, completed leak checking of hydrogen lines at an operating pressure of 2000 psi, and turned the heat on. The furnace reached the deposition temperature of 1100°C at 1:40 p.m. and Brian Bullard purged the lines, then began introducing the hydrogen from the tanks. While he was opening the third tank, a flexible stainless steel line ruptured, permitting the high pressure hydrogen to escape. Both men were knocked to the floor by the escaping gas, but were able to recover and alert others to evacuate the area. Hy-

Extinguishing Fires

drogen was ignited after about 15 seconds at the pilot of the building's heating furnace, causing extensive damage and a rapid fuel-controlled fire. Evacuation was successful and no one was injured. Soon after the explosion, however, HCl gas was identified hampering firefighting procedures and H_2S and H_2Se were also released.

The damage at MPI was mainly confined to the area where the ignition and explosion occurred. Sampling taken in various areas, however, showed the presence of selenuim compounds in the walls, carpeting and even the ceiling tiles. The TLV of H_2Se is 0.05 ppm, recommended to prevent pulmonary irritation and to prevent onset of systemic disease. In addition, H_2Se is extremely flammable; H_2S is also flammable and toxic, with a TLV of 10 ppm. The combustion of these chemicals in the explosion and fire produced toxic compounds which settled into the buildings.

The risk associated with these toxic compounds necessitated removal of all internal walls within the research laboratory. The cost of this exercise was in addition to the direct damage costs and those incurred through loss of business.

Loss control insurers have since evaluated the procedures and policies for chemical handling at MPI. Their analysis did not find fault with the procedures of the experiment being conducted, although they did recommend redesign of the manifold system using rigid plumbing. They were, however, extremely critical of the storage practice at MPI, with the hazardous H_2Se and H_2S chemicals exposed to the experiment instead of being isolated. Safe storage must consider the possibility of rupture and the mixing of incompatible materials under fire conditions. ■

EARLY DETECTION

Life safety is the major concern in fire protection. When examining how fires develop (Fig. 10.2), it is clear that we can best assure life safety through early detection. Detection permits fires to be extinguished easier and results in less damage to property. There are many available types of detection systems, operating on the principles outlined in Table 10.5. What we have to remember, though, is not how these detection systems work or how to maintain them, but how they can be best employed. Here is another area where a little innovative thought might prevent catastrophic results. For example, in Case History 10.3, what would have happened if a hard-wired smoke detector or rate-of-rise detector had been located in the auditorium space which was covered? Or even in Room D? Basically, detection also requires good management skills.

EXTINGUISHING FIRES

Detection is only the earliest aspect of a fire protection program—extinguishing the fire follows that closely. Virtually all fires are small at first and could be easily extinguished if the proper type and amount of extinguishing agents were promptly

TABLE 10.5 Common Fire Detection Devices

Detection Type	Principle of Operation	Application
Heat	Fusible link melts at predetermined temperature	Automatic sprinklers, thermal alarms
	Rate-of-rise-pressure caused by temperature rise converted to mechanical action by diaphragm	Alarms
	Continuous-line heat-sensitive insulation melts, causing short circuit	Alarms
	Bi-metallic – strips of metal having differing thermal expansion bend and make contact at present temperature	Alarms
Smoke	Photoelectric – smoke particles obscure photoelectric beam	Alarms
	Ionization – smoke particles decrease conductance of ionized air	Alarms
Flame	Infrared or ultraviolet detectors respond to radiation from flame	Fail-safe devices on gas furnaces, alarms in fuel loading areas

applied. Therefore, it is important not only to ascertain the possible types of fire that may be caused, but also to supply the right extinguishing agents to fight those types of fire. The types of extinguishing agents, how they function and how they can be distributed are summarized in Table 10.6.

The success of automatic sprinkler systems in extinguishing fires is so good (in excess of 90% satisfactory performance in all industries), that we can benefit by examining how they work. Their success is even better when we consider that the leading cause for their "failure" was that the sprinklers were shut off. A sprinkler system consists of piping—usually for water, but also for dry powder or halons—with sprinkler heads positioned to deliver the extinguishing agent. Sprinklers operate on the principle of thermal detection, based upon links fused together by a thin layer of solder which has a predetermined low melting point. Fusible links such as that shown in Fig. 10.6 both reduce the time for the solder to melt and minimize the force on the solder under ordinary circumstances. When fire does melt the solder, the separation of the link permits water to flow from the pipe through the orifice. Distribution is based upon spacing between the sprinkler heads, the water pressure, orifice size and on the deflector design. The purpose of the deflector is to break up the water stream into droplets, which cover a larger area and are more effective coolants. Optimum water distribution from automatic sprinklers is depicted in Fig. 10.7.

Sprinkler heads and their location are not the only parts of an automatic sprinkler system, however. Alarms, maintenance tests and even external Siamese connections for fire departments to connect hoses to and use the pipe system are integral parts of the system. Figure 10.8 depicts an automatic wet sprinkler system; water to the system is controlled by an OS & Y valve. When the system is inactive,

Extinguishing Fires

TABLE 10.6 Extinguishing Agents

Extinguishing Agents	Type of Fire Used For (most common underlined)	Mechanism for Estinguishment	Methods of Distribution	Problems
Water	<u>A</u>, B, D	Cooling (heat absorbed in vaporization); smothering; emulsification; dilution	Automatic sprinklers, hose systems	Freezing and pipe bursts in unheated areas
Carbon dioxide	<u>C</u>, A, B, D	Smothering; cooling	Portable extinguishers, total flooding	Reignition after dissipation
Halogenated agents (HALONS)	<u>C</u>, A, B, D	Chain breaking (chemical reaction which interferes with combustion)	Automatic sprinklers, portable extinguishers	Toxic
Dry chemicals	<u>A</u>, B, <u>C</u>	Chain breaking; smothering; cooling; radiation shielding	Portable extinguishers, automatic sprinkler	
Foams (aqueous, fluoroprotein)	<u>B</u>, A, D	Smothering	Portable extinguishers, hose systems	Horizontal fires only
Combustible metal agents (graphite, G-1 Met-L-x, Nace, Lice)	<u>D</u>	Cooling (graphite); smothering	Spread by hand portable extinguishers	Expensive

the clapper valve is closed and no water flows. However, when a fusible link melts, water flows through that sprinkler head and through the system. Alarms are sounded by the vane alarm, the pressure switch and the water motor gong. This automatic sprinkler system can be converted to a dry system where freezing conditions are encountered by pressurizing the piping beyond the clapper valve with air at a higher pressure than that of the water.

Case History 10.5 *A Case of Liability*

Tom Dennis is an entrepreneur, both with respect to business and safety. A few years ago, when Lowell was a dying city, Tom purchased an old mill with an eye to renovating it and attracting new, small businesses to the area. In the early years, renovations were strictly cosmetic—giving them a facelift with new partitions, new paint and updated electrical wiring. His efforts were successful, and he attracted a manufacturer of automobile seat covers to the largest area of

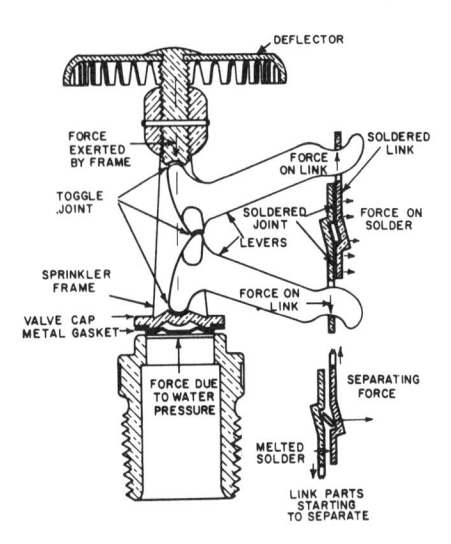

Figure 10.6 Upright fusible link sprinkler head. (Reprinted with permission from NFPA, *Fire Protection Handbook,* 16th Edition, copyright, 1986, National Fire Protection Association, Quincy, MA 02269.)

one building and a state-sponsored packaging firm which employed mentally retarded workers in a smaller wing of the building.

The building was equipped with an ancient automatic wet sprinkler system. Because the new occupants would not be using or heating all of the areas serviced by the sprinkler system, Tom had the system converted to a dry system in the Fall season. Conversion involved draining the system and installing air pressure lines and a compressor to maintain the air pressure. The compressor was located near the water inlet, OS & Y valve and clapper valve where the area was heated. In December of the first year of occupancy, the first cold weather led to disastrous flooding, mostly in the assembly room of the packaging firm. Its cause was readily discovered, for the electrician who had installed the compressor motor had not aligned the pulley properly, causing an overload which tripped the circuit breaker. Tom's problems seemed simple enough because his insurance company covered the damages, intending to subrogate against the electrical contracting firm.

After the clean-up and repair of the piping caused by freezing, the system was to be reactivated. In order to do so, the OS Y valve had to be closed, all residual water drained, and the empty piping pressurized before reopening the

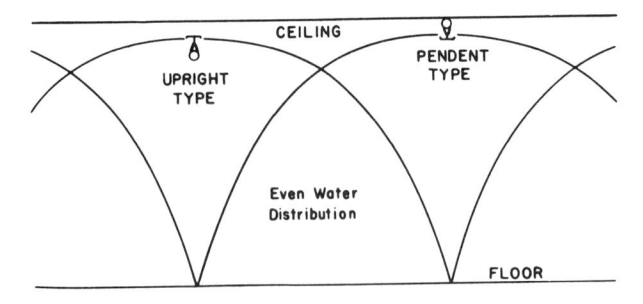

Figure 10.7 Distribution pattern from upright sprinkler system. (Reprinted with permission from NFPA, *Fire Protection Handbook,* 16th Ed.)

Extinguishing Fires

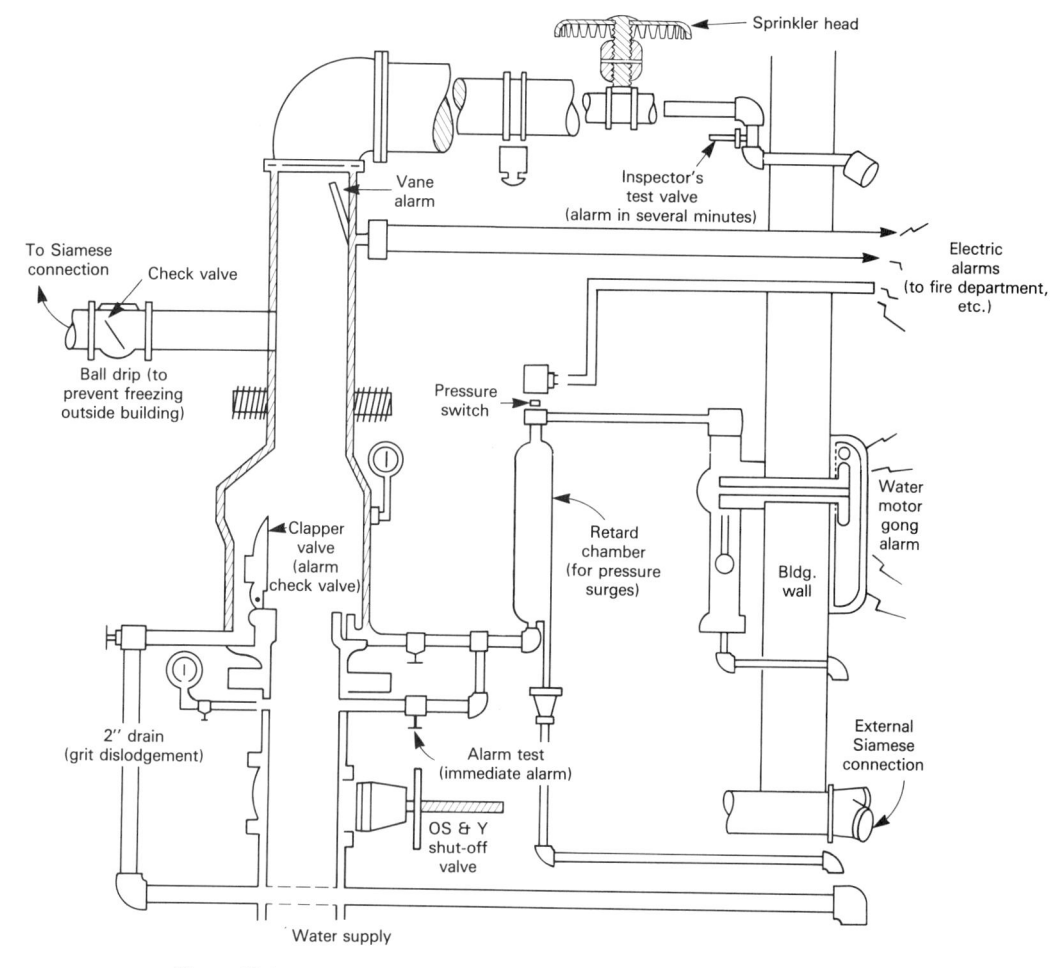

Figure 10.8 Schematic diagram of automatic wet sprinkler system.

OS & Y valve. Tom's maintenance crew selected December 24th for this task because of the disruption and time involved for reactivation. The repressurizing should have been completed by 8:00 P.M. However, drunken revellers ignited an incendiary fire in an unconverted part of the mill complex which rapidly developed into an inferno, causing a general alarm in Lowell and requiring assistance from as far away as Nashua, New Hampshire, and Boston. Firefighting was hampered by a very low water pressure, cold temperatures and fatigue. The fire still stands as the single worst fire in Lowell's history, destroying all of the buildings in the complex.

The seat cover manufacturer's business was ruined, and there was insufficient insurance coverage to cover the loss of materials and equipment, much less business interruption. The State is self-insured, and therefore both parties sued

Tom for their damages independently, citing that their losses would have been prevented if the automatic sprinkler system had been operable when the incendiary fire was ignited.

Owing to the success of automatic sprinklers in extinguishing industrial fires, Tom's chances of defending the charge of negligence was slim. When it was pointed out that the fire department had not been notified that the system was down for repairs as required by law, those chances were dented further, causing Tom to feel victimized. Tom's lawyers were diligent, however, and requested all of the automobile seat cover manufacturer's records pertaining to material purchases, sales and safety. These records indicated an OSHA citation for blocking sprinkler heads by piling materials too high, and a much higher fuel loading than usual because of the differences in incoming vinyl fabrics and finished seat covers shipped in the 3-month period preceding the fire. Because of this, Tom was cleared of negligence, but he still has the psychological scars suffered after 4 years of intensive litigation. ■

DISASTER PREPAREDNESS

If fire goes unchecked, there is the possibility of a disaster, with destruction of property, business opportunity and even loss of life. Emergency or disaster preparedness has to be part of safety management goals in order to prepare response to emergencies and disasters. Although these terms are frequently interchanged, an emergency is simply any situation which requires immediate response, whereas disaster is a sudden, unexpected and undesired event which can bring great damage, loss or destruction.

There are two types of disaster: natural (such as hurricanes, floods, winter storms or earthquakes) or man-made (created by our inability to handle technology safely). We are faced with an anomaly that overall management of natural emergencies appears to be more coordinated than is the management of the technological hazards. Examples such as Chernobyl, Bhopal, the Challenger and Valdez, which are most frequently cited, are also unfortunately recent. Media attention has focused on these and other disasters, causing public insecurity with companies even remotely involved with a potentially disastrous situation. Thus we have to prove good faith as well as prepare for coping with disasters on merit alone.

In effect, preparedness is only one phase of a risk management program which includes prevention, preparedness, response and recovery. Although we are already familiar with prevention techniques, the other phases represent management concepts. Preparedness implies planning, response is implementation of appropriate activities during the emergency, and recovery includes all activities designed to return the organization to normal, routine operations.

The methods used in developing disaster preparedness plans are not significantly different from those in developing an integrated systems safety program. However, disaster preparedness plans do involve more interaction with public safety groups, e.g. fire departments, police, hospitals and others. Communication is of primary importance, with initial response groups, leadership and command post responsibilities clearly established.

In fire preparedness programs, we must first learn about the company's fire hazards and how to combat them. Many safety engineers have chosen to hand pick and train a fire brigade in firefighting, rescue, evacuation and salvage. The brigade is the initial response group who will tailor the fire protection equipment and preventative maintenance necessary for a plant and play a key role in educating the workers on preparedness programs.

ELECTRICAL SAFETY

If we examine what ignites most fires, electrical faults rank highly, along with careless smoking, heating and cooking. Electrical causes include distribution wiring, appliances and portable tools and even static electricity which can directly ignite fires by producing sparks. They also include electrical processes which produce other ignition sources, such as arc welding where molten metal and/or slag formed by the arc is the source of ignition. This was the ignition source in Case History 10.2, the cold storage warehouse fire.

Electrical safety involves much more than concern for igniting fires. Accidental contact with energized circuits can cause shock, burns and even death. We cannot neglect the element of surprise, either, because many a fall from a ladder has been caused by a surprise shock while working on electrical repairs. But what do we have to know about electrical distributions and electronics to assure safe working conditions? Most people know little more than how to turn switches on or off, plug in an electrical device and maybe where the fuse box or circuit breakers are. Fortunately, electrical circuits may be complex and electronics even mystifying, but the hazards associated with them are not. In most cases, competent electrical engineers are needed only to ensure functional design.

If we examine the consequences of electrical hazards—fire ignition, contact and surprise—we need only to determine the aspects of electrical circuits which are related to these phenomena. Fire can be ignited by overheating or by sparks created by arcing. Contact injuries arise if we accidentally become part of the circuitry and surprise can be caused by contact or by arcing. These conditions all relate to the most fundamental laws of electricity—those which describe flow, power and resistance to flow.

Electrical equipment operates because electrical energy flows through a circuit. The potential energy of the circuit is described as the voltage, the flow is described as the current and how much flow occurs is determined by the resistance. These parameters are related by Ohm's Law, which states:

$$E = IR \qquad (10.1)$$

where E is the voltage or potential (in volts), I is the current, measured in amperes, and R is the resistance, measured in ohms. The power P, represented in any circuit is measured in watts and is given by:

$$P = IE = I^2R \tag{10.2}$$

Heat is generated within the wires of a circuit when current flows and can be described by:

$$H = I^2Rt \tag{10.3}$$

where H is the heat and t is time.

These relationships show that our safety concerns, whether they are contact, surprise or fire, all involve electrical resistance. Resistance results both from geometric considerations and from fundamental properties. This relationship for a wire is:

$$R = \rho l/A \tag{10.4}$$

where ρ is material resistivity (usually in ohm-cm), l is the length in cm, and A is the area in cm^2. Resistivity is the inverse of conductivity, and therefore it is a simple matter for us to think in terms of high conductivity and low resistivity for conducting materials and low conductivity and high resistivity for insulating materials. The resistivity values of common materials used in electrical applications are given in Table 10.7.

All materials will conduct electricity if the conditions are right. There has to be sufficient potential difference and sufficient conductivity of the path. Lightning strikes, for example, when the potential difference between the clouds and earth is high and the air path becomes sufficiently conductive to conduct the electricity between them. This uncontrolled flow of electricity between two conductors is a discharge or arc which occurs whenever the potential difference

TABLE 10.7 Electrical Resistance of Common Materials Used in Electrical Applications

Material	Typical Application	Resistivity (ohm-cm)
Conductors		
Copper	Wire conductor	1.67×10^{-6}
Aluminum	Wire conductor	2.65×10^{-6}
Gold	Microelectronics	2.35×10^{-6}
NiChrome	Heating element	116.0
Lead-tin solder	Printed circuit boards	15.0
Graphite	Susceptor	1000
Insulators		
PVC	Wire insulation	10^{14}
Glass	Light bulbs, transistors	10^7
Porcelain	Power lines	10^{14}
Mica	Power circuits	10^{13}–10^{17}

exceeds a value known as the breakdown voltage where the medium between them will conduct the electricity. Arcing is most common for movable parts of electrical circuits, such as switches and relays, which come together and separate, and in static electricity where a charge is accumulated and then discharged suddenly when nearing a conductor.

Resistance is also important when determining how damaging accidental contact with electricity can be. The hazards of using electric appliances in the bathroom relate to the decrease ($\simeq$100-fold) in resistance of our skin when wet as opposed to dry and to the conducting characteristics of water. It is the current, however, which injures people in contact with electricity. Once skin resistance is overcome, current flows readily through our blood and body tissues. The effects can vary from slight sensations through painful muscle contractions, burns, ventricular fibrillation and death. Death can result from asphyxiation caused by respiratory interference, either from muscular contraction or paralysis of the central nervous system and from ventricular fibrillation, affecting the heart directly. Electrical burns arising from arcs are usually deep, very painful and slow to heal.

Safe Electrical Circuits

The consequences of electrical hazards are fires and injuries—injuries which are quite different to crushing, lacerations or those needing an amputation. The engineering methods with which we can combat electrical consequences are also quite different. We can protect equipment and, to a lesser extent, eliminate fires and protect people by overload protection and by grounding. Overload protection includes fuses and/or circuit breakers which trip at a predetermined current load, de-energizing the circuit before damage can occur. A ground is a conducting connection to the earth which provides safe equalization of large voltage differences, such as lightning surges. Grounding applies to both equipment and the electrical system. Equipment is grounded by joining all conducting materials which normally do not carry current, but enclose current-carrying parts, such as the casing of a portable drill. Joining prevents a difference in voltage between the enclosure and the system ground. If a short-circuit between the energized portion and the ground occur, the overload protection will trip.

There are many instances where the grounds are improperly installed or the fault does not lead to current overload, thus leading to enclosures being raised to live voltage. Such situations are extremely hazardous to personnel. Although grounding helps protect workers and equipment, it is not foolproof. For these reasons, ground fault circuit interruptors (GFCIs) are required in many applications. A ground fault circuit interrupter, such as that shown in Fig. 10.9, provides better personnel protection. The interrupter is designed to sense currents leaking to the ground of as little as 5 mA, which will trip the circuit, preventing any possible injury to personnel.

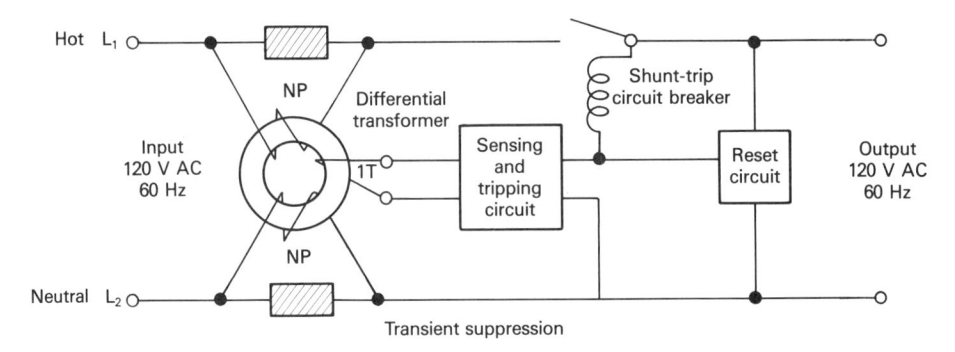

Figure 10.9 A ground fault circuit interrupter. If differential transformer senses any current flow to ground, the circuit breaker is tripped.

Management of Electrical Safety

The technical requirements for providing electrical safety can be ascertained from such sources as NFPA 70, the National Electrical Code. There is much more, however, required to ensure electrical safety in any plant operation. All of the factors which make for good management also apply here to good management of electrical safety. Because of the technical nature, training must be well thought out, and must include maintenance as well as operating personnel, covering topics such as lock-out procedures and inspection procedures. Just how far-reaching management and hazard identification processes are in electrical safety can be ascertained from the last two case histories which end this chapter.

Case History 10.6 What Can You Suspect?

Walkinshaw Shoe Company, like many shoe manufacturers, has had to reexamine their product line because of the increase in foreign competition. The company decided in 1983 to reduce the product line to specialty outdoor sporting boots. As part of the reorganization, Walkinshaw consolidated their manufacturing area to an area used principally for storage beforehand, vacating the large section of the building which was expected to be leased. Extensive rewiring was required for the move and work was contracted out to a reputable local firm.

In 1987, construction plans for a new high-rise office building adjacent to the Walkinshaw plant were approved and the one-story brick garage located only 12 feet from the Walkinshaw plant was razed. The property up to the Walkinshaw line, 3 feet from the plant, was to be converted to a parking area, and guard-rails were constructed right along the property line; these guard rails were 30 inches from ground level.

Although Walkinshaw management were aware of the construction next door, they paid no heed to its progress, simply because it was not on their property. Unfortunately, Walkinshaw Shoe Company has been named as a defendant in the wrongful death suit of a young 9-year-old boy who had lived in the neighborhood. The boy, along with two friends, had been playing ball in the new parking area until dusk. When it was too dark to play anymore, but before they disbanded to go home, they began walking along the guard rail. The unfortunate victim lost his balance and fell toward the Walkinshaw plant. He grabbed at anything to prevent his fall, unfortunately grabbing and holding the metal conduit for the electrical service into the building. His weight bent the conduit causing it to bend and break through the insulation, thus energizing the conduit and electrocuting the victim.

A post-accident inspection showed the conduit brought power from the corner of the building, with a horizontal run of 48 feet located 8 ft above the ground, where it entered the building. There were only two support straps in the 48-foot horizontal length, in direct violation of the NFPA 70 which requires a maximum distance of 10 feet between rigid metal supports. The tragedy of the incident and accidental death were bad enough, but to learn that their safety program had been lax enough to permit the obvious violation to go unnoticed for almost 5 years devastated Walkinshaw Shoe Company. ■

Case History 10.7 The Wrong Wire

In the Northeast, mountains and lakes lend themselves to boating, swimming, fishing, hiking and skiing at different times of the year. This has led to the development of weekend homes, cottages and camps. It gets very cold in January and February and weekend residents like to be able to heat their recreational homes fast and efficiently. This has led to the need for radiant heating, the sole product of Radiant Panels, Inc.

Radiant Panels began operations in 1972, making 4 × 8 foot gypsum panels which were inlaid with resistive heating wires connected electrically in series to a 220-volt source. The concept of the product, being installed mainly in ceilings, but occasionally in walls, caught on gradually and then grew rapidly as the building of new units flourished. The panels were made by cutting slots lengthwise in standard gypsum board, inserting insulated Nichrome wires in these slots, filling with gypsum slurry and adhesively bonding a new surface to the board. Nichrome wire was selected because it was a known resistance heating wire; purchase conditions specified only a certain resistance per foot.

Radiant Panels were conscious of their production costs, because the cost of purchasing and installing their product hindered the development of their market share. When the metal supplier sales representative met with Joe Daigle, the President of Radiant Panels, and recommended he buy a copper-nickel alloy wire with the same resistance per foot, which was half the price of Nichrome, Joe was delighted. Tests conducted with the new alloy wires implanted in panels were satisfactory, and therefore a limited production run was made using the new copper-nickel alloy wire.

The first hints of a problem appeared innocuous because small loops of the wire popped out of the paneling, usually near the end of the panel. Little attention was initially paid to this problem because it was only of minor, cosmetic consequence. Four months after the first installation of these special panels, however, two fires were reported, both demonstrating that fires could be initiated at the panels. Radiant Panels worked closely with fire investigators examining the remnants of panels from both fires. Evidence of several exposed small loops were found near where each fire started.

These observations led to an extensive study of the heat transfer process in the radiant heating panels. It was the metallurgist from the wire manufacturer who finally solved the problem, pointing out that the copper-nickel alloy wire was not suitable for the application. The reasoning was simple:

1. Radiant heating panels heat the room by radiation, but the panels are heated in turn by conduction from the implanted wires. Heat transfer from the wires by conduction is a function of the surface area of the wire.
2. Nichrome wires were replaced by copper-nickel alloy wires having the same resistance per foot. The resistivity of Nichrome, however, is twice that of the copper-nickel alloy. The specification could only be satisfied if the copper-nickel alloy wire diameter had been 30% smaller than the Nichrome wire.
3. Heat transfer from the smaller diameter copper-nickel alloy wire is therefore about 50% less than from the equivalent resistance Nichrome wire. Therefore, the internal temperature of the copper-nickel alloy wire was higher.
4. All metals expand when heated, with the expansion proportional to the temperature difference. When constrained at the ends, a metal which expands can fail by buckling. The copper-nickel alloy wires, expanding more because of a higher temperature, probably buckled near voids in the gypsum, forming the exposed loops.
5. The hot, exposed loops probably ignited combustible dusts which in turn ignited more substantial fuel in the room areas.

Radiant Panels has recalled all of the paneling which used the copper-nickel alloy wire, replacing the product and allowing for installation. ∎

FURTHER READING

Publications of the National Fire Protection Agency, Batterymarch Park, Quincy, Mass.
    Fire Protection Handbook, 16th Edition, 1983.
    Industrial Fire Hazards Handbook, 1st Edition, 1979.
    R. L. TUVE, *Principles of Fire Protection Chemistry,* 1976.
    NFPA 70, The National Electrical Code, 1987.
    Life Safety Code.
NATIONAL SAFETY COUNCIL, *Accident Prevention Manual for Industrial Operations,* En-

gineering and Technology, 8th Edition, Chapters 13 through 17, National Safety Council, Chicago, Ill., 1980.

Fire Investigation Handbook, National Bureau of Standards Handbook 134, U.S. Government Printing Office, Washington, D.C., 1980.

QUESTIONS

10.1. What are the four classifications of fire? List the two *best* extinguishing materials for each.

10.2. All of us have experienced extinguishing a small fire. Share that experience with us now and explain the mechanism you used in extinguishment.

10.3. Explain how simple knowledge of the fire triangle can lead to an effective fire *prevention* program.

10.4. Explain why heat of combustion is an important property of fuels.

10.5. Describe two common methods of ignition and outline a program to eliminate them.

10.6. Explain why detection is the first line of defense in a fire *protection* program. Describe the type and location of detectors you would recommend to prevent the occurrence in Case History 10.3.

10.7. Should Halon be used in an automatic sprinkler system in a computer room? Explain.

10.8. Describe the fire hazards posed in a wood furniture manufacturing firm.

10.9. Assume that you have been assigned the responsibility of a fire reconstruction. How will your knowledge of heat transfer assist in your analysis?

11

Industrial Hygiene

Safety professionals have always been interested in preventing accidents which can cause injury or damage to property. Recent trends have shifted the emphasis from this traditional safety viewpoint to a more comprehensive program which includes long-term occupational health issues as well. The impetus for this change has been fostered by legislation such as the OSHAct, which specifically established the National Institute for Occupational Safety and Health (NIOSH) to conduct research on occupational health issues, the Right-to-Know Laws, the Toxic Substances Control Act, and even products liability cases, plus increased public awareness and knowledge of industrially related diseases. Occupational health programs have predominantly been centered in medical departments, supported by industrial physicians, industrial hygienists, nurses, toxicologists and health physicists. Most safety professionals are already involved in some aspects of industrial hygiene, through ergonomic studies or through control administration, but we should be trained more thoroughly to satisfy our expanded roles in today's safety programs.

Industrial hygiene is the science and art of identifying and evaluating environmental hazards present in the workplace which may cause sickness, impaired health and well being, discomfort and inefficiency among workers or the community, and devising means to control or eliminate these hazards. This chapter looks at basic anatomy and physiology to show how hazards can be introduced to and affect our bodies, and then examines one example of each of the industrial hygiene hazard classifications: chemical, physical, biological and ergonomic.

An estimated 500,000 new work-related illnesses occur annually, nearly half of which are skin disorders, a third of which are cumulative trauma, hearing or respiratory disorders, and the balance more serious diseases such as cancer, neurologic disorders and coronary artery heart disease. The relative importance of skin disorders is exaggerated because they are easy to recognize. Many more serious health problems are not recognized as work-related, often because of the long time between exposure and diagnosis.

ROUTES OF ENTRY OF FOREIGN SUBSTANCES

The Respiratory System

Toxic substances in the workplace take many forms (Table 11.1) and can be introduced to the body by three main routes: inhalation through the respiratory system, absorption through the skin and ingestion by means of the gastrointestinal tract. The most important method, however, is inhalation. The respiratory system (Fig. 11.1) involves the delivery of oxygen to the body and removal of carbon dioxide from the body. It includes air passages such as the nose, mouth, throat, trachea and bronchi, the lungs where oxygen and carbon dioxide exchange takes place, and the diaphragm and chest muscles which provide normal respiratory movements. Air enters the nose through the nostrils where it is filtered, warmed and moistened before it enters the pharynx or throat on its way to the trachea or windpipe. The trachea divides into two bronchi which lead to the two lungs. In doing so, the bronchi divide numerous times into bronchioles which lead into small air sacs called alveoli. Exchange of gases is made with tissue blood across the cell membranes which make up the walls of the alveoli.

The respiratory tract is very efficient at removing particles from the air we breathe. Millions of tiny hairs called cilia line the tract, but how and where deposition occurs is controlled by inertia, gravity and kinetic motion. When these

TABLE 11.1 Substances Commonly Inhaled

Dusts	Airborne solid particles generated by grinding, crushing, handling, etc., which are 0.1–25 μm in size. Most dusts are larger than 5 μm and normally precipitate rapidly.
Fumes	Condensate of vaporized solids, usually oxidized, produced by welding, melting and other operations. Particle size is usually less than 1 μm.
Smoke	Solid carbon or soot particles, less than 0.1 μm in size.
Aerosols	Liquid droplets or solid particles fine enough to remain in air for extended periods without precipitation.
Mists	Suspended liquid droplets generated by atomizing or by condensation of liquid vapors.
Gases	The highest energy state of matter which can exist with no restrictions to form or shape.
Vapors	The gaseous form of substances which are normally solid or liquid at room temperature. The vapor pressure increases as temperature nears the boiling point.

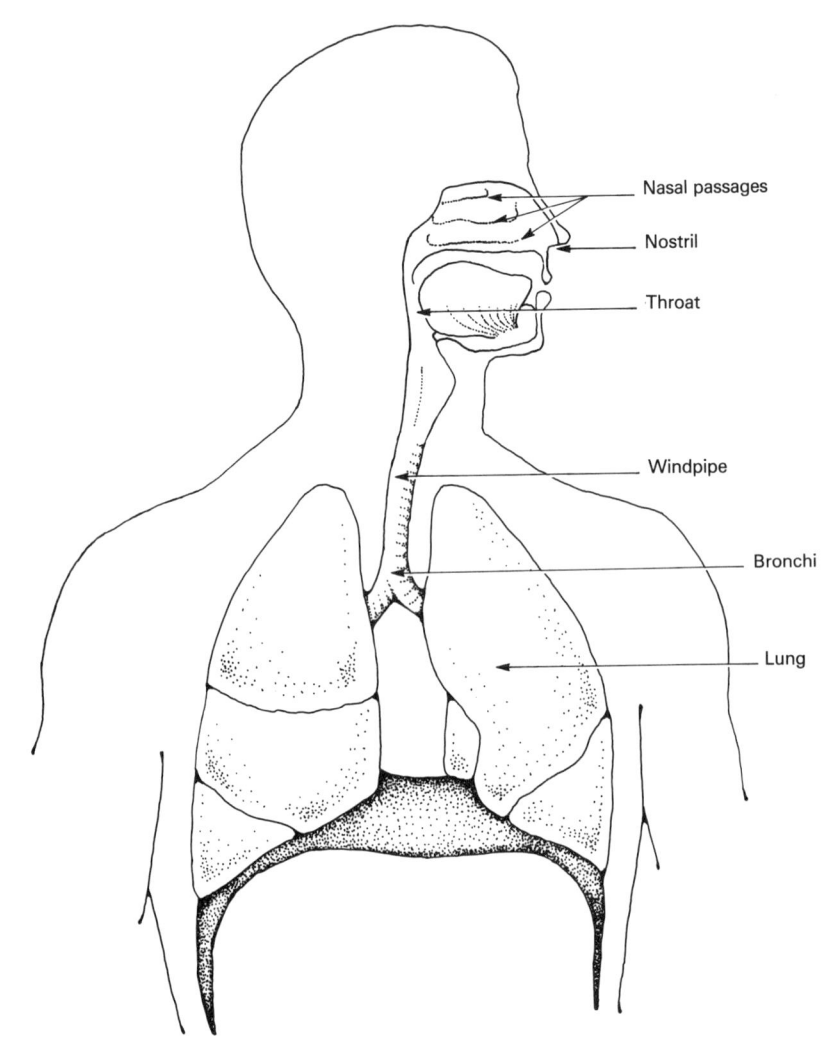

Figure 11.1 Respiratory tract.

forces cause a particle to strike the cilia or alveoli surface, it is deposited. Three factors affect the location and extent of particulate deposition: particle size and shape, the branching of the respiratory tract, and breathing pattern. Figure 11.2 compares the location of particle deposition with the types of particles deposited. Most large particles are deposited in the upper respiratory tract, with only smaller particles, less than 5 μm in size, reaching the alveoli. Those particles which are deposited in the ciliated tract are carried upward by the wave action of mucous-coated cilia, known as the mucociliary escalator. The smallest particles are cleared by the bloodstream, but those in the range 0.5–5.0 μm which deposit in the alveoli

Routes of Entry of Foreign Substances 227

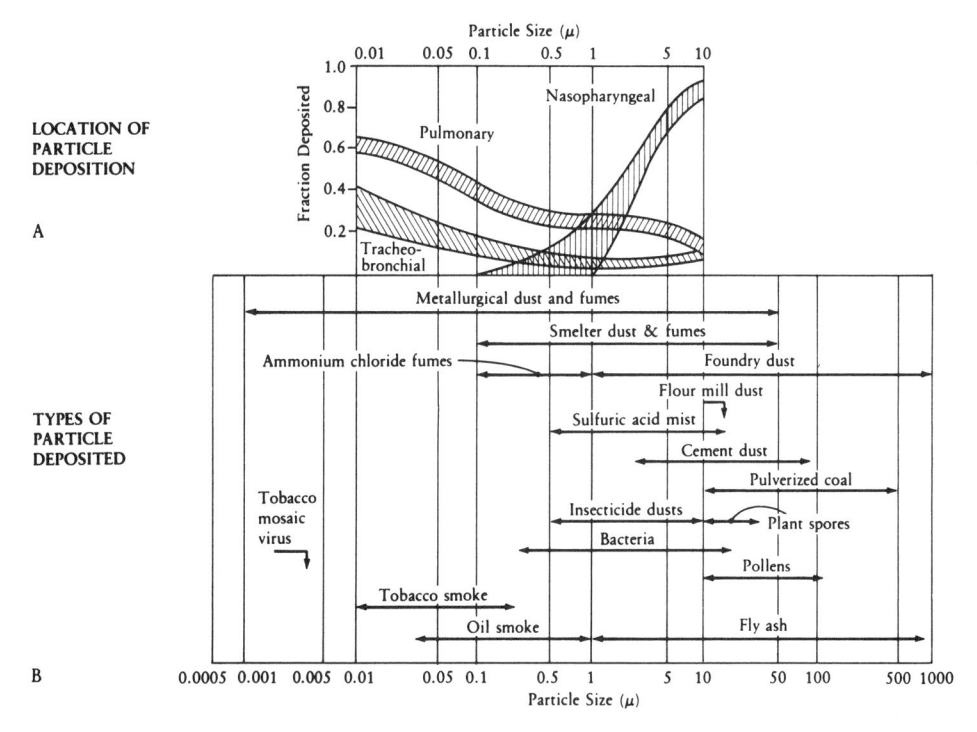

Figure 11.2 Particle size of commonly inhaled matter and location of particle deposition for various particle sizes. (Reprinted by permission of Little, Brown and Co., adapted from J. D. Brain and P. A. Valberg, *Am. Rev. Resp.. Dis., 120*, 1325, 1979.)

or bronchioles can persist, the first step in the development of pneumoconiosis. Pneumoconiosis, like asbestosis, is restrictive respiratory disease which reduces lung volume. Occupational asthma, on the other hand, is considered an obstructive respiratory disease because the air flow is interfered with, but there is no change in lung volume.

Particles are not the only substances inhaled which can cause work-related respiratory problems. Gases, vapors and even mists can damage the respiratory tract or can pass from the lungs to other parts of the body by means of the bloodstream. Irritant gases, such as ammonia, can dissolve in the moist lining of the upper respiratory tract and produce immediate irritation and tissue damage, while less soluble irritant gases such as nitrogen dioxide reach the bronchioles and alveoli where they dissolve slowly and may cause pneumonitis, and even emphysema after continuous exposure. Other gases which are not irritants, and are not soluble in water but are highly fat soluble, can cross from the alveoli into the bloodstream and to other parts of the body. This is an important means of introducing toxins in the workplace because of the variety of substances involved and the spectrum of acute and chronic disorders that may result.

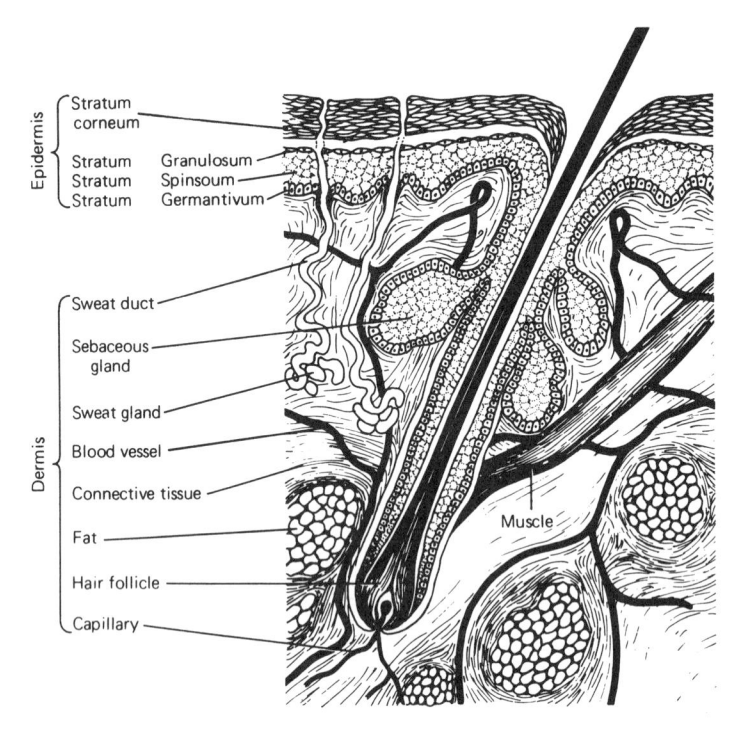

Figure 11.3 Cross-section of human skin. (From J. Doull, C. D. Klaasen, and M. O. Amdur, *Cassavette and Doull's Toxicology: The Basic Science of Poisons*, reprinted with permission of Macmillan Publishing Co., New York.)

Skin Absorption

The second most important method by which toxins can be introduced into our bodies is by absorption through the skin. The skin (Fig. 11.3) consists of three layers—the epidermis, dermis and subcutaneous fat—plus glandular structures. It is the epidermis which provides most resistance to absorption, but because of its large surface area, transepidermal absorption remains the method by which most absorption takes place. Percutaneous absorption is affected by a number of factors, such as fat solubility of the toxin, skin wetness, and prior damage. Of course, absorption through the skin can occur much more rapidly if the skin is cut or abraded.

Occupational skin diseases include contact dermatitis (inflammation), infections, pigment disorders and neoplasms, both benign and malignant. At least 200,000 working days are lost each year because of occupational skin disorders, with the high-risk industries being food preparation, leather finishing, chemicals and rubber. The most prevalent skin disorder is contact dermatitis, accountable for about 90% of all industrial cases. This disorder is characterized by redness, swelling, exudation and scabbing, and can result from primary irritant exposure or to sensitization brought about by long-term continued exposure. The toxins which cause contact dermatitis can be classified as follows:

- chemical, e.g. soaps, solvents, plastics, acids, alkalis;
- physical, e.g. humidity, heat, sunlight;
- mechanical, e.g. abrasion, pressure;
- plant, e.g. poison ivy, poison oak.

Gastrointestinal Introduction of Toxins

The ingestion of hazardous substances is not usual in the workplace. This is true not only because of personal hygiene, such as not eating in the workplace, but also because dilution of any ingested toxins by gastric and pancreatic juices tend to detoxify the substances. The most serious exposures, therefore, are slow accumulative actions of heavy metal poisons.

LONG-TERM MEDICAL DISORDERS AND EPIDEMIOLOGY

Some toxins produce an immediate reaction in our bodies (e.g. a chemical burn), but many toxins are taken up by the bloodstream and transported to other organs of the body where they may react. The destination of a particular toxin is controlled by how it can be distributed, largely determined by its ability to cross various membranes. Many of the toxins are transformed metabolically, often in the liver, and excreted with no effects.

This normal course of introduction, distribution, metabolic transformation and excretion can be interfered with by storage or immediate biologic reactions. It is at those sites where toxins have a high affinity and accumulation occurs that toxic action may occur. For example, carbon monoxide has a very high affinity for hemoglobin and reacts with it to form carboxy hemoglobin, which reduces the oxygen-carrying capacity of the blood. More commonly, however, a toxin is stored at a site where it is slowly released into the bloodstream long after exposure has ceased. Lead, for example, is stored in bone, but exerts its toxic effect on various soft tissues.

The complicated cycle of toxic entry, distribution, metabolism and storage or excretion makes hazard identification extremely difficult for industrial safety programs. The problem is exacerbated by the wide variation of susceptibility of individuals to particular toxins. Many disorders, such as contact dermatitis, are easily recognized, and normal hazard identification procedures can ascertain the hazard and means to control it. Others, such as asbestosis, are identifiable by clinical evaluations combined with normal hazard identification procedures.

However, there are a number of work-related diseases which are not recognized because physicians are unable to assess the cause–effect relationships due to the long time between exposure and onset of symptoms. In such cases, data collection systems cannot provide assistance to guide an industrial safety and health program. Fortunately, clinical medicine can be effectively combined with epidemiology to address occupational health problems. Epidemiology is the statistical study of the occurrence and distribution of disease in a human population. Although epidemiologic studies are susceptible to attack because they depend upon the quality and nature of the exposure data as well as the information collected, they have identified many industrial toxins. For example, the epidemiological data in Table 11.2 identified vinyl chloride as a significant carcinogen. Those workers exposed to vinyl chloride experienced very high levels of liver cancer, and significantly higher levels of brain and lung cancers than the average.

Epidemiology has also been used to identify such toxins as carbon disulfide, a solvent which can lead to coronary artery heart disease as a result of reduced blood flow to the coronary arteries; coronary artery heart disease is characterized by angina and irregular heart beats. Neurologic disorders such as numbness, motor weakness and muscular atrophy have also been identified by combining epidemiology with clinical studies.

How can we incorporate all this into a *safety and health* program? Much of the manifestation of occupational illness is new to safety professionals, but this should not prevent us from communicating with medical and industrial hygiene

TABLE 11.2 Observed and Expected Deaths in Vinyl Chloride Workers

Cause of Death	Observed	Expected[a]	Ratio Observed/Expected
All causes	161	161.0	1.0
All cancer	41	27.9	1.5
Digestive	13	8.3	1.6
liver and biliary tract	8	0.7	11.4
Lung	13	7.9	1.6
Brain	5	1.2	4.2
Lymphatic and hematopoietic	5	3.4	1.5
Other	5	7.1	0.7
CNS vascular	8	9.5	0.8
Circulatory	66	68.5	1.0
External	22	24.5	0.9
Suicide	10	5.3	1.9
All other	24	30.5	0.8

Source: R. R. Monson, J. M. Peters, and M. N. Johnson. Proportional mortality among vinyl chloride workers. *Environ. Health Perspect.* **11**, 75, 1975.

[a] Expected numbers based on age-time-cause-specific proportional mortality ratios for U.S. white males.

specialists or from contacting NIOSH to discuss any potential exposures within the workplace. Incorporating industrial hygiene into safety programs is possible if hazard identification procedures are followed, and management is innovative and committed.

Singer has proposed that the nervous system should be used for the early detection of work-related toxic exposures. Many chemicals which are carcinogenic or related to coronary artery heart disease are also neurotoxic. The rapid onset of neurotoxicity is actually beneficial for the detection of a disease before it becomes chronic or even fatal. Singer's recommendations include a pre-employment neuropsychological examination to determine the baseline nervous system function, periodic re-examination, and the education and treatment of workers with neurotoxicity symptoms.

Industrial hygiene hazards have been classified as chemical—used to exemplify how occupational disease is manifested—physical, biological and ergonomic. The remainder of this chapter will focus on an example of each of these categories to illustrate their importance to industrial health and safety programs:

- Chemical hazards include all the particulates, vapors, gases and mists which can be introduced into the body. The widespread use of chemical solvents makes them suitable for study.
- Ergonomic hazards include improperly designed tools, improper lifting, repeated motions while in awkward positions and psychological stress. Although cumulative trauma disorders such as carpal tunnel syndrome account for about 15% of occupational illnesses, psychological stress will be developed because it relates to all levels of workers.
- Physical hazards include temperature, pressure, electromagnetic and ionizing radiation and noise. Because of the extensive exposure to noise in many industries, this will be the topic to be discussed.
- Biological hazards include insects, molds, fungi and bacterial contamination, including food handling, sanitation, personal hygiene, sewage and hazardous industrial waste. Although hazardous waste may be considered by some to be more chemical than biological, it will be developed because of its topical importance.

INDUSTRIAL SOLVENTS

A solvent is any material which is used to dissolve another material—the solute—to form a solution. We usually think of liquid solvents for industrial purposes, but solid and gaseous solutions also exist. Liquid solvents are used widely in the home as well as in industry for cleaning and polishing. However, in industry, they also assist in extraction processes, degreasing and in the manufacture and use of such items as paints, waxes, textiles, electronics, adhesives and plastics.

Aqueous solvents, which are water-based, generally have a low vapor pressure, and therefore they are not readily inhaled, and long-term health problems do not usually result. Acids, alkalis and detergents are common aqueous solutions, and the major hygiene hazard associated with them is skin disorders, such as common "dishpan" hands, and the more serious problems of acid or caustic burns.

Organic solvents present a more difficult hazard potential because their vapor pressures are significantly higher and they can be introduced into the body by inhalation as well as by ingestion or absorption. Organic solvents are carbonaceous linear or branched molecular compounds, e.g. alcohol, turpentine, acetone, 1,1,1-trichlorethane and freon. The systemic toxicity of these solvents depends on their molecular structure. For example, refined petroleum products used as solvents are linear or branched chains of hydrogen and carbon atoms and typically cause only dermatitis. When a halogen replaces hydrogen in the chain or when a benzene ring is in the molecular structure, a different picture emerges. Carbon tetrachloride damages the liver and kidneys and benzene is known to cause bone-marrow cancer.

However, it should be recognized that there are solvents (as well as gaseous toxins) whose vapor concentrations do not cause any discomfort, ill-effects or systemic damage to most workers. These limits have been termed Threshold Limit Values (TLVs) by the American Conference of Governmental Industrial Hygienists who review and update the latest information on an annual basis. TLVs have been established as guidelines and do not mean that workers exposed to higher levels will always suffer adverse health effects. What can be assumed, however, is that continued exposure to levels *below* the TLV will probably not cause any ill-effects to most workers.

There are three categories of TLVs which can be specified: the time-weighted average (TLV-TWA), the short-term exposure limit (TLV-STEL) and the ceiling exposure limit (TLV-C). For some substances, such as irritant gases, only the ceiling value (i.e. the concentration which should never be exceeded) is important. The time-weighted average TLV is defined as the concentration to which nearly all workers may be exposed without adverse effect for a normal 40-hour week. The short-term exposure limit TLV is the maximum concentration to which workers can be exposed continuously for 15 min without suffering from irritation, chronic or irreversible tissue change or narcosis.

Solvents also present fire hazards and hazards associated with chemical reactivity. Chemical reactivity can include such effects as decomposition, corrosivity and chemical incompatibility. For example, oxidizers release oxygen upon reaction with hydrogen or chlorinated hydrocarbons form hydrochloric acid when heated. Table 11.3 lists some of the important properties and hazards of some typical organic solvents.

Industrial Solvents

TABLE 11.3 Typical Solvents

Solvent	Boiling Point (°F)	Flash Point (°F)	Health-Flammability Reactivity	Flammable Limits Lower	Flammable Limits Upper	Evap. Rate (Ether=1)	Specific Gravity	TLV TWA	TLV STEL
Acetone	133	−4	1-3-0	2.6	12.8	1.9	0.79	750	100
Benzene	176	12	2-3-0	1.3	7.1	2.8	0.88	10	—
Carbon disulfide	115	−22	2-3-0	1.3	50	1.6	1.26	10	—
Chloroform	142	n.a.	2-0-0	—	—	2.2	1.50	10	—
Ethylene glycol	387	232	1-1-0	3.2	—	—	1.12	100	12
Isopropyl alcohol	180	53	1-3-0	2.0	12.0	7.7	0.80	400	50
Methanol	147	52	1-3-0	6.7	36.0	5.2	0.80	200	25
MEK	176	16	1-3-0	1.7	11.4	2.7	0.80	200	30
Methyl chloroform 1,1,1-trichlorethene	165	n.a.	—	—	—	2.7	1.32	350	45
Tetrahydrofuran	147	6	2-3-0	2	11.8	2.0	0.90	200	25
Toluene	232	40	2-3-0	1.2	7.1	4.5	0.90	100	15
Trichloro-trifluoro ethane (Freon TF)	1/8	n.a.						1000	125
Turpentine	325	95	1-3-0	0.8	—	3.5	0.86	100	15

Hazard Assessment

There are four major sources of contaminants in any industrial operation. In most instances, they are a raw material or process chemical, but they can also be intermediate or final products. These sources cannot be eliminated but the other sources, undesirable chemical reactions and inadvertent breakdown of chemicals, can be eliminated through proper processing procedures. The identification of contaminants is not easy, however. For example, material safety data sheets have to be filed and made available for all hazardous chemicals which are used. Such records are kept on computer by many companies because literally thousands of hazardous chemicals have to be controlled. Nevertheless, contaminants from raw materials and final products are easier to control.

Hazard identification in contamination cases where an undesirable chemical reaction or the inadvertent breakdown of chemicals occurs is much more difficult to diagnose. Common contaminants from undesirable chemical reactions include carbon monoxide, which is formed by incomplete combustion, and toxic zinc oxide fumes generated by welding galvanized steel. Perhaps the best known example of chemical breakdown, however, is the production of phosgene gas from chlorinated hydrocarbon solvents by ultraviolet light from arc welding in proximity to the solvent.

Hazard identification must begin with our experiences, familiarity with testing procedures and good reference materials. However, we shall look at testing procedures first because they are quite different from those we have been working with to date. The concentration of gases or vapors in air can be determined readily by direct-reading instruments or by more sophisticated laboratory analysis when time permits and accuracy is necessary. The main requirements for direct-reading equipment are that they are accurate, particularly at small concentrations, portable and capable of measuring instantaneous and cumulative dosages. The principles upon which most instruments are based include colorimetry, electrochemical oxidation, galvanic sensing and catalytic oxidation. Each type depends upon the introduction of a certain volume of gas, its reaction and the results. Instruments are specific, i.e. they only measure a particular gas, and they are expensive, and so it is important to find further monitoring functions for them than just hazard identification.

One of the common methods for identifying unknown contaminants is colorimetry. Colorimetric devices are hermetically sealed tubes containing porous solid granules impregnated with a reagent which reacts with a specific gas and changes color. The ends of the tube are broken and a specific volume of gas is pumped through the tube, and the length of stain formed in this time is related to the concentration of the gas. These are inexpensive and not re-usable. When combined with an analysis of the process, it can point to problem areas. Let us look at a case which illustrates how difficult hazard identification is when the source of contamination is an undesirable chemical reaction.

Case History 11.1 A Case of Worker Comfort

Gunnison Fabrications, Inc. is a family-owned and -operated business which produces alloy steel shear blades for textile customers. Most of the 12 employees shape and form ¼-inch alloy steel plate by heating it with an oxyacetylene torch, bending it, and then grinding the edges. Each shear is normalized in a gas-fired furnace, and quenched into a gas-fired salt pot for tempering.

Mr. Gunnison, the sole proprietor, had incorporated many safety features. The plant (Fig. 11.4) was heated by a hot-water system and radiators were located on the outer walls. In the cold New England weather, though, the work environment was still not very warm. To compensate for this, Mr Gunnison had the ductwork above the heat-treating furnaces extended to blow the heated air into the work area *without the benefit of a heat exchanger*. Once the heat-treating furnaces had reached their working temperature, the external flue was closed and warm air was blown into the work area.

Occasionally in New England, there are very cold spells. During one of these, shortly after the ductwork was installed, four workers were taken ill, complaining of headaches and dizziness; a fifth collapsed in the rest room and had to be hospitalized for what was diagnosed as a heart attack. The symptoms suggested potential carbon monoxide poisoning and colorimetric CO analysis tubes indicated CO levels approximately 4–5 times the TLV (50 ppm).

Carbon monoxide is one of the most insidious and most feared toxic gases encountered in the industrial environment. Mr. Gunnison immediately discon-

Industrial Solvents

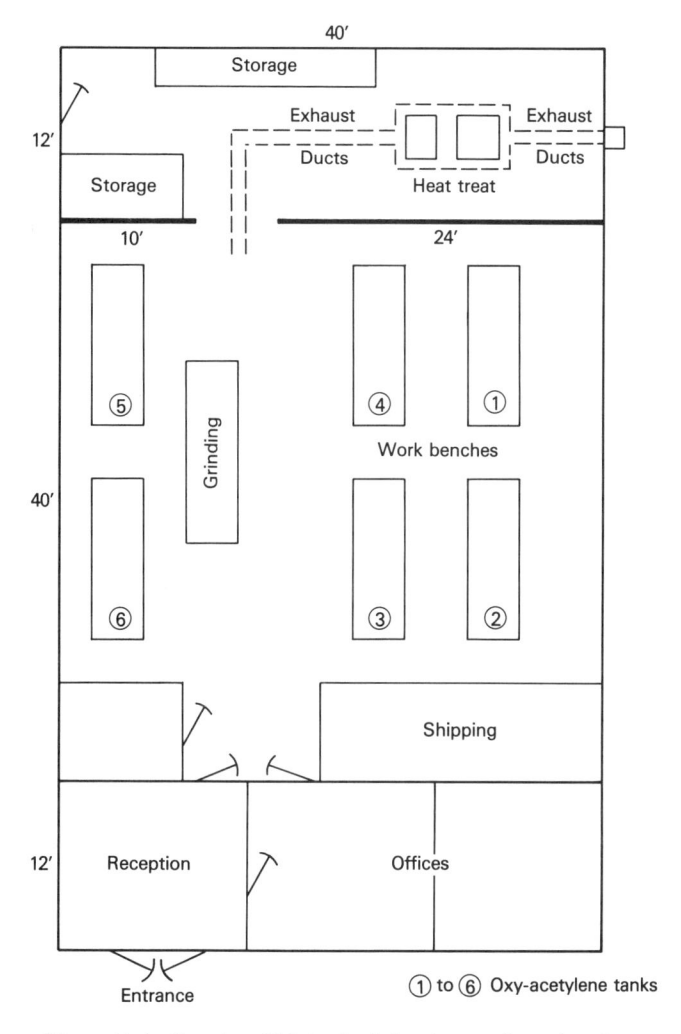

Figure 11.4 Gunnison Fabrication's Inc.'s manufacturing plant.

nected the ductwork and shut down operations until the source of the CO was identified. Evacuated air sampling devices were used to obtain representative instantaneous samples at work stations, in the heat treat area and at the natural gas heat source for the heat-treating furnaces. These were later analyzed in a mass spectrometer. The results were surprising, for carbon monoxide levels at all positions were below the TLV and no CO was identified in many of them. However, many of the samples showed oxygen depletion, the lowest being 18.5%. It was realized therefore, that dizziness and headaches are symptoms of oxygen depletion as well as carbon monoxide poisoning.

These test results led to a re-examination of the plant lay-out and processes and the problem became clearly defined. First, make-up air to the work area is limited, particularly in cold weather. Secondly, the workers all use oxyacetylene torches to heat the stock—two-thirds of the oxygen used in oxyacetylene torch applications comes from the atmosphere. Thirdly, the heated air blown from the heat-treating area was oxygen-depleted because of the combustion process. ■

Control Remedies

Once hazard identification has been made and the exposure potential has been evaluated, control remedies need to be selected and applied. The most effective methods, of course, involve source control. Substitution of a less hazardous solvent, for example, is the easiest and most effective form of hazard control. Process control, such as wet grinding, can also reduce the amount of dust contaminants which are released.

Administrative controls have special significance, not only from the viewpoint of job rotation, but in the matter of training—training in safe processing, the proper use of personal protective equipment, and right-to-know training. The American Welding Society, for example, conducts periodic seminars on all aspects of welding safety, yet too few companies have their welders participate. More than one welder has died from complications due to fume exposure, caused unnecessarily by using a torch which was too short or not wearing a respiratory mask, although heat and humidity exacerbate fume exposure.

The most common hazard controls for airborne contaminants and gases or vapors involve the pathway and personal protective equipment. Isolation, barriers and enclosures are pathway controls which are important but have limited application. Ventilation is a popular method for reducing or eliminating worker exposure, two types of which we should be familiar with—general or dilution ventilation and local exhaust ventilation.

When contaminants are released in the workroom, they diffuse according to their specific gravity and become mixed with the air, thereby becoming diluted. Either natural air make-up in rooms or fans can be used, but for dilution ventilation to be an effective contaminant control, the following criteria must be met:

- Only small quantities of contaminants are released and at uniform rates.
- Workers must be far enough from the source so that dilution is kept at safe levels.
- Contaminants must be of low toxicity and not be a fire hazard.
- Exhaust to the community cannot be hazardous.
- Dilution should not lead to corrosion or other problems.

The major disadvantages of dilution ventilation is that large amounts of make-up air might be necessary and that worker exposure near the source before dilution is sufficient may be unavoidable.

Industrial Solvents

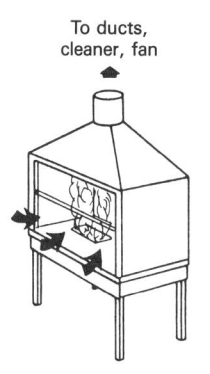

(a) Enclosures - contain contaminants released inside the hood

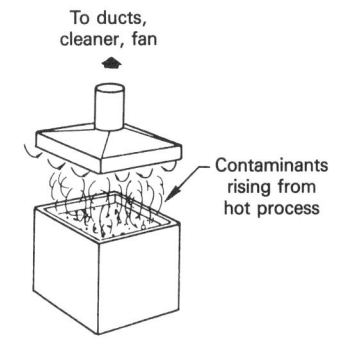

(b) Receiving hoods - catch contaminants that rise or are thrown into them

(c) Capturing hoods - reach out to draw in contaminants

Figure 11.5 Three major hood types. (Reproduced, with permission, from H. J. McDermott, *Handbook of Ventilation for Contaminant Control*, Stoneham, Mass.: Butterworth Publishers, 1976.)

Local exhaust ventilation systems capture air contaminants before they are released into the environment and remove them. This is in contrast to dilution ventilation where there is no reduction in the amount of the contaminants which are released. Local exhaust systems are used for removing and collecting particulates in grinding and woodworking operations, for example, or for removing and controlling acid fumes or solvent vapors. It is important that systems are properly designed, installed, and maintained so as to ensure that contamination is properly eliminated. Local exhaust systems consist of four parts:

- Hoods (Fig. 11.5) are the most important part of the system, for this is where the contaminants are captured or retained by flowing air currents. Enclosure hoods contain contaminants released within the hood, whereas receiving hoods catch rising vapors and capturing hoods reach out to draw contaminants into them.

- Ducts are the network of pipes which carry the contaminants through the system.
- Air cleaners or scrubbers remove particulate matter which cannot be discharged into the atmosphere. Cleaners for gaseous and vapor contaminants are also available.
- Fans provide the energy to provide movement of contaminants through the system.

Local exhaust ventilation systems work because there is a pressure differential between the high pressure at the hood entry and the low pressure generated by the fan within the ducts. The fan has to develop enough pressure to pull the specified amount of air into the hood and through the system to the point of exhaust. The total pressure in the system is made up of both static and velocity pressures and differences in either can cause air to flow. Of concern to designers, though, is a proper consideration of factors such as friction, changes in direction and design factors which cause pressure losses. When considering particulate collection, we have to remember that higher velocity pressures are needed compared with vapor or gaseous systems in order to move the contaminants. Case History 11.2 demonstrates how small oversights in design can create unexpected havoc in a plant.

Case History 11.2 A Matter of Intake

KC Industries assembles printed circuit boards and some hardware components which are used in computer numerically-controlled machining centers and in other computer-aided manufacturing equipment. After experiencing appreciable growth, they moved into much larger quarters. The building which they leased is 80,000 ft^2, open to the 30-foot ceiling in the manufacturing area. After only 8 months of occupancy, extensive corrosion was observed in the boiler room where the hot water tank had to be replaced, at the furnace which was the main heat source for the manufacturing area, and at two auxiliary gas heaters in the shipping area. The exhaust vents at roof level for the hot water heater and the main furnace were also corroded. Extensive replacement and repairs were necessary for all of these units. The plan and elevation views of the plant are shown in Fig. 11.6.

KC Industries requested Dalzell Associates to investigate the cause of corrosion. They identified potential sources of corrosive vapors at the paint spray booth, a wet grinding operation and a vapor degreaser which used 1,1,1-trichloroethane as the solvent. Containers of this solvent were used throughout the manufacturing area as well. Qualitative determination testing for chloride in the rust was conducted on-site and positive results indicated the solvent to be the most probable corrosive agent. Dalzell Associates had previous experience with heated chlorinated solvents in a dry cleaning store that released chlorine upon heating to form hydrochloric acid, which caused rapid corrosive deterioration.

Industrial Solvents

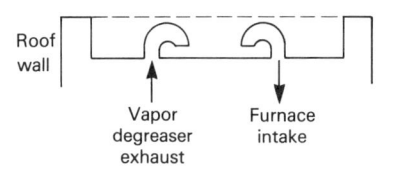

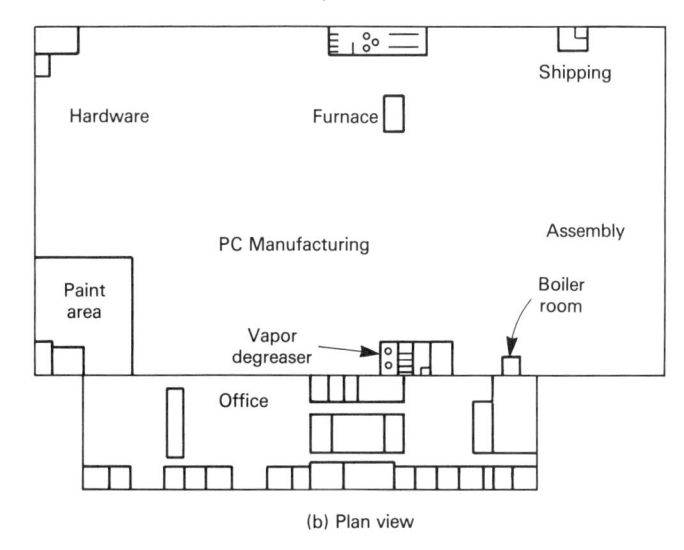

Figure 11.6 KC Industries's plant layout.

Gas samples were taken at various locations within the plant and at the roof level. Chlorides were only found in the exhaust vent above the boiler room. Pressure measurements indicated there was a slightly negative pressure in the boiler room compared to the main plant areas. Proximity and the high vapor density of the degreaser solvent indicated that some solvent vapors probably escaped from the degreaser, diffusing rapidly both because of gravity and the pressure differential to the boiler room. The problem was solved by installing a dilution ventilation intake fan in the ceiling of the boiler room. The vapors from the open solvent cans were identified as the source of corrosion of the auxiliary heaters. This was easily corrected by controlling solvent use outside of the vapor degreaser.

The source of corrosion for the main furnace was more difficult to identify because the intake air came from the vents above roof level. Only the exhaust vents, and not the intake vents, were corroded. Additional gas samples were taken and this time trichloroethane vapors were detected at the furnace intake vent, at levels only somewhat more diluted than at the vapor degreaser exhaust vent. Careful consideration of the roof design (Fig. 11.6), vapor densities and the airflow conditions around the building led to a simple, but totally effective solution. The intake vents were extended so that intake air was captured above the roof wall line. ■

Personal Protective Equipment

The control of respiratory hazards at their source, through administrative controls or by means of ventilation, is not always possible and anyone involved with such control has to be familiar with personal protective equipment. Such equipment is available for oxygen deficiency, and particulate and gaseous contaminants. Supplied air devices, of course, are necessary for oxygen-deficient environments, but air purifying devices remove contaminants from the atmosphere. Chemical respirators that contain chemical cartridges which absorb specific gases can be used for concentrations up to 1000 p.p.m., but they are not recommended for a host of gases which cannot be stopped effectively. Mechanical filters are used against airborne solid or liquid particulates which are physically trapped as air is inhaled; the particulate size is critically important when selecting the appropriate filter. To ensure that the desired protection is achieved, the correct hazard information is necessary before purchasing personal protective equipment.

STRESS AND THE WORKPLACE

One of the most important, but least understood and controlled, ergonomic industrial hygiene factors encountered is stress. Stress in this context is defined as "a non-specific body response to any demand made upon it." In other words, stress is the automatic response of the body, including the mind, to the changes, challenges and other demands which we all encounter as a part everyday life. Not all stress is bad, however, as many people believe. There are in fact, two types of stress—positive stress (eustress) and negative stress (distress). Eustress is usually experienced around changes which we look forward to, such as marriage, moving or promotion, but can also include such experiences as increased alertness and awareness of stressful situations. Distress, on the other hand, is of more concern to us because it affects both the health and well-being as well as the safety of all workers.

The effects of stress are manifested in both organic and behavioral changes. Hormones and other body chemicals are released in response to stressors, yet the symptoms exhibited might be anger, depression or simply frustration, but might also include psychosomatic symptoms, headache or stomach distress. In the extreme, stress has been linked to coronary heart disease, ulcers and premature death. We must not forget that distressed persons are also at a greater risk of accidents and injuries.

Some effects, such as the release of adrenalin or insulin, are specific body reactions to stress, but the increased flow in any case creates a demand for readjustment, a non-specific increase in the need to restore normal conditions no matter what the stressor. Dr Hans Selye, regarded as the foremost authority on stress, demonstrated that many different stressors can provoke identical biochemical reactions in the body. Certain reactions were common to all stressors, embodied in Selye's General Adaptation Syndrome (GAS) which he first described in the 1930s. GAS does not necessarily cause harmful effects; rather, the effects depend on the intensity of the stressor made on the adaptive capacity of the body. Three stages were identified as the adaptation evolves:

1. Alarm, when bodily defensive forces are mobilized.
2. Resistance, when the body fully adapts to the stressor.
3. Exhaustion, when the stressor is severe enough or lasts long enough to negate the body's coping mechanisms.

McLean has viewed the complex interactions of stress in terms of the environment or context in which interaction takes place, the vulnerability of the individual or host factor at the time of interaction, and the magnitude and type of the stressor. These three factors are dynamic, varying both in size and in time. When all interact, the symptomatic response, whatever it might be, is manifested. When vulnerability or context are absent, interaction is avoided and symptoms do not develop. In other words, any one of us can stand up to otherwise destructive stressors if the context is supportive and the vulnerability is low.

Context includes many aspects of the environment, including social, ethnic and political as well as the many aspects of the work environment. Ergonomic factors such as noise and illumination, the perception of safety programs and management attitudes toward employees, the involvement of employees and many more are important factors in the working context. We generally accept that anxiety, perceptions of job stress, and low morale are correlated. In fact, job satisfaction is the leading determinant of perceived work stress.

Work context has another very significant role in stress because work represents one of the major means by which we feel useful in life; not only do we find personal identity in work, but it serves as a prime source for social contacts. It is important to recognize that work can and often does represent a refuge from nonwork-related stress while coping defenses are being developed. Nevertheless, we traditionally tend to focus on work as a major source of stress. Studies of mortality ratios have shown that the most stressful occupations are dangerous, have conflicting demands, are subordinate or are routine, but not self-paced. Least stressful occupations, on the other hand, allow workers to control the pace of their work, have no conflict, and provide a sense of challenge and achievement.

Stressors which are present both at work and in our everyday life can be defined objectively (e.g. by outside researchers) or subjectively (as perceived by the person under stress). Objectively defined stressors which have been studied include physical noise, deadlines, responsibility, promotion or demotion, divorce and/or reconciliation. Subjectively defined stressors include role ambiguity, role conflict, responsibility, participation and relationships. Ambiguity arises when we have poorly defined task assignments, whereas conflict arises when either job demands conflict—often the case for collateral personnel who wear more than one hat—or when individual ethics conflict with organizational demands.

We should pay more attention to studies of subjectively defined stress because it is the perception of stress that is important, combined with vulnerability to stressful events and conditions. Although we are susceptible to stress regardless of age or sex, our vulnerability to stressful events continually fluctuates. However, each stage of life has its own particular problems and crises which evoke particular coping patterns. Childhood, adolescence, early adult transition, the milestone birthdays of 30 and 40, middle age, retirement and senior citizenship are stages which we all go through.

Changes in life are frequently stressful in themselves, but when several important stressors have been encountered within a short time, that person's vulnerability to illness is increased. Researchers in the 1960s asked people to assign a value to life changes, normalizing to marriage. The top 15 events appear in Table 11.4. Numerous subsequent studies using these values as predictors have been remarkable. If the annual total of the Life Change Units (LCUs) exceeded 150, almost 40% of those persons recorded an illness in the following year; when the total exceeded 300 LCUs, the percentage increased to more than 70%.

Personality characteristics also affect vulnerability to stress. Although the reactions to stress may be manifested in many ways, such as anger or withdrawal, the most definitive studies have examined the relationship between personality and the likelihood of coronary heart disease. Although an oversimplification, the early studies of Friedman and Rosenman categorized subjects as Type A—a competitive, aggressive individual who is characteristically hypercritical, both of themselves and others—or Type B—a more laid-back unflappable person. Of course, we have bits of both in us at different times and the simplistic view has

TABLE 11.4 Rank of 15 Top Life Events

Rank	Event	Life Change Units (LCUs)
1	Death of spouse	100
2	Divorce	73
3	Marital separation	65
4	Jail term	63
5	Death of close family member	63
6	Personal injury or illness	53
7	Marriage	50
8	Fired from work	47
9	Marital reconciliation	45
10	Retirement	45
11	Change in family member's health	44
12	Pregnancy	40
13	Sex difficulties	39
14	Addition to family	39
15	Business readjustment	39

been controversial as a result, but studies do consistently show that those with severe Type A behavior are at risk of coronary heart disease.

One of the perplexing problems of stress, vulnerability and the workplace involves the process of innovation, which is valued for safety management, while our natural inclination, both individually and as an organization, is to resist change. Thus the very thing that is needed within an organization to ensure success may very well be a very stressful event in itself. Careful preparation is required to minimize the stress which accompanies change. In addition, many changes at work involve loss, and whether or not any loss is real or only perceived, the reactions are likely to be the same.

Coping with Stress

There are many ways for coping or managing stress. Those of us who need help most can enter into therapy with trained psychiatrists or psychologists. Many organizations have Employee Assistance Programs (EAPs) to help employees (confidentially) under stress from chemical dependency, personal or work-related issues. When needed, EAPs can provide referrals to professionals. Many companies encourage discussion in quality or safety circles, which is helpful in coping with the stress of change in the work context. Involvement and participation, both in work planning and in after-work activities such as sports, can also be helpful.

Social support systems have always been recognized as effective coping techniques. Positive, supportive relationships with family, friends and fellow workers can provide the context in which levels of perceived job stress are reduced. But relationships and social support systems are not available to all of us and sometimes relationships we have built up are neither positive nor supportive. What then can we do, particularly in the workplace? Some safety professionals advocate education or training programs which provide a forum for discussing stress, the danger signs and how to cope with them. Discussion, in itself, is a stress-reducing technique, and it promotes self-awareness of stress. Self-awareness of trouble signs is sufficient for us to seek help when we need it. Training should also be directed toward supervisors so that they can recognize the symptoms of emotional distress. Supervisors are often the catalyst for employees seeking help in EAPs.

Other work-related stress reducers involve participation in decision making about stressors. Whenever stress can be eliminated, it should be so immediately. For example, job descriptions should include clearly defined demands and objectives. More effective planning, particularly concerning human factors, should be practiced. Emotional as well as technical or physical characteristics should be considered in people decisions whenever possible. Reward systems and participation in sports or social clubs at work, designed to make workers feel they are important members of the team, should be considered. When stress symptoms are detected early, most work-related stress problems can be resolved at the levels where they occur when listening and participative decisions are involved.

Many physical and mental exercises have been developed to increase the adaptability of essentially healthy persons to stress. One of the popular techniques in stress management programs is meditation where we sit quietly, in a comfortable position, and repeat a word known as a "mantra." The mantra can be any word, but it is used to clear the mind and tune out distracting thoughts. Progressive relaxation is a technique we can use at any time or any place. Starting at the toes, we sense a body part and then consciously release the tension from that part; we continue all the way to the scalp, remembering to keep a passive attitude. A 10–15 min routine can be very invigorating. Proper techniques, of course, must be learned, and therefore training programs are necessary.

The physiological response of our bodies to stressors has led to the development of biofeedback processes to monitor these physiological signs. By observing these signals under all conditions including stress, we develop a conscious or subconscious desire to return to normal; most of the time, we attribute this return to normal to simply thinking relaxing thoughts. The most popular biofeedback devices monitor blood pressure, but others include feedback thermometers, electromyographs to measure muscle tension, and dermographs, which measure galvanic skin response.

Whatever the method, stress relaxation or stress management techniques should become an important part of our safety programs. They can fill a wide variety of needs, from extending longevity of valued employees to just promoting self-esteem; in any event, we reap benefits in improved performance.

Case History 11.3   Distress at Evaluation Time

SuperMag is one of the many high-tech small industries spawned by scientific advances which have had financial difficulties when trying to establish themselves. SuperMag lost its backing after several years causing nearly 50% cutbacks in personnel and was forced into Chapter 11 bankruptcy. Its sponsor also removed all wire-drawing equipment used to manufacture multifilamentary superconducting wire. SuperMag received new support, however, and was awarded a large contract for multifilamentary wire for magnets used in particle physics research. In order to meet production deadlines, however, used equipment had to be located, transferred and installed. Work-related stresses were compounded when the foreman, who had over 20 years experience, decided to quit.

The combined experience of the three employees in the wire-drawing shop was less than 5 years. Plans called for three additional personnel to meet production requirements, and a decision had to be made on a new foreman.

Because of the previous foreman's work experience and the inexperience of the company, a job description for the foreman's position had not been formalized. The tasks were known, but they had not been standardized. They included chemical cleaning and clean-room handling conditions for preparation of billets, shipping and supervising extrusion of billets and initial wire-drawing off-site under contract, in-house wire-drawing to final size and solder coating the final product. His other duties were assumed rather than spelled out, such as inventory control, training, leadership, job assignments and maintenance.

With no time available to search for a foreman with experience, Don, one of the other wire-drawing shop employees, was promoted. Needless to say, he was pleased with the promotion and did perform well for some time. However, his poor preparation and lack of training, combined with role ambiguity, finally began to erode the performance of his group. Problems with equipment, then material defects, led to production delays and overtime, then to low morale and increased stress.

At review time, Don met with management to determine how to dispense available monies among the various members of his group. Because his men had fared well and he had had two substantial raises since his promotion, Don also expected a salary increase. However, he was confronted with the many problems his group had experienced instead, such as low production and his poor leadership skills. His review took nearly 2 hours, and he did not get a raise.

After lunch, which he ate alone, Don began working on the draw bench. He was observed by one manager to be angry, working hurriedly and carelessly. The manager knew the source of Don's behavior, so did not intervene, thinking the work was therapeutic. Within minutes, however, Don made a critical error, and he crushed two fingers on his right hand between the grip and die block. He eventually lost both fingers.

What could have prevented Don's accident and injury should have stimulated extensive discussion. However, in reality, there was not much change in the wire-drawing shop, except that morale was generally low. An examination of the draw bench concluded that it was not defective and that the safety microswitches had not been overridden. Don did return to work on light duty, but eventually left the company. ∎

INDUSTRIAL NOISE

The workplace is a very noisy environment and problems of communication in the presence of noise have long been recognized. Noise is nothing more than sound, albeit unwanted sound, and therefore a study of sound, its characteristics and how we hear sounds needs to be made in order to understand and control noise. Sound travels by means of electromagnetic waves, just like light waves, except that their frequency is in the audible range; testing is performed in the range 250–8000 Hertz (1 Hz is equivalent to 1 cycle per second). The speed at which sound travels is the product of the frequency and wavelength of the sound wave and is constant for the medium it travels through. For most considerations, then, we can think in terms of frequency only. Frequency determines the pitch, with low frequencies typical for vibrational sound and high frequencies typical for squeaky, high-pitch sound.

In order for sound to be heard, the intensity of the electromagnetic wave is

TABLE 11.5 The Level of Intensity of Some Typical Sounds

Sound Level (dB)	Sound
0	Threshold of hearing
20	Sound studio background
30	Soft whisper at 5 feet
40	Quiet office
50	Average home
60	Conversational speech (3 feet)
70	Freight train
74	Normal traffic (30 feet)
80	Noisy restaurant, typical factory
90	Subway
100	Textile mills
110	Punch presses, riveter
120	50 horsepower siren
140	Jet plane

also important. The sound level is expressed in decibels (dB), a dimensionless number which is a logarithmic relation of the ratio of sound wave intensity to a standard intensity:

$$dB = 10 \log I/I_0$$

The definition of dB was selected so that the weakest sound detectable by a human would be 0, yet the range could extend to the threshold of pain, i.e. 140 dB. Examples of sounds within this range appears in Table 11.5.

Hearing is a very complicated process which does not only depend on the intensity of the sound wave, but on the frequency and the hearing ability of the receiver. We hear and we perceive because we can collect sound which is converted to nerve impulses and then interpreted. Figure 11.7 illustrates the cross-section of an ear. Sound waves are gathered by the auricle and channeled through the auditory canal to the eardrum which vibrates in response to the pressure and frequency of the sound wave. The vibrations are transmitted through the middle ear by means of the hammer, anvil and stirrup that make up the ossicles to the cochlea of the inner ear. Fluid in the cochlea moves in conjunction with the stirrup vibrations, generating nerve impulses which are transmitted to the brain by means of the cochlear nerve. Pressure on either side of the eardrum is equilibrated because the middle ear is connected to the pharynx by the Eustachian tube. Semicircular canals and the vestibular nerve relate to balance and sense of motion, other functions of the ear.

We do not respond to the same sound levels for all frequencies; this is true because nerve cells in the cochlea located nearest to the middle ear are stimulated by the highest frequencies and those farthest from the middle ear are stimulated

Industrial Noise

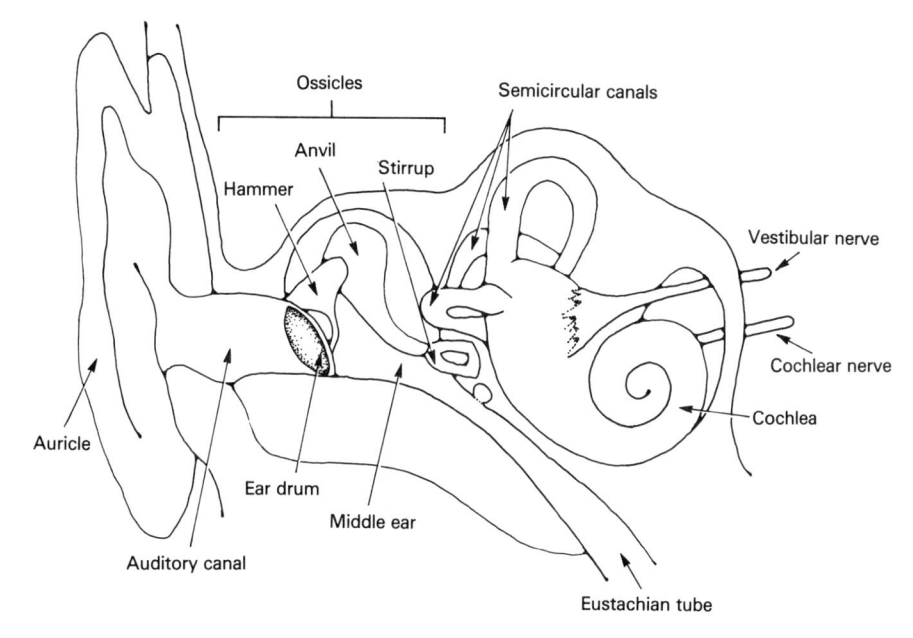

Figure 11.7 Cross-section of human ear.

by the lowest frequencies. We actually hear high frequencies better than low frequencies, and therefore a low-frequency sound must be louder to obtain the same response as a high-frequency one. Sound level meters which detect sound intensity have been electronically modified to simulate this human dependence of sound level on frequency. The A-weighted sound level measurement (dBA) has generally been adopted, since it responds in a manner reasonably similar to the human ear.

The concerns with sound in industrial safety and health programs are twofold. First, noise creates difficulties in communication and hearing can be adversely affected by long-term exposure to noise. At the same time, we must also appreciate some of the benefits of sound, such as the change in pitch which occurs just before separation in an abrasive sawing action.

Vocal instructions which are unheard or misunderstood can lead to human error and injury-causing accidents in noisy work areas. Early research on the effects of noise on vocal communication resulted in the relationships demonstrated in Fig. 11.8. This type of information is required to determine whether other forms of communication other than vocal should be developed.

Research has also been conducted to determine the effects of noise on hearing loss. For example, audiometer testing of the same sawmill worker exposed to loud work noise with no protection for 25 years showed extensive hearing loss, first noticeable at high frequencies. This behavior, shown in the composite au-

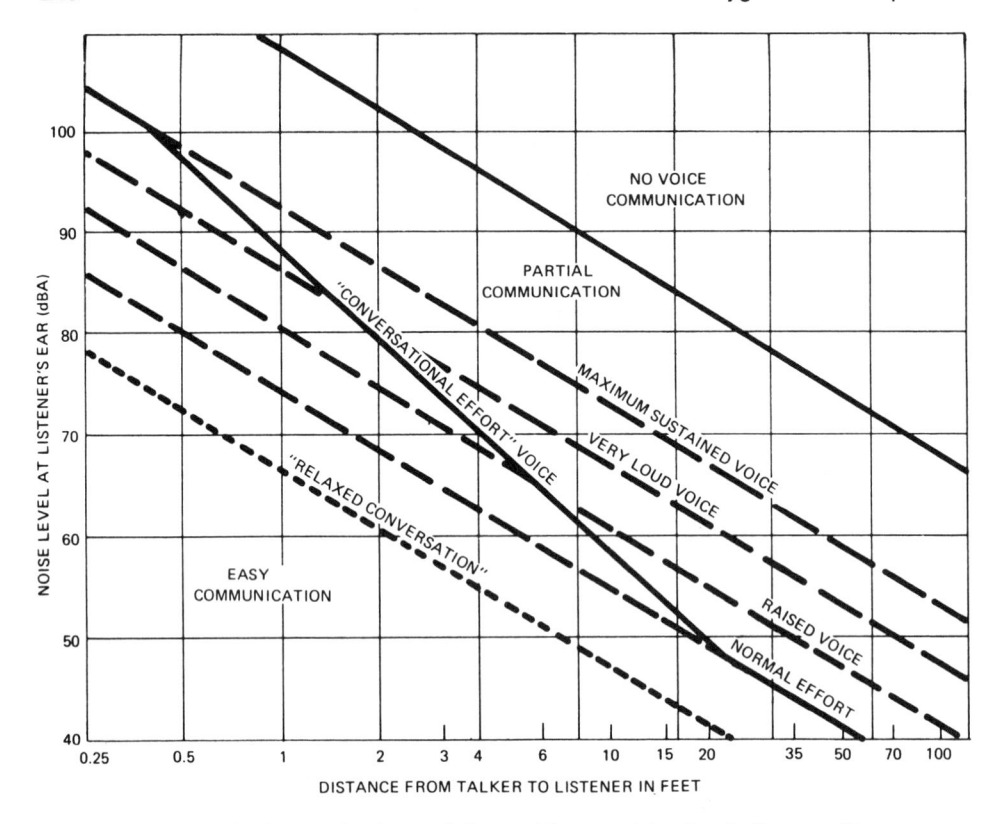

Figure 11.8 Communication as influenced by sound level and distance. (Reprinted by permission of National Safety Council, from J. B. Olishifski, *Fundamentals of Industrial Hygiene*, 1979.)

diogram of Fig. 11.9, has been well documented in all industries. Hearing loss caused by acoustic trauma is termed sensorineural because it is associated with damage to the inner ear or brain nerves. Hearing loss affecting the outer or middle ear is termed conductive hearing loss because sound does not reach the inner ear and can easily be distinguished by physicians using a tuning fork placed on the mastoid bone (behind the auricle) for bone conduction. With normal hearing or sensorineural hearing loss, air conduction is louder than bone conduction, whereas the opposite is true for conductive hearing loss.

Acoustic trauma is not the only cause of sensorineural hearing loss, e.g. aging naturally causes it. Atrophy of the brain cells and neurons of the cochlea cause an irreversible loss in the ability to transmit signals to the brain, and other physiological factors such as decreased blood supply contribute to hearing loss in the elderly. When the effects of aging in a noise-free environment and the effects of noise exposure are superimposed (Fig. 11.10), it is clear that noise exposure,

Industrial Noise

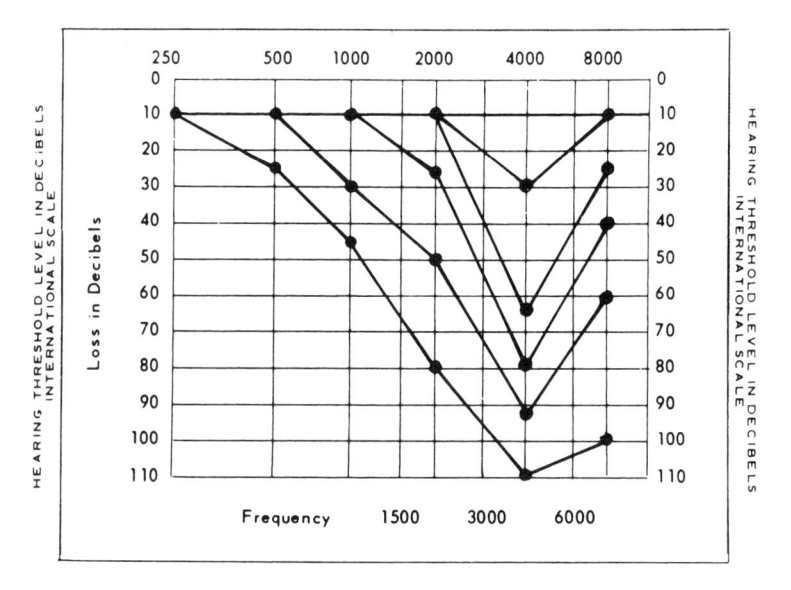

Figure 11.9 Composite audiogram of sawmill worker exposed to loud noise with no hearing protection over 25-year period. (Reprinted by permission of Year Book Medical Publishers, from C. Zenz (ed.), *Occupational Medicine: Principles and Practical Applications,* 1988.)

at least at levels of 90 dB or greater, is the major contributor to hearing impairment and that remedies must be included in safety and health programs.

OSHA Conformance Procedures

The Occupational Safety and Health Act has established limits to the noise employees can be exposed to. These regulations state that occupational noise exposure should not exceed 90 dBA for any 8-hour work period. At higher noise levels, permissible exposure time is reduced, as shown in Table 11.6.

Noise level and exposure time are both critical, but the noise level is also variable with time: many machines are not run continuously, workers move about and even tasks can change during the workshift. In order to determine the noise exposure under such variable conditions, we must determine the time of exposure to each sound level and then calculate the dosage, which is:

$$\frac{\text{Time actually spent at the sound level}}{\text{Time of permissible exposure to the sound level}}$$

Total exposure, D, for a workday is the sum of all partial doses:

$$D = \frac{C_1}{T_1} + \frac{C_2}{T_2} + \frac{C_3}{T_3} + \cdots \frac{C_n}{T_n}$$

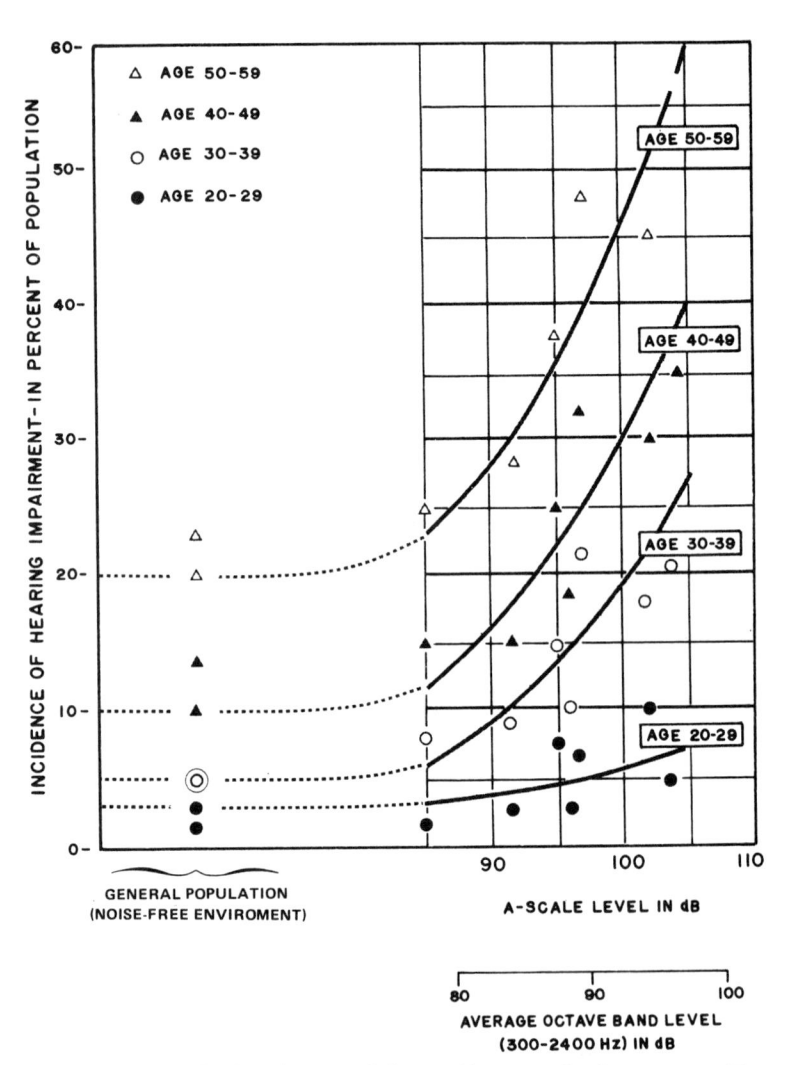

Figure 11.10 Hearing impairment as influenced by age and noise exposure. (Reprinted by permission of American Industrial Hygiene Association.)

where C is the actual exposure time and T is the permissible exposure time. Total dose should not exceed unity. If it does, steps must be taken to reduce the total exposure to acceptable levels (1 or below).

OSHA also states that action be taken when the dosage exceeds 0.5, but does not necessarily exceed unity. When the action level is reached, a hearing conservation program, noise surveys, training and recordkeeping procedures must be undertaken.

Industrial Noise

TABLE 11.6 OSHA Noise Exposure Limitations

Sound Level (dBA)	Duration Permitted (hours)
90	8
92	6
95	4
97	3
100	2
105	1
110	1/2
115	1/4

Identification of Noise Problems

Noise is most commonly measured with a sound level meter, which should be made in accordance with ANSI S1.4, specification for sound level meters, and calibrated. To conduct a survey, we should consider several factors: the layout of the area, the height at which a worker is normally positioned and the number of machines which are operational. In theory, it is possible to determine the noise emitted from a unit and calculate the noise level at a position a known distance from the unit, and combine the information from other units by using the information in Table 11.7 to obtain the combined noise level from two noise sources (easily reproduced by the definition of dB and recognition of the fact that intensity levels, not dB, can be added). Then a layout can be made based upon the emission

TABLE 11.7 Method for Determined Combined Noise Level of Two Sources of Noise

Difference Between Two Noise Sources (dBA)	Amount to Be Added to *Larger* Level to Obtain Combined Noise Level
0	3.0
1	2.6
2	2.1
3	1.8
4	1.4
5	1.2
6	1.0
7	0.8
8	0.6
9	0.5
10	0.4
11	0.3
12	0.2

of noise and dissipation as a function of distance from the source. For an ideal point source, the noise level L_p at a point r feet from the noise source L_w, is:

$$L_p = L_w - 10 \log 4 \pi r^2 - 10 \text{ dB}$$

It is more practical to survey the actual noise levels in an area because calculations do not easily take into account noise reflections from walls, ceilings, etc. Machines should be turned on and off to determine major and minor noise sources; knowledge of how to use Table 11.7 can be helpful in this task. More extensive sound measurements can be made in close proximity to noise sources *and at all frequencies* to study the internal details of potential sources (we have to remember that the A-scale is used to compensate for our hearing response, not the source emission). We must pay close attention to normal operator positions, and to changing conditions during daily routines. It is also important to consider the paths noise take, including reflections as well as direct paths from a noise source. Quite often, we can use our ears to assist in these tasks.

Noise Control

When our working environment requires us to take measures to reduce exposure to noise, there are a number of remedies that can be incorporated into a noise control program. But, first, an audiometric testing program must be implemented to establish baseline hearing for new employees and periodic re-evaluations of hearing for all employees. All managerial skills are required to prevent administrative pitfalls which can be associated with any testing program. The remedies which are available to reduce noise exposure can be categorized as source, path and receiver remedies, plus administrative remedies such as worker rotation to keep dosage of each below unity.

The best way to reduce noise is at its source. This can sometimes be accomplished by simple maintenance procedures, such as lubrication, adjusting the tension on belts or chains, or by tightening loose casings or guards. It is sometimes possible to move equipment to a more isolated location where attenuation is obtained simply by distance (sound level can be reduced by as much as 6 dBA by doubling the distance between the source and receiver). Vibrations, which are low-frequency noises, can often be damped by using shock mounts or neoprene strips. Acoustical shields can often be mounted on machines which absorb sound in the direct path between source and receiver and deflect sound in directions where reflected noise is minimized. Thus shields are more effective when the worker is close to the noise source and when the frequency of the noise is higher, i.e. absorption is greater. Sometimes, however, redesign is necessary, or other measures must be taken.

Noise reduction in the path between the source and receiver is probably the least effective remedy because of the possibility of leaks and reflections. Windows, for example, are acoustical weak spots in barriers, and ceiling reflections can become new sound paths. Generally, we have to have total enclosure if more than a 12–15 dBA reduction in noise is necessary.

Personal protective equipment is probably the most common noise remedy, although it is not the most dependable. Ear muffs, which cover the entire ear, and ear plugs, which fit into the ear canal, can provide 25 dB or more of attenuation at frequencies above 500 Hz; ear muffs are more effective than ear plugs. These hearing protectors make vocal communication more difficult in a low-noise environment, but do not impair communication in high-noise environments. The largest problem associated with hearing protection aids is getting workers to use them, because they are uncomfortable. Unfortunately, when they are more loosely fitting and more comfortable, their performance is reduced, so that only 10 dB or less of attenuation is realized.

Case History 11.4   A Punch Press Noise Problem

Brook Industries manufactures metal cabinetry for electronic applications, using large numerically controlled punch presses and smaller pneumatic punch presses to provide flat stock which is then formed in press brakes and assembled into the cabinetry. Within the punch press room are light units, an annealing furnace and a central supervisor's station. With all of the equipment turned off, the sound level at this station was less than 60 dBA, indicating no external noise sources. With the furnaces turned on, the noise level increased by about 10 dBA, which is still quite low for industrial environments. With five presses operating, however, the sound level at the station was 97 dBA and, close to the machines, it reached 110 dBA. Detailed frequency analysis from the operator's position of a pneumatic press (98 dBA) was determined and is shown in Fig. 11.11. Other pneumatic presses were found to be similar, showing increased noise at all frequencies associated with the task (stock actually being stamped) and higher frequency noises attributable to air ejection jets.

As a result of this survey, Brook personnel decided to reduce the noise level of each pneumatic press to 85 dBA at frequencies above 500 Hz. Their approach was to use a shield to reduce air noise in the path between the ejection jets and the operator and to reduce the air ejection noise at the source. Working with the manufacturer of the presses and latest methods for nozzle construction, Brook engineers devised a quiet nozzle cover and added an adjustable reducing valve to lower the air pressure to the minimum which could still enable ejection. A three-sided barrier plexiglass shield with polyurethane foam lining was assembled to fit around the die without interfering with stock progression.

When incorporated on the pneumatic punch presses, attenuations of 17–20 dBA were measured. At the supervisor's station, the noise level was reduced to 85 dBA. However, this success was short-lived, because further monitoring 2 months later showed that noise levels had increased to unacceptable levels again. An investigation revealed that the shield had become dirty and that the operators had increased the air pressure to the nozzles. Operator training and maintenance procedures were implemented and noise attenuation is now successful. ∎

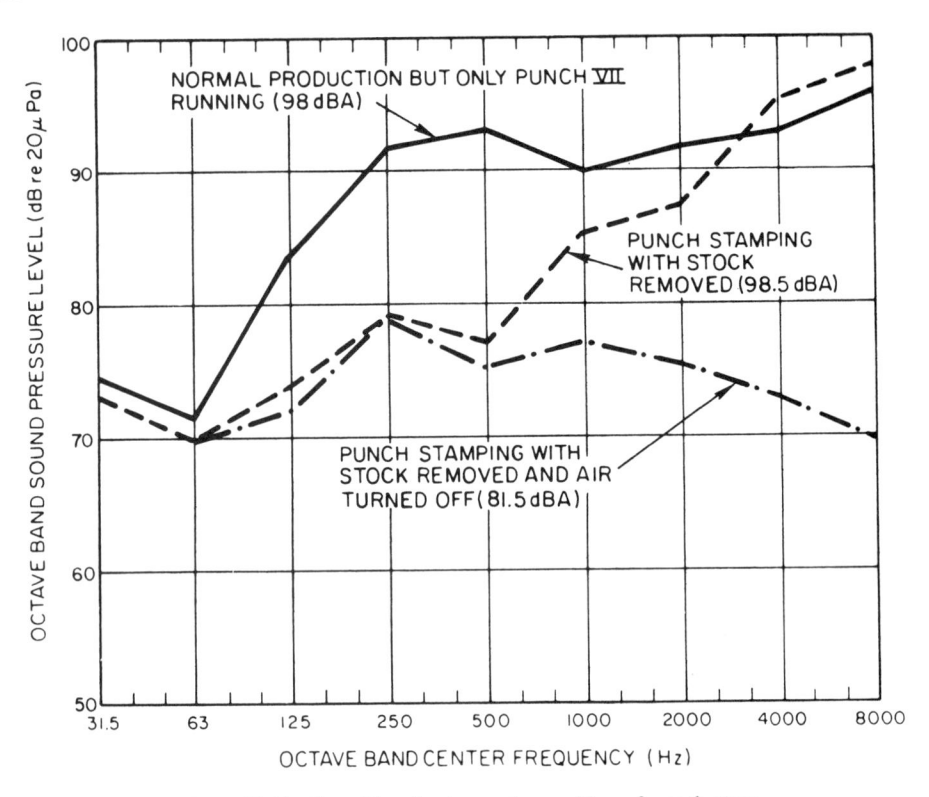

Figure 11.11 Sound levels at operator position of punch press.

HAZARDOUS WASTE

If safety professionals are to include occupational health in safety programs, they cannot ignore hazardous waste, perhaps the most significant disadvantage which accompanies today's technology. Hazardous waste is literally a by-product of the industrial world, evoking the negative effects of industrial processes. Nobody wants it, but society looks for the products which leave the waste behind. Ultimately, solutions will be political and social, but safety professionals must accept the challenge and opportunity to participate in finding the best solutions.

The term *waste* refers to material which is to be discarded because it is no longer useful, because of degradation or contamination or because it is a nonusable by-product of some primary process. Economics are important in determining what waste is, however, because reclamation can be favorable or unfavorable; significant fluctuations can occur over time. Historically, hazardous waste management has been fostered by conservationist groups interested in protecting the quality of our air and water. Controls, therefore, have focused on atmospheric emissions and wastewater discharges, thus creating incentives for land disposal, which has become associated with the major component, i.e. solid

Hazardous Waste

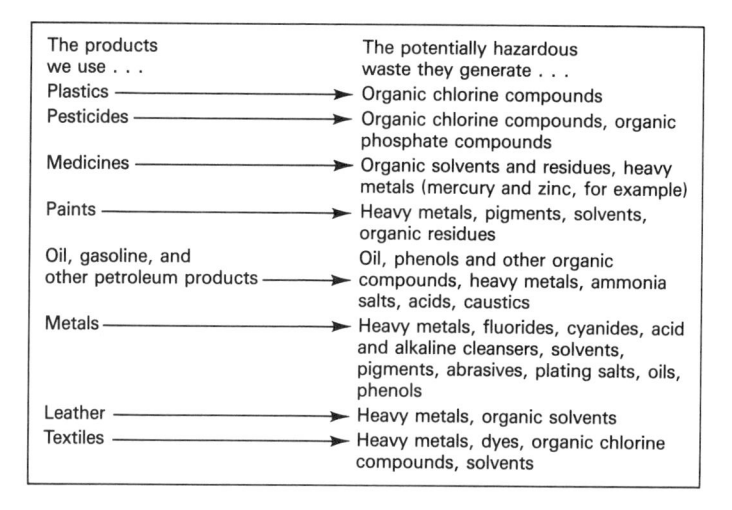

The products we use...	The potentially hazardous waste they generate...
Plastics	Organic chlorine compounds
Pesticides	Organic chlorine compounds, organic phosphate compounds
Medicines	Organic solvents and residues, heavy metals (mercury and zinc, for example)
Paints	Heavy metals, pigments, solvents, organic residues
Oil, gasoline, and other petroleum products	Oil, phenols and other organic compounds, heavy metals, ammonia salts, acids, caustics
Metals	Heavy metals, fluorides, cyanides, acid and alkaline cleansers, solvents, pigments, abrasives, plating salts, oils, phenols
Leather	Heavy metals, organic solvents
Textiles	Heavy metals, dyes, organic chlorine compounds, solvents

Figure 11.12 Some sources of potentially hazardous wastes.

waste. Thus, the accepted definition, given in the Resource Conservation and Recovery Act (RCRA) of 1976 is:

Hazardous Waste—a solid waste that may cause or significantly contribute to serious illness or death, or that poses a substantial threat to human health or the environment when improperly managed. Solid waste means any garbage, refuse, sludge and other discarded material, including solid, liquid, semi-solid or contained gaseous material.

There are two main categories of hazardous waste: listed wastes and characteristic wastes. Listed wastes are those wastes from specific or non-specific sources, unacceptable chemical formulations or acutely hazardous wastes, e.g. the sludge after treatment of spent electroplating baths. Characteristic wastes are those which are identifiable by their flammability, corrosiveness, reactivity or toxicity. Any material with a flash point less than 140°F, a pH < 2 or pH > 12.5, reacts spontaneously or violently with water or air, or has a concentration of toxic compounds above the TLV, are characteristic hazardous wastes. Figure 11.12 illustrates some of the potentially hazardous wastes generated in the manufacture or use of products we use.

Hazardous Waste Management

Regulation. Waste management is the term used to describe the full range of activities that go with custodianship and disposition of wastes from the point of origin to that of final disposal. The first major federal legislation which was directed specifically at hazardous waste management was the Solid Waste Disposal Act of 1965. This Act recognized that a national effort was needed to coordinate an effective waste management program, but that continued local or state regulations, supported by federal grants, funds and assistance were still necessary. Federal involvement was expanded in the Resource Recovery Act of 1970, but the shift to federal control of solid waste management was clearly defined by the Resource Conservation and Recovery Act (RCRA) of 1976. RCRA established methods for the identification and listing of hazardous waste, and standards applicable to generators, transporters and owners and operators of hazardous waste treatment, storage and disposal facilities.

In 1980, the Comprehensive Environmental Response, Compensation and Liability Act (CERCLA) established superfund, a pool of money generated by taxes to insure for remedial action in response to the release of hazardous materials. CERCLA also had a second fund for remedial action and damages that might occur from sites operated and closed *in compliance* with RCRA. Of course, there are many overlapping laws which interface with hazardous waste management, such as the Toxic Substance Control Act, Clean Air Act, Hazardous Materials Transportation Act, and others. These laws together are intended to achieve comprehensive national environmental control. This means "cradle-to-grave" control of hazardous materials.

Figure 11.13 summarizes the relation of hazardous and toxic materials defined by legislation. Toxic and hazardous are not synonymous—toxic is an intrinsic property, whereas hazardous can refer to intrinsic or extrinsic properties. Hazardous may include the potential for injury resulting from toxic *or other* re-

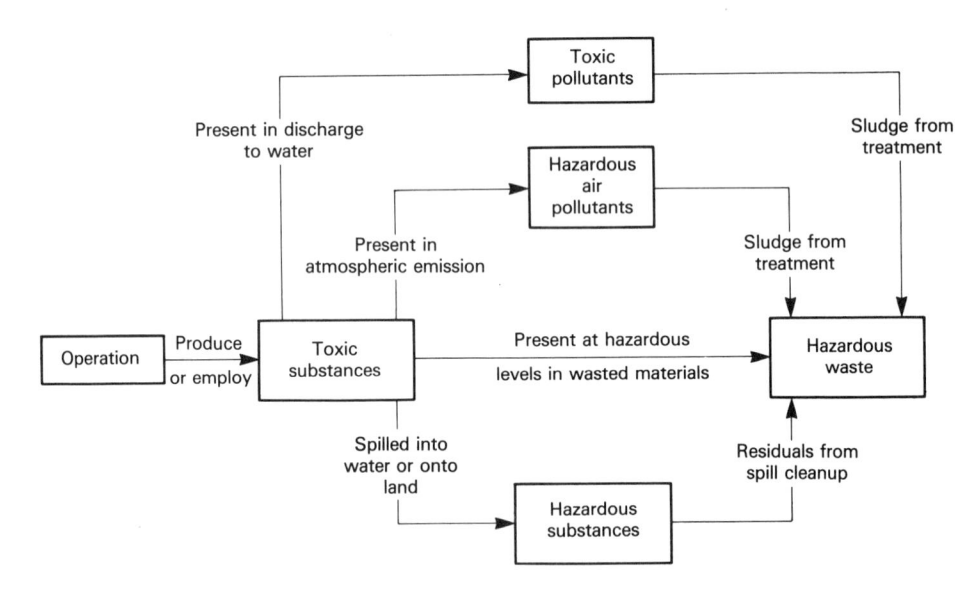

Figure 11.13 Relation of various hazardous and toxic materials. (Reprinted by permission of John Wiley and Sons, from G. W. Dawson and B. W. Mercer, *Hazardous Waste Management,* 1986.)

actions. Because hazardous can convey both meanings, a complete definition of hazardous wastes should answer questions of what, why, to what extent, at what time and under what conditions a material is hazardous.

The hazardous waste management industry. The hazardous waste management industry first emerged as a recognizable industry in about 1969. Generators, transporters, recovery specialists and cleaning operations represent some of the diverse backgrounds which form it. In the early 1970s, there were only about 10 regional plants for the treatment and disposal of hazardous wastes, typically limited to serving only a radius of 500 miles. The number of firms and the technology utilized has increased, with some facilities maintaining a full range of capabilities, including secure landfills, chemical treatment, deep well injection, incineration and more. The biggest problem which has plagued the industry and hampered the comprehensive national program has been organized public opposition to siting new facilities.

The success of firms in the hazardous waste management industry has been attributable to several factors. Location has a two-fold importance because state and local regulations have created cost benefits and economic transportation costs are frequently pivotal. Capital investment for the industry is high and funding has often been difficult to obtain; those companies with greater access to corporate financing have emerged as leaders. The ability to gain public confidence for siting, to obtain long-term contracts with large generators and to provide full service capabilities (transportation, storage, treatment and disposal) have proved keys to success.

Technology of Hazardous Waste Management

The effective control of industrial and municipal waste has not kept pace with the manufacturing processes which are responsible for that waste. Ideally, the generators of waste can help the most by developing processes that eliminate or minimize the generation of hazardous wastes. A reduction in waste can be accomplished by substitution, reformulation processes, modification, equipment redesign and recovery of waste materials. Until such efforts succeed, however, the problems of waste management will continue.

The five most popular methods presently used for managing hazardous waste are:

1. Secure landfill.
2. Waste treatment.
3. Deep well injection.
4. Incineration.
5. Recovery/recycling.

Secure landfill. Landfill disposal has the advantage of low costs, both in construction and operation, plus they can be quickly established and can handle a large variety of wastes. Landfills have been, and are expected to remain, the dominant method for waste disposal in the near future.

Landfill designs have evolved from studies of how contaminants have migrated from storage. Surface run-off, evaporation, infiltration and even wind-activated sediment suspension have been encountered. Principal design features contain:

1. A layer of topsoil.
2. A drainage system at the edges of the cover.
3. A permeable layer of sand or gravel.
4. A sealing layer of clay or polymer liners to prevent infiltration into the waste.
5. Buried waste (segregated into cells).
6. An underlayer of fine sand to support the sealing layer.
7. A venting system.
8. A drainage layer to collect leachate where it can be removed.
9. A bottom sealing layer.

In many areas where the level of the groundwater is high, a double liner/drain-layer system is used. The key to the effectiveness of landfills is the ability to monitor leachate production. Liner effectiveness is dependent on the waste, but polypropylene and polyvinylchloride membranes have performed well with most wastes.

One method which is used in conjunction with secure landfills is fixation or solidification. This process, however, is compatible with only a few wastes generated today. Inorganic wastes which contain heavy metals are of interest because we can prevent leachate formation. Solidification can be categorized as silicate or cementitious, lime-based, thermoplastic-based, organic polymer-based or encapsulation. If the fixation method succeeds in reducing the mobility of contaminants to the point where no serious stresses are exerted on the environment, then the wastes may even be disposed of in a conventional sanitary landfill since they can be considered nonhazardous.

Waste treatment. Hazardous waste treatment includes a variety of biological (activated sludge, composting, etc.) and physical/chemical (flocculation, coagulation, etc.) processes. Organic waste streams are amenable to biologic treatment. Activated carbon powders, added to chemical manufacturing waste streams, remove organic and inorganic contaminants, producing a high-quality affluent from a variable waste stream. Waste sludge is dewatered and then placed in a secure landfill. Many times, electroplating wastes are pH adjusted to a value of about 9, thus making heavy metals insoluble; precipitation produces a gelatinous sludge which is dewatered and placed in a secure landfill. In some instances, sludges with a high concentration of elemental metals can be produced and used for recovery.

Deep well injection. Waste which can be disposed of by underground injection should be low in volume and high in concentration, difficult to treat by other methods, non-reactive with the soil that it is injected into, biologically inactive and noncorrosive. Suitability depends alike on the physical and chemical characteristics of the soil as well as those of the waste. Because the potential for contamination of underground drinking water is so great, siting is carefully controlled by the EPA who administer the Safe Drinking Water Act of 1974.

A typical well consists of an outer surface pipe which extends about 180 feet below any groundwater and is cemented in place. A protective casing is inserted through the surface pipe and beyond its end into the disposal zone. It also is cemented in place. Actual waste is injected through a tubing which is sealed both at the well head and above the disposal zone.

Incineration. Incineration is a controlled high-temperature oxidation which converts most organic compounds into CO_2 and H_2O. Because the toxic or hazardous nature of the organic waste is due to the structure of the molecule, incineration destroys not only the molecule but the toxic or hazardous property. The process is most effective for flammable waste such as spent solvents or paint sludges, but can also be used for solid wastes.

There are two main types of incinerators. Liquid-injection incinerators operate by atomized waste being injected into a chamber where oxidation takes place. Atomization provides a larger surface area for incineration in the same way that fogging provides better cooling when fighting a fire. Rotary-kiln incinerators can accept any type of combustible waste, but they are used primarily to incinerate solid waste and tars which cannot be processed in the liquid-injection units. Material is combusted at high temperatures and tumbled down the inclined ramp which effectively provides the oxygen necessary for combustion. Non-combustible materials accumulate at the bottom and can be separated. Because of the high temperatures and the hazardous waste, corrosion problems abound and capital equipment costs are high because of frequent repair and replacement.

Incineration is able to handle large volumes of waste and produce an inert, detoxified residue, but the high construction costs, high operating costs and questionable stack emissions have all detracted from its widespread application. The method is, however, an attractive alternative to secure landfills which will become obsolete in the future.

Case History 11.5 Know Your Waste

Many small companies learned about hazardous waste disposal procedures only because of conformance to OSHA's Right-to-Know Laws concerning hazardous chemicals. Walnut Motors, a full-service dealership representing several foreign automotive manufacturers learned of these procedures in this manner. Most chemicals they dealt with were paint lacquers used in the body shop. Although safety was considered and practiced in the paint spray area, excess paints were collected and stored in drums in a remote area, along with old tires, waste oil and scrap automotive parts. Since implementation of the OSHA requirement, Walnut Motors has segregated lacquers and waste oils, contracting with an approved firm for their proper disposal.

Their efforts did not go far enough, however. A young independent contractor was removing old tires for Walnut Motors and asked if he could take the metal as well. Unaware that a partly filled lacquer drum remained in the tire area, permission was granted. The contractor used an oxyacetylene torch to burn a vent hole in the top, then proceeded to begin cutting the bottom with the torch. The small amount of lacquer ignited, and the gases within expanded too rapidly for venting through the small hole to be possible. As a result, the drum exploded, severely injuring the contractor and causing extensive damage to the property.

Walnut Motors, despite its apparent concern for safety, was cited for violation of NFPA storage regulations for flammable materials and was sued for negligence in causing the contractor's injuries. Although the company had viewed hazardous waste disposal as a matter of conformance, they learned that it was reflective on their safety program and complacency had no place in that program. ■

FURTHER READING

R. A. ALKOV, "Psychological Stress, Health and Human Error," *Professional Safety,* **26,** (8), 12, 1981.

K. ANDERSON and R. SCOTT, *Fundamentals of Industrial Toxicology,* Ann Arbor Science, Ann Arbor, 1981.

G. W. DAWSON and B. W. MERCER, *Hazardous Waste Management,* John Wiley, New York, 1986.

M. L. CAPELL, "The Stress-Loss Connection," *Professional Safety,* 20, (3), 33, 1985.

H. H. FAWCETT, *Hazardous and Toxic Materials: Safe Handling and Disposal,* John Wiley, New York, 1984.

M. FRIEDMAN and R. H. ROSENMAN, *Type A: Your Behavior and Your Heart*, Knopf, New York, 1974.

Handbook of Organic Industrial Solvents, **Alliance,** 1980.

R. HOLF, "Occupational Stress," in *Handbook of Stress: Theoretical and Clinical Aspects* (ed. L. Goldberger and S. Breshitz), Free Press, New York, 1982.

B. LEVY and D. WEGMAN, *Occupational Health,* Little, Brown and Co., Boston, 1988.

H. J. MCDERMOTT, *Handbook of Ventilation for Contaminant Control,* Ann Arbor Science, Ann Arbor, 1976.

A. MCLEAN, *Work Stress,* Addison-Wesley, Reading, Mass., 1979.

A. McLean, "The Corporate Environment and Stress," in *Stress in Health and Disease* (ed. M. Zales), Brunner/Mazel, 1985.

J. B. Olishifski, *Fundamentals of Industrial Hygiene,* National Safety Council, Chicago, Ill., 1979.

R. M. Singer, "Nervous System: Early Detection for Chemical Hazards," *Professional Safety,* **32** (3), 37, 1987.

M. J. Smith, "Recognition and Control of Psychosocial Job Stress," *Professional Safety,* **26,** (8), 20, 1981.

QUESTIONS

11.1. Describe, with examples, the main routes by which toxins can enter our bodies.

11.2. Industrial hygiene issues are classified by the hazards posed. What are the classifications? Give an example of each (do not use an *example* that has already been used in the Chapter).

11.3. What is epidemiology? Explain whether or not the epidemiology model of accident causation is realistic in light of application of epidemiology in understanding industrial hygiene issues.

11.4. Define TLV and explain how you would incorporate it into your safety program.

11.5. Name some major sources of contaminants in industrial operations. What problems do they pose with respect to hazard identification?

11.6. Explain why local ventilation is a better remedy than dilution ventilation in most workplace environments.

11.7. Describe a time when you have been under eustress and when you have been under distress. How did you cope with them?

11.8. How can management help reduce work-related stress? Nonwork-related stress?

11.9. Why is noise a major concern of industrial safety programs?

11.10. Are management remedies as effective in noise control as engineering remedies? Explain.

11.11. Define hazardous waste and summarize proper disposal methods.

11.12. Explain why social and political solutions to the hazardous waste problem are expected.

12

Product Safety

Product safety is frequently associated with consumers only, but third-party action suits are important and they relate directly to industrial safety programs. This importance is derived from the Workmen's Compensation Laws, which bar injured employees from suing their employers. Litigation against the manufacturer of a machine on which an employee is injured is not prohibited, though.

This chapter examines the circumstances in which injured employees are able to sue, how litigation proceeds, and what is involved in the lengthy process leading up to a trial. Although as safety professionals we may not be directly involved, time will be lost in accident reconstruction, rumors, and lost morale. However, a separate action could be brought against us by the manufacturer who is the defendant in the first case.

Until the beginning of this century, anyone hurt while using a product had no basis to sue for injuries sustained. In 1916, however, the first theory for recovery was established on the basis of negligent manufacture. Much later, in the 1960s, breach of warranty and strict liability in tort were added as means for recovery from a manufacturer. In the last two decades, the potential for recovery from manufacturers combined with inflation and outdated Workmen's Compensation awards, have escalated the number of claims.

Once the groundwork for products liability is established, this chapter will continue by prescribing how manufacturers can protect themselves from litigation claims. Effective products liability prevention programs are really quite similar to the principles established earlier for effective safety programs.

THE LEGAL SYSTEM

The legal system in the United States is known as common law and has been structured upon decisions and judgments of past court cases. If we follow this process historically, its origins are rooted in the Magna Carta in Britain. Another facet of our legal system is legislated laws, known as statutes. However, statutes are subject to interpretation by the courts.

For the purposes of studying products liability law, we do not need a comprehensive understanding of all legal aspects. Products liability involves only civil law, i.e. non-criminal in nature, where a person brings a civil action intended to enforce his private rights; the outcome usually involves only monetary damages. The court system in which products cases are tried can be either Federal Courts or State Courts and are heard first in trial courts, followed in many cases by appellate courts.

A trial court hears and decides controversies between the person(s) bringing the suit (the plaintiff) against a manufacturer of a product (the defendant) by determining facts and applying rules. This can be done by a judge only or may involve a jury. Trial by jury is a right unless waived by all parties. Appellate courts review the findings of trial courts, based only on potential or alleged errors made in the lower court. Decisions are based solely on whether the lower court acted within the law. Of course, the highest appellate court is the Supreme Court, either in the federal or state systems.

A products liability case takes a long time to be decided. The first step in an industrial injury might be taken either by the injured worker directly or by the Workmen's Compensation insuror on behalf of the injured party. The first step the plaintiff's lawyer takes is the filing of a formal complaint; this can be done either in a state court if all parties reside or do business in the state, or in the federal court under certain conditions such as diversity of citizenship.

A complaint identifies the plaintiff and defendant(s), the cause for the complaint (i.e. the injury), the relationship of the manufacturer to the product and the common law theory for recovery of damages. Once proper notification has been made to all parties, certain questions—interrogatories—are exchanged and answered. This period of time is known as discovery and follows certain rules which involve, among other issues, evidence and how information is acquired. The discovery period includes product examination, questioning of witnesses by means of deposition testimony and other preparations for trial. It is during the discovery period that corporate safety engineers oversee a host of legal and expert examinations of plant machinery where a worker was injured.

Frequently, many delays are encountered in the discovery period and trials are not scheduled by the courts for numerous reasons; in many cases, settlement negotiations are conducted, usually much more earnestly as trial dates approach. Typically, however, the time between the original complaint and the trial dates often exceed 5–8 years.

THE ORIGINS OF PRODUCTS LIABILITY LAWS

Although all products cases are brought in civil courts, the origins stem from both tort law, involving wrongdoing, and breach of warranty, involving contract law. As cases have been tried, these have remained and evolved as well to form the doctrine of strict liability in tort. Historically, manufacturers were not liable for injuries suffered with their products. This general rule of non-liability was established in 1842 in England in the case of *Winterbottom* v. *Wright*. Winterbottom was injured when he was thrown from a defective stagecoach built by Wright. Because Winterbottom had no contractual agreement with Wright as the driver of the stagecoach, he was denied recovery for damage.

The general rule for non-liability was followed up until the beginning of this century, with only a few exceptions involving imminently dangerous products, when negligence was first established as cause for recovery in the case of *MacPherson* v. *Buick Motor Co.* MacPherson was driving his car when a defective wheel caused him to lose control. The court decided that his injuries were caused by the negligent manufacture of the product which was sold without customer inspection, and that in such cases, a liability of the manufacturer follows.

What would have happened if MacPherson had not been able to prove negligent manufacture, however? In such cases, the general rule of non-liability persisted until the landmark case of *Henningsen* v. *Bloomfield Motors, Inc.* in 1960. In that case, Mr. Henningsen bought a new car as a gift for his wife. Mrs. Henningsen was injured within 2 weeks of the purchase when the steering suddenly failed at low speed causing the vehicle to veer and strike a wall. The car was demolished, and therefore defective manufacture could not be ascertained. The case was tried on breach of warranty, based upon contract law (implied warranties are based upon fitness of the product for ordinary purposes for which the product is used, and express warranties are used to induce purchase of a product). In finding for the plaintiff, the courts established breach of warranty as a recovery process and, of almost equal importance, extended products liability to users not in contractual agreement with the manufacturer (Mr. Henningsen purchased the car, not Mrs. Henningsen), and limited the effectiveness of disclaimers in products liability cases.

Strict liability in tort was established soon after the Henningsen decision. Two 1964 California Supreme Court cases, one which affirmed a lower court decision and one which reversed a lower court decision, are credited for establishing this theory. In the case of *Greenman* v. *Yuba Products, Inc.*, Greenman was injured while using a Shopsmith lathe in the intended manner, and negligent design as well as breach of warranty was argued by the plaintiff's attorneys. This decision was upheld, adding that it was sufficient for the plaintiff to prove that he was injured while using the product as intended, but because of a defective design the product was made unsafe for its intended use. In the other case, *van der Mark* v. *Ford Motor Co.*, van der Mark complained to the dealer that his new car veered when the brakes were applied and then released. The problem recurred

The Origins of Products Liability Laws

at about 1500 miles on the odometer, causing total demolition of the car. The lower court decision was reversed by the Supreme Court, stating that manufacturer's liability is non-delegable to a third party—in this case the retail dealership—and that strict liability on the manufacturer and retailer alike affords maximum protection for the injured plaintiff and works no injustice to the defendants since they can adjust the costs of such protection between them in the course of their continuing relationship.

The language describing strict liability in tort was established by the American Law Institute in the Second Restatement of Torts in 1965. Paragraph 402A has been widely adopted as a description of the rules for strict liability:

1) One who sells any product in a defective condition unreasonably dangerous to the user or consumer or to his property is subject to liability for physical harm thereby caused to the ultimate consumer or user, or to his property, if

    (a) the seller is engaged in the business of selling such a product, and
    (b) it is expected to and does reach the user or consumer without substantial change in the condition in which it is sold.

2) The rule stated in Subsection (1) applies although

    (a) the seller has exercised all possible care in the preparation and sale of his product, and
    (b) the user or consumer has not bought the product from or entered into any contractual relation with the seller.

Strict liability in tort as described by this Second Restatement of Torts has been adopted in some form in all but four states—Massachusetts has rejected the theory, only because legislative action preceded its consideration by the courts. Other exceptions still hold to a negligence doctrine.

The Second Restatement of Torts also defined the duty to warn with respect to products liability (Paragraph 388). Three factors are used to determine whether a duty to warn exists:

1. The likelihood of an accident occurring when a product is put to a foreseeable use without warning.
2. Probable seriousness of an injury if an accident does occur.
3. The feasibility of an effective warning.

There are many interesting products liability cases which demonstrate the various theories for recovery or the duty to warn, but few have the impact of a landmark case. Some of the issues raised include continuing duty to warn cases which have led to recall programs, and enhanced injury or crashworthiness cases where the product did not cause the accident, but injuries were greater because of the unsafe design of the product.

STATUS OF PRODUCTS LIABILITY LAWS

Refinements of the theories for recovery in products liability litigation have established certain proofs which must be made by a plaintiff. In *any* product liability case, the plaintiff must show:

1. The product was defective or unreasonably unsafe as made by the manufacturer.
2. The defective or unreasonably unsafe condition existed at the time the product left the manufacturer's care.
3. The defective or unreasonably unsafe condition was the cause of the plaintiff's loss.
4. The nature of the plaintiff's loss is related to the defective or unreasonably unsafe product.

Determination of what is a defective product is not always an easy matter. Defects can be classified as latent, which appear only in a single unit of a product line, or patent, which appear in all units of a product line. An example of a latent defect would be a stress-concentrating void in a metal casting which led to premature failure. The best example of a patent defect is a design defect case, such as not providing a safeguard for a known machine hazard.

Most of the relevant considerations of defectiveness concern risk associated with use of the product. The best rule to follow is the prudent man concept, whereby reasonable risk and unreasonable risk are defined as:

- *Reasonable Risk*. Risks of bodily harm to users are not unreasonable when consumers can:
  1. Understand that risks exist.
  2. Appraise their probability and severity.
  3. Know how to cope with them.
  4. Voluntarily accept them to get benefits that could not be obtained in less risky ways.
- *Unreasonable Risk*. Preventable risk is not reasonable when:
  1. Consumers do not know that it exists.
  2. Consumers are unable to estimate frequency and severity of the risk even though they are aware of it.
  3. Consumers do not know how to cope with it.
  4. Risk is unnecessary in that it could be reduced or eliminated at a cost in money or performance of the product that consumers would willingly accept if they knew the facts and were given the choice.

STATUTORY LAW AND PRODUCTS LIABILITY

As products liability law was developing, a parallel pattern of increasing public concern for safety was also developing. In the workplace, we saw the advent of OSHAct and its impact on safety in the workplace. The growing consumer movement, spearheaded by people such as Ralph Nader, led to passage of the Consumer Product Safety Act in 1972. The purposes of the Act are:

1. To protect the public against unreasonable risks of injury associated with consumer products.
2. To assist consumers in evaluating the comparative safety of consumer products.
3. To develop uniform safety standards for consumer products and to minimize conflicting state and local regulations.
4. To promote research and investigation into the causes and prevention of product-related deaths, illnesses and injuries.

The Act established a Consumer Products Safety Commission, which became responsible for administration of the CPSAct and other consumer-related acts. Table 12.1 lists these laws along with others which have an impact on products liability.

Although the influence of the CPSC has been restricted by budgetary constraints, the commission righteously deserves our credit for the safety match book cover, childproof caps for medications and other hazardous chemicals, and the accumulation of information on the safety of consumer products. One of their functions, specified in the CPSAct, is maintenance of records on safety of consumer products, which is accomplished by the National Electronic Information Surveillance System (NEISS), a computerized coordination of hospital emergency records involving injuries from consumer goods.

TABLE 12.1 Major Laws Which Impact on Products Liability

Title	Administrative Agency
Consumer Product Safety Act	CPSC
Flammable Fabrics Act	CPSC
Food, Drug and Cosmetics Act	FDA
Hazardous Substances Act	CPSC
Mine Health and Safety Act	OSHA
National Traffic and Motor Vehicle Safety Act	DOT
Occupation Health and Safety Act	OSHA
Poison Prevention Packaging Act	CPSC
Refrigerator Safety Act	CPSC
Toxic Substances Control Act	EPA
Workmen's Compensation Act	State Industrial Accident Boards

Our focus in this book, however, is not on consumer product safety, but on industrial safety. With that in mind, we have to focus on the effects of OSHA and Workmen's Compensation Laws on products liability because these laws deal directly with workplace safety. Workmen's Compensation Laws will compensate workers who are injured in the workplace *regardless of fault,* but the workers are barred from suing the employers directly for their injuries. In many instances, the compensation is not only inadequate, but requires legal representation to obtain it. In those instances where injury occurs on a machine, there is no bar to suing the manufacturer of the machine under products liability law. In fact, the Workmen's Compensation Insuror can participate in the litigation by attaching a lien on any judgment for compensation paid *regardless of fault* when the machine is found to be defective and/or unreasonably dangerous.

As we might expect, the application of products liability law to industrial accidents has caused a great deal of controversy, because most accidents have multiple causation and the law can place liability on only one cause, i.e. the defective or unreasonably unsafe machine. We must remember, however, that each products case must be tried on its own merits and that many cases are won by defendant manufacturers. Nevertheless, complaints filed against a manufacturer must be defended and products liability insurance costs have escalated, causing what has been referred to as a products liability crisis for manufacturers of products.

It is with the products liability cases arising from industrial accidents where OSHAct has an impact on products liability. Remember, OSHAct was originally passed to eliminate hazards in the workplace and placed responsibilities only on the employer and employee, remaining silent on any role of machinery manufacturers. Although OSHAct cannot expand a manufacturer's duty or liability, OSHA regulations are standards for safety, and can be used as evidence in products liability cases just as ANSI or NFPA standards are used. As a matter of fact, both the plaintiff's and the defendant's attorneys use OSHA regulations for different purposes.

Plaintiffs use OSHA for evidence of defendants' negligence, evidence of the unreasonable danger of a machine or as the basis for unbiased expert testimony. We must remember that a jury can believe whomever they wish and some jurors might believe that expert testimony is swayed by fees being paid them. Pre-existing OSHA standards established in the public interest, however, are neutral to either party and therefore unimpeachable. If a machine is lacking in some safety characteristics which are defined by OSHA, the burden of proof of defectiveness or unreasonable danger is reduced considerably for the plaintiff.

The defendant manufacturer's attorneys can also use OSHA regulations to advantage. First of all, compliance to regulations can be demonstrated if applicable, but a more common application is to point out improper or unsafe behavior on the part of the plaintiff or the employer. Unlike all previous safety codes, OSHA explicitly created duties for employees as well as employers; thus, if conduct which should have been undertaken by the plaintiff (e.g. safety glasses supplied by employer) to prevent injury was not undertaken, then contributory negligence is evident. Such behavior might sway a jury to the defendant manufacturer's position.

THE EXPERT WITNESS IN PRODUCTS LIABILITY

The descriptive adjective, expert, is much more imposing than it should be. An expert is simply someone who, on the basis of training or experience in the matter involved in a products liability case, can be qualified in a court of law. An expert has to be qualified for each and every trial according to rules of civil procedure in effect for the particular court. Thus, experts include those persons with years of experience in a field as well as consultants, academic professionals and even responsible employees. The expert's role can be varied, from providing guidance to counsel on technical issues during discovery to being perceived by a jury as a believable, unbiased witness. For these reasons, expert witnesses cannot accept payment on a contingency basis and employees of defendants do not make good witnesses because of their loyalty bias. Such employees, however, serve well behind the scenes in a guidance role.

Although whether a jury believes a witness or not is paramount, selecting an expert witness is best made on professional qualifications, such as education, certification in the professional field, publications pertinent to case issues and an ability to present the issues as simply and honestly as possible. In addition, the expert should have the ability to stand by his/her convictions under cross-examination which sometimes takes the form of personal attack.

Safety professionals are interested only in workplace injuries because they impact safety programs. In order to be an effective expert in these third party actions, he/she must be aware of the purposes of the action, the theory for recovery, and what has to be proved *by the plaintiff* (see p. 266). These proofs cannot be overlooked; any honest expert has informed plaintiff's counsel numerous times that a product was not defective or unreasonably unsafe. The same honest expert has also advised defendant's counsel that a claim is legitimate. Moreover, many claims against manufacturers have been dropped after the expert pointed out that the hazard resulted because of modifications made to the machine.

In most cases, however, a defective or unreasonably unsafe condition is not readily apparent. Those instances where a latent defect which appeared only in a single product and caused an accident and injury are quickly settled. An expert, therefore, must be able to recognize the defective condition which renders the product unreasonably unsafe, with emphasis placed on the "unreasonably." In workplace accidents and injuries, the issues of reasonable v. unreasonable can become very complicated, because most industrial accidents have multiple causes.

How does an expert proceed? His or her job is similar to that of a safety professional, but with one important exception; he or she already knows that an injury has occurred which makes identification of the hazard much more simple. Perhaps we can best describe the methods used by an expert as failure analysis or accident reconstruction methods. Their tools are no different from the ones we have studied in hazard identification, including interviews, ergonomic analysis of tasks and movements, application of a few principles of mechanics, testing of design and materials, and even literature reviews to determine state-of-the-art or standards applicable at the time of manufacture. These same tools are used by experts for both sides of the litigation. Using these tools, combined with knowledge of the theories for recovery and the proofs required, permit the expert to form an opinion in a particular litigation.

Let us now look at a few case histories which demonstrate the proofs required by liability law and how the lawyers and experts acted in each case.

Case History 12.1   Kales v. Gem Wheels, Inc.

Philip Kales is a high school graduate with 1 year of trade school. His work history is as a laborer for the town highway department, and foundry and metal finishing industries. For 10 years he had been working as a grinder for Victor Industries, a producer of hardened steel tools (e.g. jackhammer bits) used in the construction industry. These tools were made by forging, heat treating to a hardness of Rockwell C 55, and then finished by grinding the work edge.

For 3 years Phil operated the 30-inch grinder manufactured by Gem Wheels, Inc. Four days before Phil's 30th birthday, he was adjusting the work rest inward in order to dress the wheel when his hands slipped off the work rest. His left arm came in contact with the grinding wheel and it had to be amputated 3 inches below the elbow as a result of the injury. He was in hospital for 6 weeks where he was treated both for the physical and psychiatric effects of the injury. Six months later he was first fitted with a prosthesis, and continued physical and psychiatric therapy at a rehabilitation center. His benefits from workmen's compensation ceased after 2 years and he was forced to apply for welfare. Phil finally was able to find employment with the U.S. Post Office nearly 5 years after the incident. During this time, his Workmen's Compensation lawyer filed a products liability complaint against the manufacturer of the grinding machine, Gem Wheels, Inc.

How should we proceed if we are hired as an expert witness on this case? Remember, it does not matter whether or not we are hired by the plaintiff or defense counsel because our honest opinions of how the injury occurred are required. What differs is whether we find fault with the other party or we reduce the fault leveled against our own party (within the confines of honest opinion), and the methods in which we can obtain necessary information. Of course, there are common sources of information such as interrogatory answers, standards and rights to inspect equipment. In this case, if we are the plaintiff's expert, we can interview him directly to judge his knowledge of the machine and the task and his personal character. However, if we are the defense expert, we have to direct our questions through the attorney either by interrogatory questions or means of deposition of the plaintiff. As defense expert, however, we have the benefit of direct questioning of the defendant with respect to the history of the machine design patents and modifications made over the years, comparison of competitors' designs, information made available to the plaintiff's expert only through a literature search, and interrogatory questions or depositions of the appropriate defendant executives or engineers.

In Phil's incident and injury, all of these methods were employed by experts. Without considering which expert first uncovered the issues, let us make a list of the issues based upon multiple causation possibilities.

1. Machine Factors
   - *Work rest design:* Held by set screw, adjustable by hand, minor modification made to extend width of rest, weight was 85 lb.
   - *Comparison to other work rest designs:* Patent held by Gem Wheels, Inc. for rotating screw adjustment of work rest, requiring flexible bellows covering to keep out debris; design used for smaller grinder. Competitor's design had moveable grinding wheel axis, which reduced number of work rest adjustments.
2. Task Factors
   - *Work rest position:* 4 inches from wheel (ANSI B7.1, the use, care and protection of abrasive wheels, states that work rests should be a maximum of $\frac{1}{8}$ inch from the wheel to prevent work from being jammed between the wheel and rest). The specific task, however, could not be performed with such a restriction.
   - *Work rest adjustment:* The wheel had to be dressed up to six times per hour because of wear associated with grinding a hardened steel. Each dressing required adjustment before and after dressing.
   - *Productivity v. Safety:* ANSI B7.1 requires the wheel to be turned off when adjusting the work rest. However, it took 3 min for the wheel to come to a stop when turned off and there was no brake. That means that up to 36 min of each hour would be lost to production just waiting for the wheel to stop before and after dressing.
3. The Man
   - Phil was experienced in using the grinder and in adjusting the work rest.
   - Phil was taking Dilantin for seizure control, but had never had a seizure at work or in the last 12 years preceding his injury.

The accident reconstruction was conducted separately by the plaintiff's and defendant's experts. Human factors analysis was employed by each, with the plaintiff's expert opinion having the feet braced at an angle to the wheel, the right hand applying an upward torque and a forward force on the work rest and the left hand a forward force. This method is depicted in Fig. 12.1. The plaintiff's expert claimed that any slip off the work rest would cause contact with the wheel

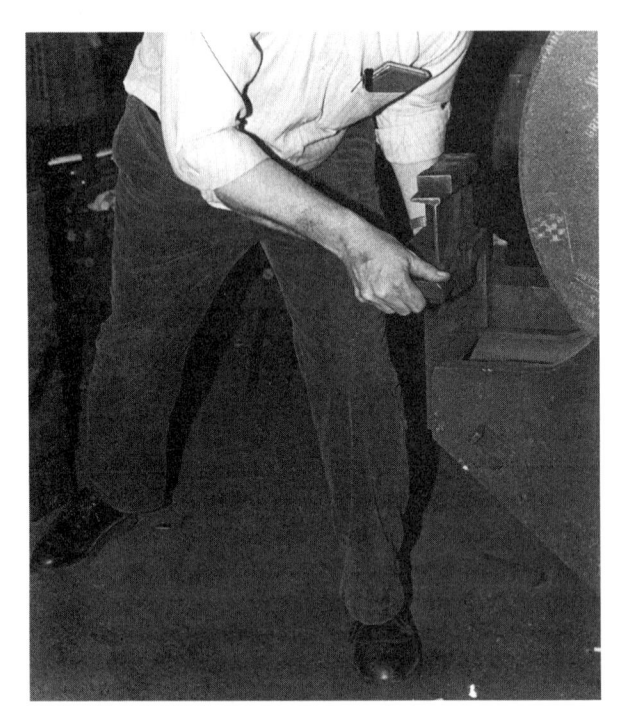

Figure 12.1 Adjusting work rest of grinder.

on the left forearm. The defendant's expert opinion varied slightly from this, with the operator motion more parallel to the wheel and the left hand on the top of the work rest when his hands slipped.

These expert opinions were presented at the time of the trial. The defendant's expert emphasized the violations of ANSI B7.1, i.e. not turning off the grinder when adjusting the work rest and working with the 4-inch space between the wheel and work rest. These he attributed to employer emphasis on productivity and further found fault with the employer for poor job–worker compatibility based on Phil's use of Dilantin. The plaintiff's expert emphasized the heavy weight of the work rest, the safer patented manufacturer's work rest design, the competitor's design which required fewer work rest adjustments, and presented evidence of the advertising of Gem Wheels, Inc. which emphasized productivity while competitors' advertisements emphasized safety.

Thus the jury had to deliberate whether to believe the plaintiff's allegation of defective and unreasonably unsafe work rest design which existed at the time of manufacture, or the defendant's opinion that the plaintiff's injury was caused by poor task design, poor worker–job compatibility and violation of safe practice with grinding wheels. After 6 hours of deliberation, the jury found in favor of the plaintiff, awarding Phil in excess of $1 million. This award was later reduced to within the limits of the Gem Wheels, Inc. products liability insurance policy by agreement. ∎

The Expert Witness in Products Liability

Case History 12.2   Mahoney v. Bernardston Chair, Inc.

Sally Mahoney is a single parent with a ninth-grade education. She had been employed for 8 years as an assembly operator at Newport Industries when she was seriously injured from a fall which required three operations on her lower back and a painful condition which will never improve. She was 28 years old at the time of the incident.

At the time, Sally was working with three other women at a 4 × 5 foot table, 36 inches above the floor. Parts were fed through a chute to the table where the three women would add a component, and Sally would inspect each and box it for shipment. All of the women were seated on polypropylene stools which would swivel, but not tilt. These stools had been purchased 2 years earlier on the basis of an advertisement which described them as heavy-duty strong stools with "super-comfortable" polypropylene bodies which provide spine relief and air vent space.

Sally described the incident which occurred about 2 hours into her shift. She turned to the left to pick up an empty box, turned back and placed it on the table, then straightened up, leaning slightly against the chair. As she did so, she and the seat fell to the concrete floor, Sally landing on her coccyx and the back of her head.

When Sally's back did not improve after the first surgery, she consulted a lawyer and a complaint was filed against Bernardston Chair, Inc. on the basis of negligence and strict liability in tort. Her lawyer retrieved the stool from Newport Industries and saw where the seat had broken from the pedestal. These breaks appear in Figs 12.2 and 12.3.

Sally's lawyer went to the University of Lowell where he asked a design instructor, Professor Gauthier, to examine the stool. After his examination, mea-

Figure 12.2 Broken welds of seat support.

Figure 12.3 Broken welds of chair pedestal.

surements and calculations, Professor Gauthier expressed his opinion that the design was adequate for the intended purpose and that some abuse must have led to the failure. Disappointed, Sally's lawyer decided to obtain a second opinion and went to an independent laboratory, Dalzell Associates, where a failure analyst (also a professional engineer) was assigned to the case. The following are excerpts from his report:

Investigation

The subject chair was thoroughly examined and photographed. The seat is identified by a stamping on the rear of the back, B.C.I. Bernardston Chair, Inc., Bernardston, MA. It is one-piece, molded black polypropylene, has numerous scratches on the seat surface and is shaped like the descriptive advertising photographs. The seat is $15\frac{7}{8}$ inches wide at its widest point, and the back tapers from 15 inches at the bottom to 12 inches wide at the top. The back height is $15\frac{1}{2}$ inches maximum and is contoured. There is also a triangular open section where the seat intersects the back.

Underneath the seat, there are molded projections to which tubular steel is riveted to the seat and back. Tubing has a 0.637 inch O.D. and is 17 inches long from rivet to rivet, being flattened of course in the rivet area. There is on each side a contoured bracket welded to the tube, such bracket in turn welded to a square tube, the outside cross-section of which is 1.014 inches square. Each side's square tubing is $4\frac{1}{16}$ inches long and is butt-welded to similar cross-section tubes which parallel the front and back of the seat, thus forming a rectangular support made of the square tubing, $4\frac{1}{16}$ inch by 9 inch in size. The 9-inch long sections of the tubing are arc-welded to a 0.140-inch thick bracket which is attached to the pedestal leg.

Where measurable, the square tubing cross-section thickness was between 0.050 inch and 0.070 inch. Most welds were covered with paint, but the four welds attaching the contoured bracket to the square tubing were unpainted and oxidized.

The failure of the seat support system occurred in one of the 9 inch long square

tubing sections remaining with the seat and one remaining with the pedestal. Examination of the section remaining with the seat revealed the following:
  A. Tube-to-tube weld failed on right hand side, displacing the tube upward and causing longitudinal cracks about $\frac{1}{16}$ inch long and a crack along the weld, 0.488 inch long.
  B. The weld between the tube and pedestal bracket failed on the right side with the weld remaining on the tube. On the left side the weld remained with the pedestal bracket and there is a hole, 0.675 inch by 0.257 inch, in the tubing.
  C. Microscopic examination showed clear shear marks on the fracture surfaces.

Examination of the 9-inch long tubing section which remained upright with the pedestal base revealed the following:
  A. All failures occurred in the tubing section with wall thickness measured between 0.042 and 0.047 inch.
  B. Weld metal from the hole in the other tubing section remains on the pedestal bracket.
  C. On the right side, the weld metal sheared, with some weld metal remaining on each part.
  D. Microscopic examination revealed clear shear marks on all fracture surfaces.

Discussion

This investigation has shown that the incident most probably occurred as described by Ms. Mahoney. As she turned and leaned back slightly, she felt a rotation of the seat; this was probably the result of the tubular corner failure, which caused the 9-inch long section to bend upward. Next, when Ms. Mahoney again leaned backward slightly, the remaining failed sections broke, causing her and the seat to fall to the floor.

It is the opinion of Dalzell Associates that the failures and the incident most probably are the result of an improper joining method used in the design. The use of arc-welding to join thin sections to thick sections is not a recommended joining process because of the potential weakening of the thin members. It was the thin members which predominantly failed in the support section.

Conclusions

The following conclusions represent the opinions of Dalzell Associates, Inc.
  A. The incident which injured Ms. Sally Mahoney occurred as she described it in her deposition; the seat support system of her stool failed, causing her and the seat to fall to the floor, leaving the pedestal upright.
  B. The subject stool is a heavy duty polypropylene stool identical to Bernardston Chair, Inc. Model 26 which was purchased by Ms. Mahoney's employer, Newport Industries.
  C. Failure of the subject stool support system probably was caused by weakening of the design by the arc-welded joints.
  D. Failure of the subject stool support was caused by shear of the weakened joints under normal stresses.
  E. Bernardston Chair, Inc. was negligent in the manufacture of the subject

stool and, as a result, it was defective and unreasonably dangerous for its intended use.

Professor Gauthier was never identified to attorneys for Bernardston Chair, Inc. On the basis of Dalzell Associates' report, the extent of Sally's injuries and the agreement of the insurer, a settlement of $16,000 was agreed upon and the litigation was never tried in court.

Case History 12.3 Berger v. Middlesex Conveyor

Edward Berger has been a maintenance laborer since graduating from high school 32 years ago. For the last 5 years, he has been employed by Claremont Plastics Corporation who manufacture rigid foam products. Three years ago, a new shear was installed and a roller conveyor system was needed to move the cut foam to the shipping department. The system was designed by Middlesex Conveyor, but the cost was prohibitive. Claremont Plastics did purchase roller conveyor sections and pulleys from Middlesex Conveyor and constructed the conveyor system in-house, in part with Ed's assistance. Last year, the belt bought from Goshen Industrial Fabrics was installed and tested. During the test, Ed noticed the belt started to slip to the side and he tried to recenter it while it was moving. His arm became caught in the in-running nip point and he was severely injured.

Ed was out of work for 3 months and his arm was badly scarred. He went to an attorney who filed a claim against Middlesex Conveyor Corporation, whose name plate was on the conveyor sections. Ed's attorney contacted Professor Harcourt at the University of Lowell and arranged to have him inspect the conveyor.

Professor Harcourt examined the conveyor, made measurements and took photographs. Figure 12.4 shows the rigid foam being transported on the conveyor, and the in-running nip point where Ed was hurt (between the belt and pulley), is shown by the arrow in Fig. 12.5. Professor Harcourt learned of the history of the conveyor system during his investigation. The following are excerpts from his report:

> The incident which injured Mr. Berger occurred at the pinch point between the pulley and belt of the power conveyor system. Production conditions would not permit barrier safeguarding of this pinch point because the conveyor system would be unusable. However, American National Safety Institute Standard ANSI B20.1 requires prominent warnings whenever guarding is not feasible; no warnings are posted at Claremont Plastics Corporation. Therefore, the power conveyor system is defective.
>
> The defect is one of warning about a pinch point between the belt and pulley. Such a pinch point did not exist at the time of delivery of the pulley by Middlesex Conveyor. They provided only a component of the conveyor system and have no responsibility for the defect in the system in the opinion of the undersigned.
>
> In addition to the existence of the defect over which Middlesex Conveyor had no control, the particular circumstances at the time of the incident should be ad-

The Expert Witness in Products Liability

Figure 12.4 Conveyor at Claremont Plastics.

dressed. Mr. Berger attempted to prevent the moving belt from sliding off the pulley when the incident occurred.

Pulleys are fixed on the bearing blocks which are bolted to the framework of the pulley system. For the belt to move around each pulley and not slip depends on friction which is controlled by tension of the belt itself. Factors which influence belt tension include the pitch of the pulleys, drag of the belt across roller conveyors and others. None of these factors were either controlled or controllable by Middlesex Conveyor when they sold components for the system.

Conclusions

The following conclusions represent the opinions of the undersigned:

    A. The incident which caused injury to Mr. Berger occurred at the pinch point

Figure 12.5 In-running nip-point of conveyor.

between the belt and pulley of a power conveyor system at Claremont Plastics Corporation.
B. The conveyor system is defective because there are no prominent warnings posted.
C. Middlesex Conveyor supplied components of the conveyor system, but did not have any control of the system design, installation or maintenance of the system.
D. None of the technical factors which might have caused the belt to slip from the pulley at the time of the incident were controllable by Middlesex Conveyor.

Ed and his attorney were disappointed with their own expert's opinion, but recognized that the defective condition did not exist when it left the manufacturer's care and therefore dropped the litigation against Middlesex Conveyor. ∎

Case History 12.4   Pottle v. Foreign Motors, Ltd

Dave Pottle drove a local delivery truck for A&J Distributors, Inc. The truck was an old one made by Foreign Motors, Ltd which "looked terrible, but ran well." After three accident-free years with A&J Distributors, Dave was about to climb into his cab one day after making a delivery when the first step collapsed and his right knee was twisted when he fell. The knee required surgery, and has never been the same since the incident.

Once Dave recovered sufficiently to recognize the economic impact of having a permanently stiff leg, he sought legal assistance. Dave's lawyer filed a products liability complaint against Foreign Motors, Inc. for negligent design of the means for entering the cab of the truck. Dave's lawyer did not hire an expert immediately because Dave could not afford the fee and the truck was traded in for a new one after the step was repaired. When an expert was finally hired, the subject truck could not be located and the mechanic who had repaired the step had passed away.

The expert, Professor Cirrito from the University of Lowell, had worked with Dave's lawyer on a number of cases, and therefore agreed to assist on Dave's case. First, he interviewed Dave to determine what happened, using the drawing of the steps provided by Foreign Motors, Ltd during discovery. This drawing, reproduced in Fig. 12.6, shows that the steps are attached to brackets which are welded to straps which hold the twin diesel fuel tanks to the tractor on each side of the cab. Dave indicated it was the right side which failed and that it was the strap which broke when he stepped up on to the lower step.

Professor Cirrito located an exemplary vehicle of the same year which was being used by a local oil company and arranged to inspect the steps of this truck. There were two steps attached to the brackets which were welded to the straps. Each step was $3\frac{1}{2}$ inches wide and 23 inches long and made of expanded metal for slip resistance. The lower step was 16 inches off the ground and there were $20\frac{1}{2}$ inches between the steps, although an intermediate step 5 × 11 inches was

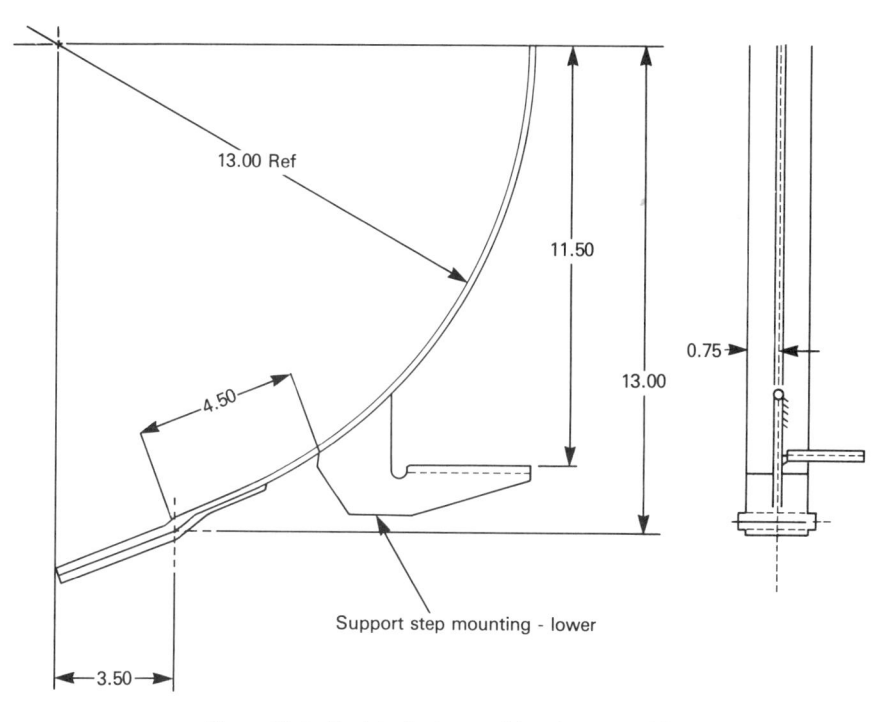

Figure 12.6 Fuel tank strap and bracket assembly.

18 inches to the side (attached to the body, not the fuel tank) and midway between the two steps. Professor Cirrito observed rust on the straps by the lower step, but none by the upper step. Such rust is shown in Fig. 12.7.

At the time of the trial, defense counsel argued vehemently to prevent Professor Cirrito's testimony on the basis that he did not evaluate the subject failure.

Figure 12.7 Rust formed at strap of lower step.

The judge overruled and Professor Cirrito was allowed to testify, despite rigorous cross-examination during qualification. Professor Cirrito's analysis of the design of the steps was based on the strap being placed in tension when someone stepped on the lower step (conversely, stepping on the upper step only compressed the strap). Repeated tensile stretching permitted moisture to get between the strap and tank, causing crevice corrosion and weakening of the strap over a period of time. It was Professor Cirrito's opinion that such corrosion was not obvious because of the location beneath the tank and that the strap failed when Dave Pottle stepped on it in its weakened condition.

Defense counsel cross-examined Professor Cirrito intensely, but did not utilize an expert witness for the defense case. Instead, he laboriously presented documentation of truck usage and maintenance through the accountant for A&J Distributors. The last witness was the maintenance clerk who was questioned about the straps, steps and corrosion. When he stated that the repair made by the late mechanic was to weld the bracket to the strap and that the brackets were always causing problems, defense rested.

The jury deliberated for 6 hours before returning with the verdict, which was that the design of the step/straps system was defective because corrosion could result from repetitive use and go unrecognized; that this defective condition, by virtue of being a design defect, existed when it left Foreign Motors, Ltd's care; but that this defective condition was *not* the proximate cause of Dave's injury. Rather, the bracket weld failure caused the incident and injury and the weld was not necessarily original. The verdict is essentially a defense verdict because Dave did not recover for his injury. ■

PRODUCTS LIABILITY PREVENTION PROGRAMS

We have not studied these case histories simply for amusement. In our role as safety professionals, we are not only interested in the safety of our workers, but have to be concerned about the safety of the users of our products as well. There are numerous examples of small companies which are no longer in business because of one incident and injury which occurred involved their product. Larger companies have survived, but their public image has suffered. For example, the McDonnell-Douglas Corporation has been scrutinized in detail as a result of the cargo door defect in the original DC-10 because 346 lives were lost in a 1974 crash attributed to the defect. Similarly, Morton-Thiokol Corporation has been adversely affected by the defective seal which led to the Challenger disaster which cost the lives of seven astronauts and worldwide publicity of the defect.

Therefore, we have to recognize our responsibility to manufacture safe products, to know the profile of our users (just as we have to know the profile of our workers), to learn or develop that vague quality of foreseeability and to learn how to market our products with safe use in mind. With only a few exceptions, the principles of a products liability prevention program, or, more positively, a products safety program, are very similar to the development of a good safety program. In other words, we should apply systems safety concepts to product safety.

There is no magic in formulating a good product safety program. All we have to do is implement information and knowledge that we already possess, then apply all of the brilliant management techniques we have learned in Chapter 4 to our product safety program. But what information do we have? Let us examine a study of 645 cases in the period 1967–73 reported in *Professional Safety* in 1981 where plaintiff verdicts resulted. The reasons were as follows:

- Negligence in manufacture — 42%
- Design defect — 25%
- Failure to properly warn — 13%
- Defective material or improper material selection — 6%
- Negligent quality control — 5%
- Packaging defects — 5%
- Misrepresentation in labeling — 2.5%
- Incomplete instructions for use, installation and maintenance — 2%

There are often overlapping features of a products case where issues of failure to warn are not differentiated from instruction issues or where failure to properly warn is raised only to emphasize a design defect or improper material selection. Regardless of overlap and judgments made in compiling this study, it points out the areas which our systems product safety programs have to emphasize. Fortunately, we do not have to reinvent the wheel, but simply introduce these concepts into our company in an environment where they can work effectively.

Traditionally, product engineering has involved strictly technical matters, such as design, strength, material selection, manufacturability and cost. Basically, our purposes are to broaden the scope, including such items as hazard analysis, design review, more attention to codes, standards, and record-keeping, consideration of warnings and instructions, quality control and packaging. Actually, the Consumer Product Safety Commission published a handbook and standard for Manufacturing Safer Consumer Products in June 1975, which very effectively fulfills its purpose "to provide guidelines to executive industrial management for establishing systems to prevent and detect safety hazards in consumer products." CPSC directed the standard to executive industrial management because only management has the resources and authority to institute sustained actions to prevent and detect product safety hazards.

MANAGEMENT OF PRODUCT SAFETY PROGRAMS

For the same reasons that a written policy for safety in an industrial operation is needed, so is there a need for a written policy for product safety. That policy should convey commitment of the corporation to product safety and the reasons for the commitment. In most cases, it pays to be frank and convey selfish interests such as legal and insurance costs as well as customer considerations. This written commitment needs to be effectively reinforced by our innovative, participative management personnel. We do not need to study management functions further, but simply state that the management of product safety adds only one more function to our safety management responsibilities.

One area of management is worthy of mention, though, particularly for organizations that have not been formally concerned with product safety in the past, i.e. training. The basic purpose of product safety training is to broaden each employee's understanding of the meaning and significance of product safety and provide them with information and tools to utilize in the program. CPSC suggests the following elements in a training program:

1. Review of the corporate program and relationships of each employee's work to the program.
2. Regulations and standards applicable to corporate products.
3. Manufacturing processes.
4. Inspection and test procedures.
5. Record-keeping and reporting requirements.

TECHNICAL REQUIREMENTS OF PRODUCT SAFETY PROGRAMS

Design Review

Product design is rarely the effort of a single individual; rather, the process of product development involves a wide range of technical and managerial skills. Many authors such as Eads and Reuter, Anderson, or Kolb and Ross have divided this product design process into three stages: the conceptual stage, the intermediate stage and the final preproduction stage. We can see what considerations have to be made by examining Table 12.2. What is important is that a design review should be conducted after the completion of each stage.

TABLE 12.2 Considerations Made in Design Stages

Stage	Tasks to be Considered
Conceptual	Functional performance features cost to produce, reliability/life expectancy, environment for use, special characteristics such as safety, instructions
Intermediate	Layouts, electrical schematics specifications, initial tooling needs
Final (preproduction)	Detailed, schematics and drawings, with tolerances, materials, manufacturing processes

The CPSC suggests that persons representing production, quality control, and consumer services in addition to management personnel comprise the design review committee. The CPSC also suggests that the committee address product design safety which takes into account hazards not only of the product, but hazards developing because of user ergonomics, record-keeping of hazard identification procedures and remedial actions, and designer defense to committee questioning. Surveys reported by Eads and Reuters indicate that about 80% of respondents had product liability individuals or committees, most of which operated only part-time. Of these, 57% exercised approval of product design. The representation of specialists on the committees is diverse, but included management, engineering, manufacturing and quality control, legal, safety and risk management, and marketing functions on a majority of committees.

The make-up of the safety design review committee does not represent the starting place for safety in any new design, however. Most executives would argue that safety is incorporated as soon as design work is begun on a new project. One designer requires new designs to satisfy six fundamental concepts—function, structure, cost, environment, safety and health, and statutory requirements. From the safety viewpoint only, the design team should conduct a preliminary hazards analysis.

Some authors suggest the first need in design safety review is to make up a hazards control checklist. Most, however, are wary of using such checklists because they tend to be depended on too heavily. From our viewpoint, such checklists are important, but if compiled at the initial design stages bias errors can be overlooked and continued, and the innovation and creativity of others performing design reviews at later stages can be stifled. In the same context, standards should not be incorporated into the early design considerations to avoid absolute dependence.

Once the preliminary hazards analysis has been completed and the initial design has been established, we should begin the formal design review process. Because of the technical nature and complicated analysis involved, several reviews are always necessary, with 1–2 weeks between them to allow for individual analysis of all aspects affecting the safety of the proposed products. Typical review agendas and the critical times for reviews are summarized in Table 12.3.

The most important aspect of design reviews that we are concerned with—hazard analysis—has not been addressed because the methods are the same as those discussed in Chapter 5.

Documentation and Change Control

Accounting is a financial documentation essential to effective management. Technical record-keeping is also necessary for effective management, but nowhere is this more evident than in products liability prevention programs. Records of product development have to be produced in litigation matters when specifically requested and can be used as evidence in a trial. For example, in the well-known

TABLE 12.3 Typical Formal Design Review Stages and Agendas

Design Review	Agenda
1. Conceptual stage	(a) Marketing needs, strategies for new product (b) Priorities—cost, reliability, performance (c) Scheduling
2. Preliminary design completion	(a) Design team presentation (b) Testing, manufacturing needs
3. Final design completion	(a) Design team presentation of hazard analysis (b) Safety hazard analysis discussion (c) Redesign, testing, warnings and instructions, packaging
4. Preproduction review	(a) Product requirements reviewed (b) Hazard analyses reviewed (c) Redesign, testing (if any) reviewed (d) Manufacturing, quality control testing, packaging and instruction manual draft reviewed (e) Final scheduling blocked

Ford Pinto case, documentation served to prove the plaintiffs' case against Ford Motor Company because the records showed that the manufacturer knew of the defective condition. Despite such examples, though, no manufacturer can feel secure without having records of product development. In most cases, documentation of safety considerations throughout design, manufacturing and marketing of a product serves to demonstrate the reasonable safe behavior of the manufacturer.

Any changes in design, production and distribution have to be controlled and incorporated into all documentation. It is particularly important that drawings, manufacturing changes, quality control tests and inspections be current with design. There is no doubt that attention to these control details alleviated some of the apprehensions caused by the discovery of cyanide-laced Tylenol capsules in 1982 because the lots affected were readily identified and traced back to the control procedures, leading to early identification of post-manufacture tampering as the cause for the presence of the cyanide. Traceability is very important and many manufacturers are complying with traceability requirements in CPSC statutes. Although this is easy for machinery which has serial numbers, many companies have found innovative methods to trace the origin of products which do not have identifying labels.

Purchase Product Controls

Many of the products we use carry the identification of the manufacturer who may, in all respects, only be an assembler of parts made by others. Yet that manufacturer is the one liable for the safety of the product. Because of this, control of the quality of purchased parts is equally important to the quality control of the final product. It is necessary, therefore, that we exercise control over vendors to a degree consistent with the potential safety impact of the parts they supply.

This control is often over the purchase specifications, which may include dimensions and tolerances or might be related to material properties. In any case, the selection of suppliers with a proven ability to provide acceptable safe parts is imperative. Even with proven suppliers, we must have meaningful specifications for our purchases. For example, Bradford Manufacturing Company wanted to purchase cadmium-plated screws for use in assembling metal shelving units. A price increase made them change to another electroplater whose price was favorable. The new vendor's product, however, was found to be brittle in Bradford's testing program of incoming products. It was found that the new supplier's screws were embrittled by hydrogen evolved in the plating process. In another case, Acworth Manufacturing Co., makers of intrusion alarms, purchased the steel stamping used for the bell ringer. When field tests showed that these parts were susceptible to fatigue, the cost of retooling was prohibitive and Acworth was forced to drop the line, rather than be faced with burglary losses because the alarm did not function.

The CPSC has established some purchase product control actions which are as applicable today as when they were first published:

1. Purchase documents should be clear and concise with respect to design, material and safety specifications.
2. Select proven vendors.
3. Verify conformance of supplies to contractual requirements.
4. Prompt corrective action when necessary.
5. Agreement on responsibilities of suppliers for reporting hazards to the manufacturer.

The last item on the CPSC action list is noteworthy, particularly when not adhered to. Sometimes, indemnification can rule, such as an automobile manufacturer indemnifying the dealership. However, battles between defendants sometimes ensue because no pre-purchase agreement was addressed. In one case study, a baler manufacturer employed a tying machine which fitted directly into the baler, requiring no separate electrical connections or modification of the rest of the baler. When a serious injury occurred involving the tying system, the two defendants argued over who was liable, much to the pleasure of plaintiff's counsel.

Manufacturing and Quality Control

Concern for manufacturing safe products is not different from safely manufacturing products. In both cases, we must consider the task, the machinery and the workers. Within the task, we should consider the purchased materials which must be identified by tags, stamps, or other means of shop identification to ensure proper use. We should also consider those operations performed with or on the materials and how they affect both worker safety and user safety (see, for example, Case History 12.2). Machinery and procedures should conform to all safety regulations and workers should be certified for critical tasks such as welding or brazing. Implications of Just In Time programs and robotics in manufacturing should also be of concern to us for any effects on user safety.

We should be equally interested in detecting any manufacturing defects to prevent user injury. Whereas design defects can affect the safety of all the units of a product, manufacturing defects usually affect only individual units. (There are exceptions to this, such as a seam-welded galvanized tube which failed prematurely because of corrosion due to zinc vaporization during processing.) In order to prevent manufacturing defects, we must incorporate quality control procedures. Such procedures include various inspection and testing procedures, selected because of the nature of the specific product, and statistical methods, based upon random sample test procedures and statistical analysis of results. While specific quality control procedures are not within the scope of this text, we must be aware of the importance of calibration of test equipment, adherence to standard procedures, documentation and proper analysis of deviations or failures.

A special caveat for product safety assurance is necessary for this critical area of quality control testing. We must analyze the test results fairly, with respect to user safety. Two examples are worthy of mention in this respect. First, in testing the DC-10 fuselage, the collapse of the cabin floor was falsely attributed to human error, permitting the cargo door defect to go undetected. Secondly, a support system for a folding hammock of continuous nylon loop through a ring and grommets was thought superior because it automatically readjusted stress. However, when static load tests showed the central grommet to elongate, no adjustment was made. Instead, a product which was unstable in comparison to its competitors was marketed.

Packaging and Marketing

Packaging is normally thought of as the protection for products during shipment. However, packaging also includes instructions for assembly (where necessary), use and maintenance. In some cases, even the packaging can be important (e.g. see Case History 7.1).

In most cases, however, we must think of instructions and warnings when packaging is considered. Remember that these closely related topics arise from a duty to warn of hazards which cannot be eliminated or reduced, but the product is desirable for societal needs. This duty is based on both negligence and strict liability theories. Failure to warn is also used as an additional argument for defective conditions by plaintiffs.

Once established, there are a number of considerations regarding warnings:

- foreseeability
- adequacy,
- warning to whom?,
- extent of duty,
- exceptions.

A supplier need not warn against abnormal, unforeseeable uses (or misuses) of his product. However, there are numerous cases where prudence should have required special warnings or instructions. For example, a deck railing around an above-ground swimming pool was fixed by long bolts and nuts. If installed in one direction, the long bolt projected toward the pool where someone could be caught on it, but if installed correctly in the opposite direction, it would not come in contact with anyone during normal use. Failure to issue proper instructions led to a quick settlement with a youth whose finger was amputated when her ring caught on the protruding bolt.

Most failure to warn cases built on adequacy have been based upon location, size and failure to warn of consequences. In cases of whom to warn, it has generally been accepted that warnings to users discharges the duty, whereas warnings to the general public would not only be difficult, but unnecessarily alarming. One expert recently chuckled when asked to determine if a seatbelt on a Subaru was defective; the request was accompanied by an article on defective seatbelts in Hondas.

Recall notices are the result of continuing duties to warn (or extent of duty). Such notices arise out of safety problems which crop up during use. These include such matters as the differential problems in the Corvair, made famous by Ralph Nader, and the product mentioned earlier which was made defective by the evaporation of zinc during seam welding.

PLANNING FOR WARNINGS AND INSTRUCTIONS

The purpose of warnings and instructions is to control or modify the behavior of the user in order to avoid bodily injury. While each warning sign and safety instruction must be closely designed with the product in mind, there are certain requirements we can apply generally. Each warning must communicate the hazard, specifying the consequence both in words and graphically; signs must be placed conspicuously and be durable. Guidelines should be developed for creating warnings by each manufacturer faced with residual risk of injury in their final product. Many manufacturers have posted signs stating "Think Safety" and incorrectly believed that it constituted a warning. Many others, in error, have used warnings to disclaim liability, with signs such as "Remember, only you can prevent an accident." Such approaches, of course, are unacceptable.

We must consider three basic elements which are contained in proper warnings. First, there must be a signal word which clearly identifies the presence of a hazard and extent of the danger. Secondly, the nature of the hazard and the consequences that could result when the warning is not heeded should be conveyed by a symbol or pictogram. The third element should be wording which describes how we can avoid the hazard. The three most commonly used signal words are usually defined as:

- *Danger:* Immediate hazard which will result in severe injury or death.
- *Warning:* Hazard which could result in severe injury or death.
- *Caution:* Hazard which could result in minor injury or property damage.

Colors are used to reinforce the message and seriousness of the hazard as well. Stereotypical colors are red for danger, orange for warning and yellow for caution. Figure 12.8 illustrates a properly designed warning label encompassing these elements.

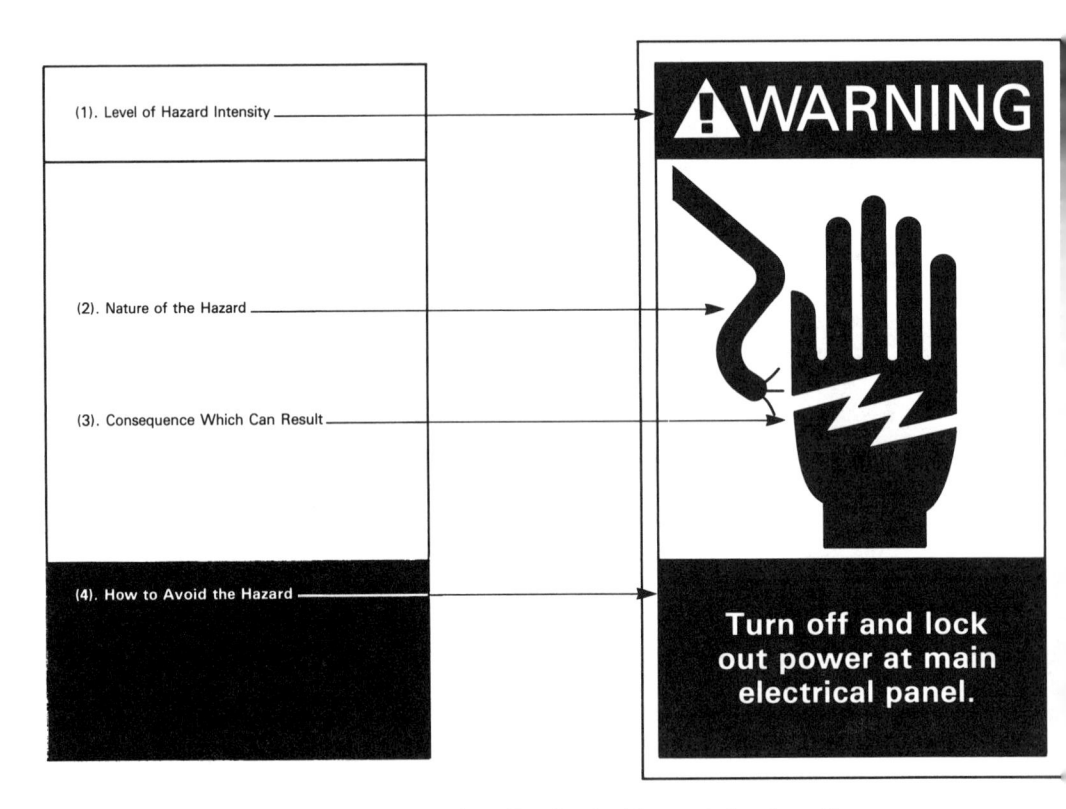

Figure 12.8 A proper warning. (Reprinted with permission from K. Ross, "Legal and Practical Considerations for the Creation of Warning Labels and Instruction Books," *J. Products Liability*, Vol. 4, 1981, Pergamon Press, Inc.)

Instructions or directions serve a purpose different from warnings which appear on the product: instructions guide the user in the proper and safe use of the product. Although they serve a different purpose, their adequacy is judged by the same standards. Most experts recommend that instructions should be written during product development and that they precede the selection of warnings. They also recommend that the front page should be dated and describe clearly the model number of the product. Although safety considerations should be covered in each instruction area, all safety matters should be compiled and emphasized on a safety page. Other elements of instructions include most of the following:

- introduction and description,
- handling and storage,
- assembly and installation,
- operation,
- inspection, maintenance and repair,
- supplemental information.

Perhaps the best advice to remember when designing warnings and instructions is that their purpose is not to avoid liability, but to prevent incidents and injuries. We must be prudent, and design warnings and instructions which communicate yet do not confuse, and limit warnings and instructions which are redundant.

RECALL PLANNING AND CUSTOMER RELATIONS

Recall programs are a direct result of the continuing liability of product manufacturers, where a defective condition is discovered after sale. We recommend that a written policy should be established in the event that recall becomes necessary. The purposes of the recall planning system are:

- to protect the assets of the corporation,
- to protect the user,
- to comply with laws,
- to remove unacceptable or questionable products from the market at minimum detriment to the corporation.

Issues which must be considered when planning include product traceability notification, the correction or cessation of further manufacturing, the cost of the program, and prevention of any recurrence.

Product recall programs are frequently integrated with the corporate consumer affairs or public relations department. These departments serve important roles in products liability prevention programs. Instead of learning about defective conditions because of litigation, proper attention to consumer complaints can often identify developing problems early. When acted upon judiciously, a caring image can be correctly projected to the public and enhance business; this is particularly true for multiple product corporations where difficulties with one product can affect the sales of others.

Some consumer departments monitor customer letters. When a new product is introduced, a large number of letters is received, but these then diminish to a steady state. As a rule of thumb, when the complaints reach a value of 40 to 50 per million products sold, a close investigation of the complaints is warranted. Unfortunately, in those companies manufacturing only a limited number of products, customer relations have to be conducted on a more personal level. In these cases, developing problems are best identified through service departments, who must also be integrated into the products liability prevention programs.

SUMMARY

Industrial safety is directly influenced by product safety because we use other's products in manufacturing our own. It is therefore our workers who benefit by having safer machines. This was not always the case, as we have learned in this chapter, and litigation based upon product safety theories for recovery (negligence, breach of warranty and strict liability in tort) has led to the need for manufacturers to produce safer products. The principles we have developed for product safety programs have been shown to be similar to those developed for a safer workplace.

FURTHER READING

R. ANDERSON, "On the Design of Products", in *Designing for Safety: Engineering Ethics in Organizational Contexts* (ed. A. Flores), RPI, Troy, NY, 1982.

S. BROBECK and A. AVERYT, *The Product Safety Book,* E. P. Dutton, New York, 1983.

G. EADS and P. REUTER, *Designing Safer Products,* Rand Publication Series, The Institute for Civil Justice, Santa Monica, Calif., 1983.

L. R. FRUMER and M. I. FRIEDMAN, *Products Liability,* Matthew Bender, 1975.

G. HALL, *The Failure to Warn Handbook,* Hanrow Press, Columbia, Md., 1986.

W. HAMMER, *Product Safety Management and Engineering,* Prentice-Hall, Englewood Cliffs, N.J., 1980.

Handbook and Standard for Manufacturing Safer Consumer Products, CPSC, 1975.

J. KOLB and S. ROSS, *Product Safety and Liability: A Desk Reference,* McGraw-Hill, New York, 1980.

D. Noel and J. Phillips, *Products Liability in a Nutshell,* West Publishing, 1974.

Product Recall Planning Guide, ASQC, Milwaukee, 1981.

K. Rodd and M. Foley, *Product Liability of Manufacturers: Prevention and Defense,* Practicing Law Institute, New York (Annual).

J. Thorpe and W. Middendorf, *What Every Engineer Should Know About PL,* Marcel Dekker, New York 1979.

A. Weinstein, A. Twerski, W. Daneher and H. Piehler, *Products Liability and the Reasonably Safe Product,* John Wiley, New York, 1978.

QUESTIONS

12.1. Explain in your own words the three theories for recovery in a products liability complaint.

12.2. How can you relate the proofs required by a plaintiff to these recovery theories?

12.3. What proofs required by the plaintiff are demonstrated in Case Histories 12.1 through 12.4?

12.4. List the representatives who should be on a design review committee and explain *why* their participation is valuable.

12.5. Examine Case History 7.1. Do you think the PET manufacturer could be held liable for Fred's incident and injury? Explain.

Appendix A

Safety Information

We live in the age of information explosion; often there is seemingly so much information that proper measures become confusing and even misleading. In our studies, we have focused our attention on people, managing people and innovation and creativity in our programs rather than on simple checklists of what to do. Nevertheless, the latest technical information is indispensable for sound safety programs.

Sources to obtain safety information are numerous, so numerous that even selecting which reference to use becomes challenging. Fortunately, how safety professionals obtain new information has been studied, thereby making our task of organizing safety information sources easier. Feldman and Beck[1] found that safety professionals most frequently consulted colleagues and reference manuals for information, suggesting that involvement in professional organizations and their activities is valuable, if only to develop information sources. We know, however, that such activity is much more enriching and that information is almost incidental. Perplexing in their study, however, was the perception of safety professionals that the need for current and new information was more important than workplace hazards. We must exercise caution to prevent the means from being mistaken for the end.

With these thoughts in mind, the sources for safety information have been organized as follows:

---

[1] R. Feldman and K. Beck, *People Who Need to Know Seek Colleagues' Advice First*, Occupational Health and Safety, **53**, No. 1 (1984), p. 61.

Appendix A: Safety Information

- To take advantage of personal contacts as well as other informational sources available in professional organizations.
- To utilize reference manuals that are flexible and contain regularly updated changes and new materials.
- To maintain the focus on where and how as well as on what by using recognized textbooks.
- To draw attention to the many excellent periodicals available in the safety profession.
- To direct attention to government agencies which frequently have the most current safety information available and deserve to be consulted more frequently by professionals.

Appendix B

Professional Organizations

Alliance for American Insurers
1501 Woodfield Road
Schaumberg, IL 60195

American Board of Industrial Hygiene
345 White Pond Drive
Akron, OH 44320

American Council of Government Industrial Hygienists
P.O. Box 1937
Cincinnati, OH 45201

American Industrial Hygiene Association
345 White Pond Drive
Akron, OH 44320

American Insurance Association
85 John Street
New York, NY 10038

American National Standards Institute
1430 Broadway
New York, NY 10018
> ANSI is the coordinator of America's voluntary standards system, including extensive standards for safe design operation, and maintenance of equipment.

American Occupational Medical Association
150 N. Wacker Drive
Chicago, IL 60606

Appendix B: Professional Organizations

American Society of Mechanical Engineers
345 East 47th Street
New York, NY 10017

American Society of Safety Engineers
1800 East Oakton Street
Des Plaines, IL 60018-2187

American Society for Testing and Materials
1916 Race Street
Philadelphia, PA 19103

Chemical Transportation Emergency Center
CHEMTREC/CMA
2501 M Street NW
Washington, DC 20037

> CHEMTREC provides information and/or assistance to those involved in or responding to chemical or hazardous material emergencies. It gives immediate advice on the nature of the product, given only the name of the product and the nature of the problem. *Emergency calls only:* 1-800-424-9300.

Human Factors Society
P.O. Box 1369
Santa Monica, CA 90406

National Fire Protection Association
Batterymarch Park
Quincy, MA 02269

National Safety Council
444 No. Michigan Avenue
Chicago, IL 60611

> The NSC Library has an in-house database of more than 75,000 documents and will conduct searches for members. It provides a bibliographic printout quickly and copies of articles upon request. (312) 527-4800, ext. 5303.

National Safety Management Society
6060 Duke Street
Alexandria, VA 22302

Society of Automotive Engineers, Inc.
400 Commonwealth Drive
Warrendale, PA 15086

System Safety Society
P.O. Box A
Newport Beach, CA 92663

Underwriters Laboratories, Inc.
333 Pfingsten Road   or   270 E. Ohio Street
Northbrook, IL 60062   Chicago, IL 60611

Appendix C

Bibliography

MANUALS AND REFERENCE BOOKS

American Institute of Chemical Engineers, *Guidelines for Hazard Evaluation Procedures,* New York, 1985.

Anderson, D., and R. Scott, *Fundamentals of Industrial Toxicology,* Ann Arbor Science, Ann Arbor, Mich., 1981.

Best's Safety Directory, A. M. Best Co., Oldwick, N.J. 08858.

Bird, F., *Management Guide to Loss Control,* Institute Press, Atlanta, 1974.

Blundell, K., *Machinery Guarding Accidents,* Hanrow Press, Columbia, Md., 1983.

Brannigan, F., R. Bright, and N. Jason, *Fire Investigation Handbook, National Bureau of Standards Handbook 134,* 1980.

Burgess, W., *Recognition of Health Hazards in Industry,* John Wiley, New York, 1981.

Dickie, D. E., *Rigging Manual,* Construction Safety Association of Ontario, 1975.

Eads, G., and P. Reuter, *Designing Safer Products: Corporate Responses to Product Liability Law and Regulation,* The Institute for Civil Justice, Rand Corp., Santa Monica, Calif., 1983.

Eastman-Kodak, *Ergonomic Design for People at Work,* 2 vols, Van Nostrand Reinhold, New York, 1983.

Fawcett, H., *Hazardous and Toxic Materials: Safe Handling and Disposal,* John Wiley, New York, 1984.

Fire Protection Handbook, NFPA.

Fulscado, A., *Laboratory Safety: Theory and Practice,* Academic Press, New York, 1980.

Goldberger, L., and S. Breshitz, *Handbook of Stress: Theoretical and Clinical Aspects,* Free Press, Division of MacMillan Publishing Co., 1982.

Appendix C: Bibliography

GORDON, H., *Hazard Control Information Handbook,* International Institute of Safety and Health, Rockville, Md. 1982.

Guards Illustrated, 4th ed. National Safety Council, Chicago, 1981.

Handbook of Organic Industrial Solvents, Alliance, Chicago, 1980.

HETZLER, D., *Industrial Fire Protection,* Fire Protection Publications, Stillwater, Oklahoma, 1982.

Industrial Fire Hazards Handbook, NFPA, Quincy, Mass., 1979.

JENSEN, P., C. JOKEL, and L. MILLER, *Industrial Noise Control Manual,* NIOSH, Washington, DC, 1979.

LEES, F., *Loss Prevention in the Process Industries: Hazard Identification Assessment and Control,* Butterworths, Stoneham, Mass., 1980.

LEVY, B., and D. WEGMAN, *Occupational Health: Recognizing and Preventing Work-Related Diseases,* Little, Brown and Co., Boston, 1988.

LEVY, P., and LEMESHOW, S. *Sampling for Health Professionals,* Van Nostrand Reinhold, 1980.

Life Safety Code Handbook, NFPA, Quincy, Mass., 1985.

"Machinery and Equipment Safeguarding Manual," Engineering and American Insurance Association, New York, 1979.

MCDERMOTT, H., *Handbook of Ventilation for Contaminant Control,* Ann Arbor Science, Ann Arbor, Mich., 1976.

MCELROY, F., *Accident Prevention Manual for Industrial Operations, Administration and Programs, Vol. 1,* National Safety Council, Chicago, 1981.

MCELROY, F., *Accident Prevention Manual for Industrial Operations, Engineering and Technology, Vol II,* National Safety Council, Chicago, 1980.

National Electrical Code Handbook, NFPA, Quincy, Mass., current.

Noise Control: A Guide for Workers and Employers, ASSE, Des Plaines, Ill., 1984.

OLISHIFSKI, J., *Fundamentals of Industrial Hygiene,* National Safety Council, Chicago, 1979.

PERGIANI, L., *Encyclopedia of Occupational Health and Safety,* International Labor Organization, Geneva, 1983.

PETERS, G., *Safety Law: A Legal Reference for the Safety Professional,* ASSE, Park Ridge, Ill., 1983.

POPE, M., FRYMOYER, J., and ANDERSSON, G., *Occupational Low Back Pain,* Praeger, New York, 1984.

ROSEN, S., *The Slip and Fall Handbook,* Hanrow Press, Inc., Columbia, Md., 1983.

SAX, N., and R. LEWIS, *Hazardous Chemicals Desk Reference,* Van Nostrand Reinhold, New York, 1987.

SCHULTZ, *Fire and Flammability Handbook,* Van Nostrand, New York.

TEPLOW, L., *Regulating Safety and Health,* ASSE, Des Plaines, Ill., 1988.

TUCKER, M., *Industrial Hygiene: A Guide to Technical Information Sources,* American Industrial Hygiene, Akron, Ohio, 1984.

TUVE, R., *Principles of Fire Protection Chemistry,* NFPA, Quincy, Mass., 1976.

W. WOODSON, *Human Factors Design Handbook: Guidelines for the Design of Systems, Facilities, Equipment and Products for Human Use,* McGraw-Hill, New York, 1981.

TEXTBOOKS

ALEXANDER, D., *The Practice and Management of Industrial Ergonomics,* Prentice Hall, 1986.

ALLEGRI, T. H., *Materials Handling, Principles and Practice,* Van Nostrand Reinhold, New York, 1984.

ALLOCCA, J., and H. LEVENSON, *Electrical and Electronic Safety,* Reston, Va., 1982.

BLANCHARD, K., and S. JOHNSON, *One Minute Manager,* William Morrow and Co., New York, 1982.

BROWNING, R., *The Loss Rate Concept in Safety Engineering,* Marcel Decker, New York, 1980.

BURGESS, J., *Designing for Humans: The Human Factor in Engineering,* Petrocelli, Princeton, N.J., 1986.

DAWSON, G., and B. MERCER, *Hazardous Waste Management,* John Wiley, New York, 1986.

P. DRUCKER, *The Frontiers of Management,* Truman Tally Books, New York, 1986.

ELLING, R., *The Struggle for Worker's Health: A Study of Six Industrialized Countries,* Baywood Publishing, Farmingdale, N.Y., 1986.

FERRY, T., *New Directions in Safety, American Society of Safety Engineers,* Des Plaines, Ill., 1985.

FERRY, T., *Safety for Engineers and Managers,* Charles C. Thomas, Springfield, Ill., 1984.

FERRY, T., *Safety Management Planning Manual,* Merritt, Santa Monica, Calif., 1982.

FINDLAY, J., *Leadership in Safety,* Institute Press, Loganville, Ga., 1980.

FIRENZE, R., *The Process of Hazard Control,* Kendall/Hunt, Dubuque, Iowa, 1978.

GLOSS, D., and M. GAYLE, *Introduction to Safety Engineering,* John Wiley, New York, 1984.

GOLDSMITH, F., and L. KERR, *Occupational Safety and Health,* Human Sciences Press, New York, 1982.

GRIMALDI, J., and R. SIMONDS, *Safety Management,* Irwin, Homewood, Ill., 1984.

GROOVER, M. P., *Automation, Production, Systems, and Computer Integrated Manufacturing,* Prentice Hall, Englewood Cliffs, N.J., 1987.

HAMMER, W., *Occupational Safety Management and Engineering,* Prentice Hall, Englewood Cliffs, N.J., 1981.

HAMMER, W., *Product Safety Management and Engineering,* Prentice Hall, Englewood Cliffs, N.J., 1981.

HEINRICH, H., D. PETERSEN, and N. ROOS, *Industrial Accident Prevention,* McGraw-Hill, New York, 1980.

KAUTOWITZ, B., and R. SORKIN, *Human Factors: Understanding People-System Relationships,* John Wiley, New York, 1983.

Appendix C: Bibliography

KIRK, P., *Fire Investigation*, John Wiley, New York, 1969.

LUBBEN, R. T., *Just In Time Manufacturing*, McGraw-Hill, New York, 1988.

MALASKY, S., *System Safety Technology and Application*, Garland STPM Press, New York, 1981.

MARSHALL, G., *Safety Engineering*, Brooks/Cole, Monterey, CA, 1982.

MCLEAN, A., *Work Stress*, Addison-Wesley, Reading, Mass., 1979.

MINTZ, B., *OSHA, History, Law and Policy*, Bureau of National Affairs, Inc., Washington, DC, 1984.

OUCHI, W., *Theory Z*, Addison-Wesley, Reading, Mass., 1981.

PETERS, T., and R. WATERMAN, *In Search of Excellence*, Harper and Row, New York, 1982.

PETERSEN, D., *Human Error Reduction and Safety Management*, Garland STPM Press, New York, 1982.

ROLAND, H., and B. MORIARTY, *System Safety*, John Wiley, New York, 1983.

SANDERS, M., and E. MCCORMICK, *Human Factors in Engineering and Design*, McGraw-Hill, 1987.

SCHERMERHORN, J., J. HUNT, and R. OSBORN, *Managing Organizational Behavior*, John Wiley, New York, 1985.

TARRANTS, W., *The Measurement of Safety Performance*, Garland STPM Press, New York, 1980.

WORRALL, J., *Safety and the Work Force*, ILR Press, Ithaca, N.Y., 1983.

ZALES, M., *Stress in Health and Disease*, Brunner/Mazel Publishers, New York, 1985.

JOURNALS

Canadian Occupational Health and Safety News, 1450 Don Mills Rd., Don Mills, Ontario, M3B 2x7, Canada, 1978.

Canadian Occupational Safety, 222 Argyle Ave., Delhi, Ontario, N4B 2Y2, Canada, 1963.

Canadian Occupational Safety and Health Law, 1450 Don Mills Rd., Don Mills Ontario, M3B 2x7, Canada, 1978.

Chilton's Industrial Safety and Hygiene News, One Chilton Way, Radnor, PA 19089, 1967.

Environmental Health and Safety News, University of Washington, Department of Environmental Health, School of Public Health and Community Medicine, F-461 Health Sciences Bldg., Department of Environmental Health Sc-34, Seattle, WA, 98195, 1951.

Ergonomics, The Manor House, Gaulby, Leicestershire, LE7 9BE, U.K.

Fire Technology, NFPA, 470 Atlantic Ave., Boston, MA 02210.

Hazard Prevention, System Safety Society, 14252 Culver Drive, Irvine, CA 92714.

Hazardous Materials Control, Hazardous Materials Control Research Institute, 9300 Columbia Blvd., Silver Spring, MD, 20910.

Hazards Review, Elsevier International Bulletins, 52 Vanderbilt Ave., New York, N.Y., 10017, 1979.

Health and Safety at Work, Maclaren Publishers Ltd., 19 Scarbrook Rd., Croyden Surrey, CR9 1QH, England, 1978.

Industrial Fire World, P.O. Box 9161, 208C Southwest Parkway East, College Station, TX 77840, 1985.

Industrial Hygiene Digest, Industrial Health Foundation, 5231 Centre Ave., Pittsburgh, PA 15232.

Journal American Industrial Hygiene Association, American Industrial Hygiene Association, 345 White Pond Drive, Akron, OH 44320.

Journal of Occupational Accidents, Elsevier Science Publishing Co., Inc., P.O. Box 1663, Grand Central Station, New York, NY 10163.

Journal of Occupational Medicine, Industrial Medical Association, 150 N. Wacker Dr., Chicago, Ill. 60606.

Material Handling Engineering, Penton Publishing, Cleveland, OH.

Modern Materials Handling, Cahners Publishing, Newton, MA 02158.

Occupational Hazards, Industrial Publishing Co., 614 Superior Ave., W. Cleveland, OH 44113.

Occupational Hazards, Industrial Safety/Security Management, 1111 Chester Ave., Cleveland, OH 44114.

Occupational Health and Safety Letter, 1331 Pennsylvania Ave., N.W., Washington, D.C. 20004.

Occupational Health and Safety, Medical Publications, Inc., 225 N. New Road, Waco, TX 76710.

Occupational Safety and Health Reporter, The Bureau of National Affairs, Inc., 1231 25th St., N.W., Washington, D.C. 20037.

Proceedings of the Human Factors Society, Human Factors Society, P.O. Box 1369, Santa Monica, CA 90406.

Professional Safety, American Society of Safety Engineers, 1800 E. Oakton St., DesPlaines, Ill 60018–2187.

Safety and Health (Washington), The Bureau of National Affairs, Inc., 1231 25th N.W., Washington, D.C. 20037.

Safety and Health Report, 951 Pershing Dr., Silver Spring, MD 20910.

Work and Stress, Department of Psychology, University of Nottingham, University Park, Nottingham, NG7 2RD, U.K.

Appendix D

Government Agencies

American Public Health Association
1015 Fifteenth Street N.W.
Washington, DC 20005

Bureau of Labor Statistics
U.S. Department of Labor
Washington, DC 20212

Bureau of National Affairs, Inc.
Occupational Safety and Health Reporter
1231 25th Street, N.W.
Washington, DC 20037

Commerce Clearing House
Employee Safety and Health Guide
4205 W. Peterson Avenue
Chicago, IL 60646

Environmental Protection Agency
401 M Street, S.W.
Washington, DC 20001

National Bureau of Standards
U.S. Department of Commerce
Washington, DC 20234

National Institute for Occupational Safety and Health (NIOSH)
4676 Columbia Parkway
Cincinnati, OH 45226

Occupational Safety and Health Administration
U.S. Department of Labor
200 Constitution Avenue
Washington, DC 20210

Superintendent of Documents
U.S. Government Printing Office
Washington, DC 20402

U.S. Consumer Product Safety Commission
Washington, DC 20207

Index

A

Accident:
 cost, 79
 definition, 27
Alcoa, 89, 90, 99
ANSI, 6, 167, 168, 169
Anthropometry, 55–61
Assumption of risk, 3
Attitudes of workers, 44
Autoignition temperature, 196, 198
Automated material handling, 157
 back tracking, 158
 straight-line flow, 158
Axioms of industrial safety, 28

B

Behavior management, 90
Biomechanics, 140
Bipedal gait, 181, 182
Brainstorming, 113, 137
Budgeting, 79

C

Chains, 145, 146
Coefficient of friction, 181, 182
Combustion, heat of, 197, 199

Commitment, 87
Common sense, 2
Communication, 82, 88
Consumer Product Safety Act, 119, 267
Consumer Product Safety Commission, 267
Contributory negligence, 3
Controls, 83, 135
Corrosion, 108, 146
Cranes, 156
Critical behavior index, 92
Cumulative trauma disorders, 71
Customer relations, 289

D

Deep well injection, 259
Design review, 282
Disaster preparedness, 216
Domino theory, 29, 30
Duty to warn, 265

E

Electrical safety, 217, 220
Employee Assistance Programs (EAPs), 243
Endurance, 63
EPA, 10

Epidemiology, 33, 140, 229, 230
Ergonomics, 55, 142
Experience, 104
Expert witness, 269
Extinguishing fires, 211

F

Facilitator, 97
Failure mode and effects analysis, 109
Falls:
  free, 192
  ladder, 189
  stairway, 187
Fault tree analysis, 116
Feedback, 91, 124, 125, 133, 135, 137, 169
Fellow servant rule, 3
Fire:
  chemistry, 196
  classes of, 198
  detection, 211, 212
  duration, 202
  extinguishing, 211
  gases, 199, 200
  heat transfer, 206
  protection, 210
  severity, 202
  triangle, 195
Flame spread, 203
Flammability limits, 197, 198, 233
Flash point, 196
Fork lifts, 153, 156
Free falls, 192
Frequency of occurrence, 118

G

Ground fault circuit interrupters, 219, 220

H

Handrails, 188, 189
Hazard, definition of, 103
Hazard analysis:
  failure mode and effects analysis, 109
  fault tree analysis, 116
  HAZOP, 110, 113, 114, 115, 116
  preliminary hazard analysis, 107
Hazard control, 124
Hazard identification, 103, 169, 171, 174
Hazardous waste, definition of, 255

HAZOP, 110, 113, 115
  combined with human error, 116
  guide words, 113
  method flow diagram, 114
Hearing, 246, 249
Heat of combustion, 197, 199
Hooks, 146
Human error, 31, 45, 54, 171
  definition of, 45
Hygiene factors, 41

I

Ignition, 202
Illumination, 51, 52
Incidence rate, definition of, 122, 133
Incineration, 259
Information display, 69, 70
Information processing, 48
Ingestion, 229
Innovation, 87, 132
Instructions, 286, 289
Intervention, 91, 93
Interviews, 104
Involvement, 88, 132

J

Job description, 49
Just-in-time manufacturing, 160, 161

K

Kaiser Aluminum, 154
Kinesiology, 63
KISS principle, 41, 42

L

Ladder falls, 189
Ladders, step, 190, 191
Landfill, secure, 258
Leadership, 82
Legal system, 263
Life safety, 202, 203
Lifters, 147
Load, 50
Lower back pain, 138, 139, 140
  rehabilitation, 143

M

Machine controls, 63, 69
Machine hazards, 165, 166

Machinery safeguarding, 164
Management:
  approaches, 75
  attributes, 87
  classical school, 86
  contingency school, 86
  controls, 124, 128
  definition of, 74
  human relations school, 86
  roles, 76
  skills, 76
  tasks, 75
Management by objectives:
  controlling, 83
  directing, 81
  organizing, 79
  planning, 77
  staffing, 80
Manual handling, 138, 140, 141
Marketing, 286
Monitoring, 132
Motivators, 41
Material safety data sheets, 10

N

National Electrical Code, 220
National Safety Council, 2, 5, 78, 119
Negligence, contributory, 3
NEISS, 119, 267
NIOSH, 6, 141, 142, 154
Nip point, 165
Noise, 245, 254

O

Operator factors, 169, 171
Organizing, 79
OSHA, 6, 164, 167, 169, 268
  record keeping, 17, 20, 26
  standards, 6, 8
OSHAct, 5
Overload, 31, 46

P

Packaging, 286
Palletizing, 148
Pallet trucks, 151

Pareto analysis, 91, 106
Perception, 40, 42, 44, 69, 137, 159
Performance, 82
Physiological criteria, 140
Pinch point, 165
Plaintiff's proofs, 266
Placement, 127
Point of operation, 165
Preliminary hazard analysis, 107
Press brake, 169, 170
Product recall, 289
Products liability laws, 264, 269
Products liability prevention, 280
Psychological stress, 240, 245
Psychophysical criteria, 140
Purchase product control, 284, 285

Q

Quality control, 285

R

Reasonable, definition of, 266
Reasoning:
  deductive, 106, 116
  inductive, 106
Record keeping, 17, 20, 26, 283
Relative humidity, 52
Remedies:
  cost, 125
  engineering, 124, 128
  management, 124, 128
  no remedy, 125, 126
  placement, 127
Respiratory system, 225, 226
Rigging, 138
Right-to-know laws, 10, 95
Risk analysis, 118
Risk management, 99
Robotics, 176, 178
Ropes, 145, 146

S

Safeguards, 167, 168
Safety circles, 97
Safety committees, 97
Safety goals and objectives, 78
Safety information, 292

Safety policy, 78, 79
SARA III, 10, 111
Self-regulation, 135
Severity of consequences, 118
Severity rate, definition of, 133
Skin absorbtion, 228
Slings, 145, 146
Slips, 181
Solvents, 231, 240
Spine, 139
Staff, 134 (*see also* Management by objectives)
Staffing, 80
Stairway falls, 187
Standards, 7
State, 53
Step ladders, 190, 191
Stereotypes, 64, 65
Stockpickers, 151
Strength, 63
Stress:
    coping with, 243, 244
    psychological, 240, 245
Systems safety, 74, 103, 110, 133, 150
    model for accident causation 33, 34

T

Task analysis, 52, 116, 142
Task, definition of, 169

Task factors, 169
Technical controls, 135
Temperature, 52
Testing, 105, 106
Tetrachorosilane (TET), 17
THERP, 54
Threshold limit value (TLV), 232, 233
Tool design, 71
Training, 16, 94
Trauma disorders, 71
Trips, 183
Trust, 87, 88
Two-factor theory, 40, 41
Typography, 69

V

Vendors, 284, 285
Ventilation, 236, 238
Video tapes, 106, 122

W

Warnings, 286, 289
Waste treatment, 258
What if analysis, 106, 114
Wholistic safety, 134, 136
Worker selection, 49
Workmen's compensation laws, 4, 168
Workplace of the future, 176